Guide to the JCT Design and Build Contract (2024 Edition)

This book provides a comprehensive guide to the 2024 edition of the JCT Design and Build Contract. It offers a straightforward reference to help practitioners understand the rights and obligations of the parties, and to administrate the Contract in an accurate and timely manner.

Part 1 provides commentary on all the key provisions of the Contract, in plain English. It adopts the same structure as the Contract and includes the relevant extracts to reduce the need for cross-referencing. There is also guidance on current market practice, and recommendations for best practice.

Part 2 contains over 300 corresponding resources for the administration of the Contract, by both the Contractor and the Employer. It includes notice and other correspondence templates, checklists, and trackers to assist record keeping.

Written to meet the needs of busy construction practitioners, this is an accessible, practical reference for those acting on behalf of the Contractor and the Employer (and Employer's Agent), including project teams, consultants, and advisors. It is also ideally suited for students of construction law.

Seán Mac Labhraí is a solicitor and the Chief Operating Officer at the Ardmore Group. He is a Fellow of the Chartered Institute of Arbitrators and a CEDR Accredited Mediator.

"*Ever wondered what to buy for the contract administrator in your life who has everything? Wonder no longer. This book provides an invaluable, step-by-step guide for how to run a contract properly, complete with example resource documents, all in line with the JCT 2024 Design & Build Contract, and cross referenced against the relevant clauses. My only concern is that people will now do this so well that they will not end up in disputes and I'll be out of a job.*"

David Pliener KC, Gatehouse Chambers

"*This admirable book is the Google Maps for the 2024 edition of the JCT Design and Build Contract. Contract administrators now have a trusty friend, and a reliable navigation system.*"

Julian Bailey, Visiting Professor at the Centre for Construction Law at The Dickson Poon School of Law, King's College London, author of *Construction Law* (Fourth Edition), former Chairman of the Society of Construction Law, and Partner, Jones Day

"*This book is timely. In the 2024 report 'Construction Adjudication in the United Kingdom: Tracing trends and guiding reform' published by King's College London and the Adjudication Society, the two leading causes of disputes in construction adjudication were inadequate contract administration and the lack of competence of project participants. It is therefore clear that there is very much a need for project participants to be kept up to date and informed about construction contracts, in order to increase efficiency and avoid disputes. The author acknowledges the importance of a comprehensive working knowledge of the contract and consistent contract administration for successful project delivery, and should be applauded for producing what I predict will be the 'go to' guide or handbook for those involved with projects which adopt the 2024 edition of the JCT Design and Build Contract. This must have been a gargantuan task for the author, who should be both congratulated on his efforts and thanked for taking the time to share his knowledge for the benefit of not just those who read it, but also the wider construction industry.*"

Matt Molloy, Adjudicator, Arbitrator, and Mediator, and co-author of *Adjudicating Construction and Engineering Disputes*

Guide to the JCT Design and Build Contract (2024 Edition)

Seán Mac Labhraí

Routledge
Taylor & Francis Group
LONDON AND NEW YORK

Designed cover image: Getty Images

First published 2025
by Routledge
4 Park Square, Milton Park, Abingdon, Oxon OX14 4RN

and by Routledge
605 Third Avenue, New York, NY 10158

Routledge is an imprint of the Taylor & Francis Group, an informa business

British Library Cataloguing-in-Publication Data
A catalogue record for this book is available from the British Library

ISBN: 978-1-032-78023-8 (hbk)
ISBN: 978-1-032-74403-2 (pbk)
ISBN: 978-1-003-48591-9 (ebk)

DOI: 10.1201/9781003485919

Typeset in Sabon
by KnowledgeWorks Global Ltd.

In memory of Peter Clarke (1945–2024)
Fellow and Tutor in Law, Jesus College Oxford (1971–2009)

Contents

Foreword

I was delighted when Seán Mac Labhraí asked me to write the foreword to this book, and for three reasons.

The first is that Seán and I obviously share a passion for JCT Contracts in general, and the JCT Design and Build form in particular. JCT Design and Build is certainly the form I come across, and indeed choose to use, most as a lawyer in practice, for anything from a new private hospital, to large-scale social housing, to an office fit out. It is telling that this is the Contract from the new 2024 Contract suite which the Joint Contracts Tribunal chose to release first. Seán's provision of a detailed guide to this newly revised contract form is therefore extremely timely.

Secondly, this impressive guide contains a wealth of useful background information, including not only the changes within the latest edition and the reasons for them, but also the statutory provisions referred to (including the Building Safety Act 2022), and the practical issues to consider in selecting options and completing the documentation. For those of us who regularly find ourselves lost in the weeds of an unhelpfully compiled index, in this book it could not be easier to locate a particular provision.

It speaks in simple, clear, and concise language, and is entirely practical in its approach. It speaks with the authentic voice of in-house counsel in a construction business, who is always very close to the process. It provides invaluable resources, such as summaries of each part of the Contract, full text excerpts from relevant legislation, and essential case law. It can equally be a banquet or a snack: read in its entirety for the purposes of study (or pleasure), or dipped into for guidance on a specific provision with which you may be having difficulty. I am sure that it will soon come to be regarded as a key work on the subject.

Finally, the third reason that I was delighted to be asked by Seán to write this foreword is far more personal. Seán and I, some considerable years apart, both had the honour and the very great pleasure of being taught our law by the late Peter Clarke, Fellow and Law Tutor at Jesus College Oxford, to whom this book is dedicated.

Karen Kirkham
Chair, Joint Contracts Tribunal
November 2024

Introduction

In April 2024, the JCT published the latest edition of its Design and Build Contract.

Like the 2016 edition, the JCT considers that the 2024 edition is 'primarily appropriate for larger works, where the Employer has defined its requirements and where the Contractor is not only to carry out the works, but also to complete the design for them in accordance with those requirements'.[1]

The updates included within the 2024 edition are to be welcomed, as they preserve and refine the carefully considered risk allocation between the Contractor and the Employer.

The 2024 edition follows the enactment of the momentous Building Safety Act 2022, which establishes a regulatory framework to improve the industry's approach to building and fire safety. This includes a greater emphasis on ensuring the competence of the key parties (or 'dutyholders') involved throughout the course of the project, and in certain cases, establishing a 'golden thread of information' relating to the project.

However, although the Contractor and the Employer will share the aim of delivering their projects safely, accurately, on programme, and within budget, the nature of the Design and Build Contract (like many other standard forms) is such that their competing commercial interests will inevitably remain, with the consequent risk of differences and disputes.

Furthermore, a 2024 report published by the Centre of Construction Law & Dispute Resolution at King's College London concluded that the two leading causes of disputes in adjudication in the previous year were 'inadequate contract administration' and 'lack of competence of contract participants'.[2]

Taken together, it is therefore suggested that both a comprehensive working knowledge of the building contract, along with its accurate and consistent administration, are now more important than ever to the successful delivery of a project.

This book aims to support these objectives.

It provides guidance and support (primarily to practitioners acting for and on behalf of both the Contractor and the Employer, but also for students of construction law) and adopts the following structure:

- **Part 1** provides concise commentary on all the key provisions of the Design and Build Contract, along with practical recommendations and guidance on current market practice.
- **Part 2** includes over 300 resources for contract administration, including notice and correspondence templates, checklists, and trackers, along with drafting guidance.

---

[1] JCT Design and Build Contract Guide 2024 (DB/G 2024), at paragraph 2

[2] Nazzini, R & Godhe, A 2024, *2024 Construction Adjudication in the United Kingdom: Tracing trends and guiding reform. King's College London.* Available at: https://doi.org/10.18742/pub01-192

DOI: 10.1201/9781003485919-1

With an emphasis on consistent, clear, documented communication, it is hoped that the Parties can reduce the potential for disagreement and uncertainty, and focus on their shared objectives, to their mutual benefit.

How to use this book

This book is primarily written as a reference or companion to those involved in the administration (and negotiation) of the 2024 edition of the JCT Design and Build Contract (afterwards simply referred to as 'the Contract').

Part 1 (Commentary) therefore adopts precisely the same structure as the Contract. It contains the relevant extracts from the Contract, along with recommendations and guidance on current market practice, and links to the relevant resources in Part 2 (Resources for Contract Administration).

Clause 1.3: Agreement etc. to be read as a whole

> 1.3 *The Agreement and these Conditions are to be read as a whole. Nothing contained in any other Contract Document or any Framework Agreement, irrespective of their terms, shall override or modify the Agreement or these Conditions.*

Clause 1.3 assists with the interpretation of the Contract Documents as a whole, by effectively establishing a hierarchy.

The Agreement and the Conditions are to be '*read as a whole*'; in other words, construed or understood together.

In the event of a conflict between:

(i) the Agreement or the Conditions; and
(ii) any other Contract Document (including the Employer's Requirements, Contractor's Proposals, Contract Sum Analysis, or (where applicable) the BIM Protocol), or any Framework Agreement,

then the Agreement or Conditions will take precedence.

This is '*irrespective of their terms*'; in other words, no matter what those other Contract Documents may state.

Recommendation: Negotiating, revising, and checking the Contract Documents

Similar to the recommendation made in respect of Article 5 (Employer's Requirements and Contractor's Proposals), in light of the effect of clause 1.3, the Parties are encouraged to leave sufficient time at the end of the contract negotiations to undertake a comprehensive review of the draft Contract Documents in their entirety, to remove any inconsistencies, conflicts, and discrepancies in or between the Agreement, the Conditions, and the remaining Contract Documents.

Resource 4 page	Notice: Contractor or Employer to the other Party	A notice of a conflict within the Contract Documents.

Market practice: Hierarchy of Contract Documents

Clause 1.3 is often amended to provide for a more prescriptive hierarchy of the Contract Documents, in the event of a conflict. For example, it may provide that the Contract Documents are to be read and construed in the following order:

1. the Agreement
2. the Conditions
3. the Employer's Requirements
4. the Contractor's Proposals
5. the Contract Sum Analysis
6. the BIM Protocol (where applicable)

Finally, the Reference section contains a detailed subject index and resource index, along with recommendations for further reading.

Key

Below is a key to the symbols used in this book:

Symbol	Description
	Resources for the Contractor
	Resources for the Employer
	Resources for both Parties

Symbol	Description
	Notice
	Other correspondence
	Template
	Instruction
	Checklist
	Request
	Consent/approval/permission given
	Consent/approval/permission withheld
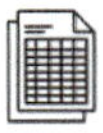	Tracker
	Record keeping
	Reminder
	Recommendation
	Urgent action required
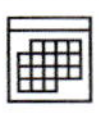	Diary management
	Further reading

Changes from the 2016 edition

The table below summarises the key changes from the 2016 edition of the Contract.

Provision	Change
Article 3 (Collaborative working)	This provision was previously included within Supplemental Provision 5 (Collaborative Working) of the 2016 edition, and has been transferred to the Articles, to reflect its increased importance. This aligns with the aims of *The Construction Playbook*, published by the UK Government.
Article 6 (CDM Regulations – Principal Designer and Principal Contractor)	This Article has been updated to provide for the role of both the Principal Designer and the Principal Contractor under the CDM Regulations.
Article 7 (Building Regulations – Principal Designer and Principal Contractor)	This Article has been introduced to provide for the role of both the Principal Designer and the Principal Contractor under Part 2A of the Building Regulations.
clause 1.1 (Definitions), definition of Building Regulations	This is a new definition and refers to the Building Regulations 2010.
clause 1.1 (Definitions), definition of Statutory Provider	This is a new definition and replaces the definition of 'Statutory Undertaker' included in the 2016 edition. This is to reflect the extent to which the utilities sector has been privatised.
clause 1.1 (Definitions), definition of Termination Payment	This is a new definition and relates to a payment under clause 8.7.6 (Termination by Employer: Consequences of termination under clauses 8.4 to 8.6), clause 8.8.3 (Termination by Employer: Employer's decision not to complete the Works), clause 8.12.5 (Consequences of Termination under clauses 8.9 to 8.11, etc.), and clause 8.13 (Termination Payment – final date, notices and amount).
clause 1.7 (Notices and other communications)	This clause has been updated to expressly permit the service of notices (including those required by the Contract to be given in accordance with clause 1.7.4) by email, to the email addresses included against the references to clause 1.7.3 and clause 1.7.4.2 in the Contract Particulars.
clause 2.1.5 (Contractor's Obligations: General obligations)	The provision previously included within paragraph 8.1 of Supplemental Provision 8 (Sustainable development and environmental considerations) of the 2016 edition has been developed and transferred to clause 2.1.5. Again, this aligns with the aims of *The Construction Playbook*, published by the UK Government.

(*Continued*)

Provision	Change
clause 2.2.2 (Materials, goods and workmanship)	The provision previously included within paragraph 8.2 of Supplemental Provision 8 (Sustainable development and environmental considerations) of the 2016 edition has been transferred to clause 2.2.2. Again, this aligns with the aims of *The Construction Playbook*, published by the UK Government.
clause 2.7.2 (Contract Documents)	This clause has been updated to refer to building information required to be provided to the Contractor under regulation 11A(4) of the Building Regulations.
clause 2.17 (Design Work – liabilities and limitation)	This clause has been updated to introduce a new standard of care in clause 2.17.1.1, namely the reasonable skill and care to be expected of a qualified and experienced architect (or other appropriate professional designer) undertaking the design of works similar in scope and character to such design of the Works. Clause 2.17.1.2 also introduces an express fitness for purpose exclusion in respect of design. Clause 2.17.2 and clause 2.17.3 have also been updated to apply to work in relation to a dwelling or dwellings to which the Defective Premises Act 1972 applies.
clause 2.24 (Notice by Contractor of delay to progress)	A new clause 2.24.4 has been introduced to the effect that if the Employer requires further information as is reasonably necessary to enable it to reach a decision under clause 2.25.1 (Fixing Completion Date), then it must notify the Contractor not later than 14 days from receipt of the relevant particulars under clause 2.24.2, or notification under clause 2.24.3.
clause 2.25 (Fixing Completion Date)	Clause 2.25.2 has been amended to reduce the period for the Employer to make its decision from 12 weeks to 8 weeks from the later of the date of receipt of (i) the particulars under clause 2.24.2, or (ii) a notification under clause 2.24.3, or (iii) further information required by the Employer under clause 2.24.4.
clause 2.26.4 (Relevant Events)	This Relevant Event has been updated to reflect the expanded clause 3.15 (Antiquities, asbestos, contaminated material and unexploded ordnance) to which it relates.
clause 2.26.7 (Relevant Events)	This is a new Relevant Event, relating to the effect of epidemics on carrying out the Works.

(*Continued*)

Provision	Change
clause 2.26.8 (Relevant Events)	Whilst it is a new Relevant Event, it represents a significant expansion on the previous Relevant Event included in clause 2.26.12 of the 2016 edition and relates to legislation, the exercise of statutory powers, and the publication of guidance (whether by the United Kingdom Government, any of the devolved administrations, any Local or Public Authority, or in the case of guidance, by the Construction Leadership Council), which affects the execution of the Works.
clause 2.29.5 (Payment or allowance of liquidated damages)	This is a new clause, to reflect the Supreme Court's judgment in *Triple Point Technology Inc v PTT Public Company Ltd* [2021] UKSC 29. If the Contractor's employment is terminated under the Contract, it provides for the Employer to deduct liquidated damages for the period between the Completion Date and the date of termination, but not afterwards, when the Employer may only bring a claim for general damages.
clause 3.4 (Conditions of sub-contracting)	This clause has been updated to reference Employer and Funder step-in rights (in clause 3.4.1) and to require compliance, as applicable, with Part 2A of the Building Regulations (in clause 3.4.2.3).
clause 3.15 (Antiquities, asbestos, contaminated material and unexploded ordnance)	This clause has been significantly expanded to provide for the discovery of asbestos, contaminated material, and unexploded ordnance, in clause 3.15.3 and clause 3.15.4.
clause 3.16 (CDM Regulations and Part 2A of the Building Regulations)	This clause has been expanded to refer to not only the CDM Regulations, but also Part 2A of the Building Regulations.
clause 4.21.3 (Relevant Matters)	Similar to clause 2.26.4 (Relevant Events) above, this Relevant Matter has been updated to reflect the expanded clause 3.15 (Antiquities, asbestos, contaminated material and unexploded ordnance) to which it relates.
clause 4.21.6 (Relevant Matters)	Similar to clause 2.26.7 (Relevant Events) above, this is a new Relevant Matter, relating to the effect of epidemics on carrying out the Works. However, unlike the corresponding Relevant Event, this Relevant Matter is optional.

(*Continued*)

Provision	Change
clause 4.21.7 (Relevant Matters)	Similar to clause 2.26.8 (Relevant Events) above, this is a new Relevant Matter, relating to legislation, the exercise of statutory powers, and the publication of guidance (whether by the United Kingdom Government, any of the devolved administrations, any Local or Public Authority, or in the case of guidance, by the Construction Leadership Council), which affects the execution of the Works. Again, unlike the corresponding Relevant Event, this Relevant Matter is optional.
clause 5.4.5 (The Valuation Rules: Measurable Work)	This is a new clause (and thereby an addition to the Valuation Rules), which requires allowance to be made for any percentage or lump sum adjustments in the Contract Sum Analysis.
clause 5.5 (The Valuation Rules: Daywork)	The references to the RICS definitions of prime cost of daywork have been updated.
clause 6.13.5.4 (Loss or damage – insurance claims and reinstatement)	This is a new clause, to the effect that reinstatement work will be treated as a Change, for the purposes of the corresponding Relevant Event contained in clause 2.26.1 (Relevant Events), but not otherwise (including the Relevant Matter contained in clause 4.21.1).
clause 6.15.2 (Professional Indemnity Insurance: Obligation to insure)	This clause has been updated to the effect that professional indemnity insurance must be maintained by the Contractor, provided that it is available at both commercially reasonable rates *and terms*.
clause 6.16 (Professional Indemnity Insurance: Increased cost and non-availability)	Similar to clause 6.15.2 above, this clause has been updated to provide for when professional indemnity insurance ceases to be available at commercially reasonable rates *and terms*.
clause 7.7.1 (Third Party Rights from Contractor: Rights for Purchasers and Tenants)	Similar to the changes to clause 1.7 (Notices and other communications), this clause has been updated to provide for the possibility of service of notices to Purchasers or Tenants by email.
clause 7.8.1 (Third Party Rights from Contractor: Rights for a Funder)	Similar to the changes to clause 1.7 (Notices and other communications) and clause 7.7.1 above, this clause has been updated to provide for the possibility of service of notices to a Funder by email.
clause 8.1.4 (Meaning of insolvency)	This clause has been updated to include two new insolvency grounds, in clause 8.1.4.3 and clause 8.1.4.4, to reflect the relevant provisions of the Corporate Insolvency and Governance Act 2020.
clause 8.7 (Termination by Employer: Consequences of termination under clauses 8.4 to 8.6)	This clause has been updated to provide for a Termination Payment, to be made in accordance with clause 8.13 (Termination Payment – final date, notices and amount).

(*Continued*)

Provision	Change
clause 8.8 (Termination by Employer: Employer's decision not to complete the Works)	Similar to clause 8.7 above, this clause has been updated to provide for a Termination Payment, to be made in accordance with clause 8.13 (Termination Payment – final date, notices and amount).
clause 8.11.1.5 (Termination by either Party and regulation 73(1)(a) of the PC Regulations)	Similar to the new Relevant Event in clause 2.26.7 (Relevant Events) and the new Relevant Matter in clause 4.21.6 (Relevant Matters), this is a new clause (and ground for termination) relating to the effect of epidemics on carrying out the Works.
clause 8.11.1.6 (Termination by either Party and regulation 73(1)(a) of the PC Regulations)	Similar to the new Relevant Event in clause 2.26.8 (Relevant Events) and the new Relevant Matter in clause 4.21.7 (Relevant Matters), this is a new clause (and ground for termination) relating to legislation, the exercise of statutory powers, and the publication of guidance (whether by the United Kingdom Government, any of the devolved administrations, any Local or Public Authority, or in the case of guidance, by the Construction Leadership Council), which affects the execution of the Works.
clause 8.12 (Termination by Contractor: Consequences of Termination under clauses 8.9 to 8.11, etc.)	Similar to clause 8.7 (Termination by Employer: Consequences of termination under clauses 8.4 to 8.6) and clause 8.8 (Termination by Employer: Employer's decision not to complete the Works) above, this clause has been updated to provide for a Termination Payment, to be made in accordance with clause 8.13 (Termination Payment – final date, notices and amount).
clause 8.13 (Termination Payment – final date, notices and amount)	This is a new clause and provides a mechanism for the Termination Payment under clause 8.7.6 (Termination by Employer: Consequences of termination under clauses 8.4 to 8.6), clause 8.8.3 (Termination by Employer: Employer's decision not to complete the Works), and clause 8.12.5 (Consequences of Termination under clauses 8.9 to 8.11, etc.). It was introduced to expressly provide for both the payment and notice requirements of the Housing Grants, Construction and Regeneration Act 1996.
clause 9.1 (Notification and negotiation of disputes)	This provision was previously included within Supplemental Provision 10 (Notification and negotiation of disputes) of the 2016 edition and has been transferred to the Articles, to reflect its increased importance. Again, this aligns with the aims of *The Construction Playbook*, published by the UK Government.

(*Continued*)

Provision	Change
clause 9.4 (Arbitration: Conduct of arbitration)	This clause has been updated to refer to the JCT 2024 edition of the Construction Industry Model Arbitration Rules (CIMAR).
Various	Several updates have been made throughout the Contract relating to gender-inclusive language.
Fluctuations Options	Schedule 7 of the 2016 edition contained the JCT Fluctuations Option A. This has been removed, but it is still available on the JCT website.

Further reading: JCT Design and Build Contract 2024 Tracked Change Document

The JCT has published a document entitled '*DB 2024 Tracked: Design and Build Contract 2024 Tracked Change Document*', which clearly highlights all of the changes from the 2016 edition.

The Building Safety Act 2022

Overview

The Contract contains a number of minor amendments which take account of the Building Safety Act 2022 and related legislation.

Whilst this legislation is expected to have an extensive and profound impact on the industry, given the very broad range of projects which adopt the JCT Design and Build Contract, and to which different aspects of this legislation will apply, more extensive revision to the Contract would likely have proved impractical.

The effects of this legislation are therefore only very briefly considered here, in relation to the law in England. (Further guidance is widely available, including from the Health and Safety Executive and the JCT.)

The key point is that the legislation falls within the definition of 'Statutory Requirements' (included in clause 1.1 (Definitions)), with which the Contractor is required to comply when carrying out the Works.

However, when negotiating and drafting their building contracts, the Parties may wish to give further consideration to the following aspects of the legislation, which may be relevant to their projects.

The dutyholder regime

The Building Regulations etc. (Amendment) (England) Regulations 2023 (made pursuant to the Building Safety Act 2022) introduced a new Part 2A to the Building Regulations 2010, entitled 'Dutyholders and Competence', and known as the 'dutyholder regime'.

The 'dutyholders' will have responsibility under the legislation for managing risks throughout the design, construction, and occupation phase of the project. During the construction phase, the dutyholders include the following:

- clients (which includes the Employer);
- the Principal Contractor (which potentially includes the Contractor);
- the Principal Designer (which potentially includes the Contractor);
- designers; and
- contractors (which includes the Contractor's sub-contractors).

These roles are similar to (albeit distinct from) those under the Construction (Design and Management) Regulations 2015 (defined in the Contract as the 'CDM Regulations').

The Contract therefore adopts the same approach for both, and the dutyholder regime is provided for in the following provisions:

- Article 7 (Building Regulations – Principal Designer and Principal Contractor); and
- clause 3.16 (CDM Regulations and Part 2A of the Building Regulations).

The dutyholders' responsibilities are greater in the context of 'higher-risk buildings', considered further below.

Higher-risk buildings

The Building Safety Act 2022 provides for a stricter regulatory regime in respect of 'higher-risk buildings', which are potentially more at risk in the context of fire safety. In general, 'higher-risk buildings' are those of at least 18 metres in height or at least seven storeys, and which comprise two or more residential units (including care homes and hospitals, albeit not hotels nor military barracks).

The Contract does not include any provisions specific to the higher-risk building regime, as it may not apply. However, if it does, the Parties may wish to consider the following elements of the regime.

The 'golden thread of information'

The Building (Higher-Risk Buildings Procedures) (England) Regulations 2023 (made pursuant to the Building Safety Act 2022) requires dutyholders to contribute to the 'golden thread of information' during the design and construction phase of the project.

This 'golden thread of information' relates to the creation, collation, sharing, and maintenance of all related documents and information from the initial design stages, through to the construction phase and eventually the occupation phase.

The aim is to allow all relevant information to be stored, managed, and updated, such that it can be accessed when necessary.

The information must also be stored, and be capable of being transferred, in an electronic format (for example, by email or a document management platform).

The following provisions of the Contract may therefore be relevant in this regard:

- clause 2.37 (As-built Drawings);
- clause 2.27 (Practical completion);

- clause 2.7 (Contract Documents); and
- clause 2.8 (Construction information).

The gateway regime

Section 33 of the Building Safety 2022 established a new 'gateway' regime for all higher-risk buildings, for the relevant dutyholders to demonstrate to the (newly established) Building Safety Regulator that they have sufficiently accounted for safety during the design, construction, and occupation phases. There are three gateways in the regime:

- **Gateway one** (introduced by the Town and Country Planning (Development Management Procedure and Section 62A Applications) (England) (Amendment) Order 2021, under the Town and Country Planning Act 1990) relates to the planning phase, where those seeking planning permission will have to demonstrate that the relevant fire safety issues have been considered and incorporated into the proposed scheme.
- **Gateway two** takes place during the pre-construction phase, and requires the dutyholders to first obtain building control approval from the Building Safety Regulator before building work can commence.
- **Gateway three** takes place at the completion of the project, when the Building Safety Regulator will assess whether the project has been properly completed in accordance with the Building Safety Act 2022 and other relevant building safety regulations. Until the Building Safety Regulator has issued the relevant completion (or partial completion) certificate, and the building has been registered with the Building Safety Regulator, the building cannot be occupied.

Whilst gateways one and two may have been completed before the Parties enter into the Contract, they may wish to give further consideration to the allocation of responsibility in respect of gateway three, including:

- liaising with the Building Safety Regulator throughout the course of the project, and obtaining their approval;
- any amendments to the requirements for practical completion, particularly in respect of the requirement to obtain a completion certificate from the Building Safety Regulator, and to register the building with the Building Safety Regulator; and
- the liability for any delays associated with obtaining the Building Safety Regulator's approval.

The following provisions of the Contract may therefore be relevant in this regard:

- clause 2.3 (Date of Possession – progress);
- clause 2.4 (Deferment of possession);
- clause 2.37 (As-built Drawings);
- clause 2.27 (Practical completion);
- clause 2.26 (Relevant Events); and
- clause 4.21 (Relevant Matters).

Part 1
Commentary

Overview

A comprehensive working knowledge of the Contract is central to its effective administration.

Part 1 therefore provides brief commentary on all the key provisions in the Contract. It adopts the same structure as the Contract, and includes both:

- the Agreement, which comprises the Recitals, the Articles, and the Contract Particulars, along with the Attestation section; and
- the Conditions, which include the clauses set out in section 1 to section 9, along with the six Schedules.

Contract Documents

The Contract is divided into a number of key documents, described as the '*Contract Documents*'.

The Contract Documents are defined in clause 1.1 (Definitions) as including the following:

- the Agreement and the Conditions; together with
- the Employer's Requirements;
- the Contractor's Proposals;
- the Contract Sum Analysis; and
- (where applicable) the BIM Protocol.

Each will be considered briefly in turn below.

The Agreement

The '*Agreement*' is divided into the following further parts (again as defined by clause 1.1):

- the Recitals;
- the Articles; and
- the Contract Particulars.

The Attestation section is included at the end of the Agreement, where the Parties execute the Contract, whether 'under hand' (as a 'simple' contract), or as a deed.

The Conditions

The '*Conditions*' are annexed to the Agreement.

DOI: 10.1201/9781003485919-3

They are divided into the following further parts (again as defined by clause 1.1):

- the clauses set out in section 1 to section 9, namely:
  - section 1: Definitions and Interpretation;
  - section 2: Carrying out the Works;
  - section 3: Control of the Works;
  - section 4: Payment;
  - section 5: Changes;
  - section 6: Injury, Damage and Insurance;
  - section 7: Assignment, Performance Bonds and Guarantees, Third Party Rights and Collateral Warranties;
  - section 8: Termination; and
  - section 9: Settlement of Disputes,
- together with and including the Schedules, namely:
  - Schedule 1: Design Submission Procedure;
  - Schedule 2: Supplemental Provisions (further divided into Part 1 and Part 2);
  - Schedule 3: Insurance Options;
  - Schedule 4: Code of Practice;
  - Schedule 5: Third Party Rights (further divided into Part 1 (Third Party Rights for Purchasers and Tenants) and Part 2 (Third Party Rights for a Funder)); and
  - Schedule 6: Forms of Bonds (further divided into Part 1 (Advance Payment Bond), Part 2 (Bond in respect of payment for off-site materials and/or goods), and Part 3 (Retention Bond)).

The Employer's Requirements

The Employer's Requirements are described in the First Recital, and they are identified in the Contract Particulars (against the reference to Article 5).

The Employer's Requirements are the documents which show, describe, or otherwise state the Employer's requirements for the design and construction of the Works. They will therefore include design information, typically in the form of drawings, specifications, schedules, reports, and surveys, along with any relevant pre-construction information and further requirements the Employer may have.

The Contractor's Proposals

The Contractor's Proposals are described in the Second Recital, and they are identified in the Contract Particulars (against the reference to Article 5).

The Contractor's Proposals are provided by the Contractor in response to the Employer's Requirements, and are the documents which show and describe the Contractor's proposals for the design and construction of the Works.

The Contract Sum Analysis

The Contract Sum Analysis is also described in the Second Recital, and is identified in the Contract Particulars (against the reference to Article 5).

Similar to the Contractor's Proposals, the Contract Sum Analysis is provided by the Contractor in response to the Employer's Requirements, and includes an analysis or breakdown of the Contract Sum.

The BIM Protocol

Very briefly, building information modelling, or 'BIM', describes a collaborative process to assist in the design and construction of the Works in a coordinated manner, where the relevant parties share the same information, in the same model or format.

Where the Parties choose to adopt a BIM Protocol, its identity will be specified in the Contract Particulars (against the reference to clause 1.1).

The Agreement

Overview

The Agreement comprises the Recitals, the Articles, and the Contract Particulars, along with the Attestation section, each of which are considered below.

Key provisions

The key provisions in the Agreement are as follows:

- the date of the Agreement;
- the identity and details of the Contractor and the Employer;
- the seven Recitals;
- the ten Articles;
- the Contract Particulars; and
- the Attestation section.

Date of the Agreement and details of the Parties

The Agreement begins with the date of the Agreement, which should only be inserted once the Contract Documents are in agreed form, and the Attestation section has been completed by both Parties.

This is followed by the identity and details of the Contractor and the Employer (which are both assumed throughout to be private limited companies, registered in England and Wales), including:

- their company names; and
- their associated company numbers and registered office addresses, both of which should be available on the Companies House website.[1]

[1] Available at: https://www.gov.uk/government/organisations/companies-house

DOI: 10.1201/9781003485919-4

The Recitals

Overview

The Recitals provide a brief background to the project, and describe the events leading up to the Parties entering into the Contract.

In summary:

- The Employer would like something designed and built (the 'Works'), and has shown the Contractor documents to that effect (the 'Employer's Requirements').
- In reply, the Contractor has shown the Employer how it would design and build the Works (the 'Contractor's Proposals'), and provided a breakdown of the cost for doing so (the 'Contract Sum Analysis').

Whilst this may not always accurately describe the tender stage in practice, it serves to introduce some of the key Contract Documents.

In completing the Recitals, the Parties will need to have regard to the following:

- the description of the Works, along with their location (First Recital);
- the preparation of the Employer's Requirements by the Employer (First Recital);
- the preparation of the Contractor's Proposals and the Contract Sum Analysis by the Contractor (Second Recital);
- whether the Employer considers that the Contractor's Proposals appear to meet the Employer's Requirements (Third Recital);
- whether the Employer is a 'contractor' for the purposes of the Construction Industry Scheme (CIS) (Fourth Recital);
- whether the Works are to be divided into Sections (Fifth Recital);
- whether there is a Framework Agreement between the Parties, and if so, whether they wish for the Contract to be supplemented by it (Sixth Recital); and
- which of the (optional) Supplemental Provision 1 to Supplemental Provision 7 are to apply (Seventh Recital).

Key provisions

- the First Recital, providing a description of the Works and introducing the Employer's Requirements
- the Second Recital, introducing the Contractor's Proposals and the Contract Sum Analysis
- the Third Recital, relating to the Employer's Requirements and the Contractor's Proposals

DOI: 10.1201/9781003485919-5

- the Fourth Recital, relating to the status of the Employer under the Construction Industry Scheme (CIS)
- the Fifth Recital, relating to the division of the Works into Sections
- the Sixth Recital, providing for a Framework Agreement
- the Seventh Recital, relating to the Supplemental Provisions

First Recital: Description of the Works and the Employer's Requirements

First Recital *the Employer wishes to have the design and construction of the following work carried out:*

___________ at ___________ ('the Works')

and the Employer has supplied to the Contractor documents showing and describing or otherwise stating its requirements ('the Employer's Requirements');

The First Recital includes the definition of both '*the Works*' and '*the Employer's Requirements*':

- it provides a brief description of the design and construction work the Employer would like the Contractor to carry out, along with their location (the Works); and
- it states that the Employer has provided the Contractor with documents setting out its requirements in that regard (the Employer's Requirements).

Second Recital: The Contractor's Proposals and the Contract Sum Analysis

Second Recital *in response to the Employer's Requirements the Contractor has supplied to the Employer:*

- *documents showing and describing the Contractor's proposals for the design and construction of the Works ('the Contractor's Proposals'); and*
- *an analysis of the Contract Sum ('the Contract Sum Analysis');*

The Second Recital includes the definition of '*the Contractor's Proposals*' and '*the Contract Sum Analysis*', both of which are provided by the Contractor to the Employer, in response to the Employer's Requirements:

- the Contractor's Proposals are the documents setting out the Contractor's strategy or plan for carrying out the design and construction of the Works, in order to meet the Employer's Requirements; and
- the Contract Sum Analysis is a breakdown or itemisation of the Contract Sum (as defined in Article 2 (Contract Sum)), which the Employer will pay to the Contractor, in exchange for the Contractor carrying out and completing the design and construction of the Works.

Third Recital: The Employer's Requirements and the Contractor's Proposals

Third Recital	*the Employer has examined the Contractor's Proposals and, subject to the Conditions, is satisfied that they appear to meet the Employer's Requirements;*

The Third Recital continues the narrative set out in the First Recital and Second Recital, confirming that, subject to the Conditions (considered below), the Employer is content that the Contractor's Proposals seem to meet the Employer's Requirements.

If it subsequently transpires that the Contractor's Proposals do not, in fact, meet the Employer's Requirements in any way, the Contract contains a number of mechanisms for resolving any such inadequacies, discrepancies, or divergences, set out in clause 2.10 to clause 2.16 (Discrepancies and Divergences).

Recommendation: Divergences between the Employer's Requirements and Contractor's Proposals

Footnote [3] of the Contract states: '*Where the Employer has accepted a divergence from its requirements in the proposals submitted by the Contractor, the divergence should be removed by amending the Employer's Requirements before the Contract is executed.*'

This is helpful advice, and if followed, means that the Parties will not subsequently need to resolve any such divergence using the clauses listed above, during the course of carrying out the design and construction of the Works.

More generally, both Parties are strongly encouraged to undertake a comprehensive review of all documents forming part of the Employer's Requirements and Contractor's Proposals prior to executing the Contract, in order to ensure that, as far as possible, any inadequacies, discrepancies or divergences are identified and removed.

This should be viewed as a worthwhile investment, which creates certainty, and reduces the risks of associated delays and additional costs at a later stage.

Market practice: Responsibility for the design in the Employer's Requirements

The Third Recital is often amended to the effect that, instead of the Employer examining the Contractor's Proposals, the Contractor instead examines the Employer's Requirements, and agrees to accept responsibility for the design contained within them. (Several other related provisions are also consequently amended, to reflect the same position.)

This represents a very significant transfer of risk from the Employer to the Contractor, particularly in circumstances where the Contractor may have had limited involvement in the creation of the Employer's Requirements, or only a brief tender stage in which to examine them.

Fourth Recital: The Construction Industry Scheme (CIS)

Fourth Recital *for the purposes of the Construction Industry Scheme (CIS) under the Finance Act 2004, the status of the Employer is, as at the Base Date, that stated in the Contract Particulars;*

The Fourth Recital departs from the narrative set out the previous three Recitals, and is primarily for tax purposes.

It relates to the Construction Industry Scheme (abbreviated to 'CIS'), which was introduced by the Finance Act 2004, and is described by the UK Government as follows:

'Under the Construction Industry Scheme (CIS), contractors deduct money from a subcontractor's payments and pass it to HM Revenue and Customs (HMRC). The deductions count as advance payments towards the subcontractor's tax and National Insurance.'[2]

The Contract Particulars (considered below) will confirm whether or not the Employer is a '*contractor*' for the purposes of the CIS, as at the Base Date (which is also stated in the Contract Particulars).

Both Parties are encouraged to consult with a tax adviser to assist with matters relating to the CIS.

Fifth Recital: Division of the Works into Sections

Fifth Recital *the division of the Works into Sections is shown in the Employer's Requirements or in such other documents as are identified in the Contract Particulars;*

The Fifth Recital provides for the possibility of dividing the Works into Sections, each with its own Date of Possession, Date for Completion, rate of liquidated damages, etc.

[2] Available at: https://www.gov.uk/what-is-the-construction-industry-scheme

Footnote [4] of the Contract advises that, if the Works are *not* divided into Sections, the Fifth Recital should be deleted.

Recommendation: Showing the division of the Works into Sections

If the Works are divided into Sections, this division should be clearly identified in documents contained within the Employer's Requirements (or elsewhere, if the Parties agree). These will often take the form of suitably annotated drawings.

For the purposes of giving possession (clause 2.3 (Date of Possession – progress)), and determining whether practical completion has been achieved (clause 2.27 (Practical Completion)), it is important to establish where one Section ends, and another begins.

Sixth Recital: Framework Agreement

Sixth Recital	*where so stated in the Contract Particulars, this Contract is supplemented by the Framework Agreement identified in those particulars;*

The Sixth Recital provides for the possibility of a related Framework Agreement between the Parties, which may be in place in circumstances where the Parties are working together on a number of projects under separate building contracts, or otherwise have a well-established commercial relationship.

It provides that the Contract is '*supplemented*' by the Framework Agreement identified in the Contract Particulars (against the reference to the Sixth Recital).

However, in accordance with clause 1.3 (Agreement etc. to be read as a whole), considered below, nothing contained in any such Framework Agreement, irrespective of its terms, will override or modify the Agreement or the Conditions.

Seventh Recital: Supplemental Provisions

Seventh Recital	*whether any of Supplemental Provisions 1 to 7 apply is stated in the Contract Particulars;*

The Seventh Recital relates to the first seven Supplemental Provisions (there are nine in total), which are included in Schedule 2 (Supplemental Provisions), considered below.

It confirms only that the Contract Particulars will state whether any of these seven Supplemental Provisions apply, as they are optional.

The Articles

Overview

The Articles set out some of the key terms of the Contract, including:

- the Contractor's obligation to carry out and complete the design and construction of the Works in accordance with the Contract Documents;
- the Employer's obligation to pay the Contract Sum to the Contractor; and
- the obligation on both Parties to work with each other in a co-operative and collaborative manner.

The Articles also introduce a number of further parties, including:

- the Employer's Agent;
- the Principal Designer and Principal Contractor, both for the purposes of the CDM Regulations; and
- the Principal Designer and Principal Contractor, both for the purposes of the Building Regulations.

Finally, the Articles provide a number of external methods of disputes resolution, namely adjudication, arbitration, and litigation.

In completing the Articles, the Parties will need to have regard to the following:

- the amount of the Contract Sum (Article 2);
- the appointment by the Employer of the Employer's Agent (Article 4);
- the identification of the Employer's Requirements, the Contractor's Proposals, and the Contract Sum Analysis (Article 5);
- the appointment by the Employer of the Principal Designer and the Principal Contractor, for the purposes of the CDM Regulations (Article 6);
- the appointment by the Employer of the Principal Designer and the Principal Contractor, for the purposes of the Building Regulations (Article 7); and
- whether or not Article 9 (Arbitration) is to apply.

Key provisions

- Article 1 (Contractor's obligations)
- Article 2 (Contract Sum)
- Article 3 (Collaborative working)
- Article 4 (Employer's Agent)
- Article 5 (Employer's Requirements and Contractor's Proposals)

DOI: 10.1201/9781003485919-6

- Article 6 (CDM Regulations – Principal Designer and Principal Contractor)
- Article 7 (Building Regulations – Principal Designer and Principal Contractor)
- Article 8 (Adjudication)
- Article 9 (Arbitration)
- Article 10 (Legal proceedings)

Article 1: Contractor's obligations

Article 1 ***Contractor's obligations***

The Contractor shall complete the design for the Works and carry out and complete the construction of the Works in accordance with the Contract Documents.

Article 1 sets out two of the Contractor's key obligations under the Contract, namely to design and build.

The Contractor must:

- complete the design for the Works; and
- carry out and complete the construction of the Works,

in both cases, in accordance with the Contract Documents.

Article 2: Contract Sum

Article 2 ***Contract Sum***

The Employer shall pay the Contractor at the times and in the manner specified in the Conditions the VAT-exclusive sum of

____________ (£____________ ('the Contract Sum') or such other sum as becomes payable under this Contract.

Article 2 sets out one of the Employer's key obligations under the Contract, namely to pay the Contractor.

In particular, the Employer is required to pay the Contract Sum, or such other sum which becomes payable under the Contract.

This is to reflect the fact that several provisions within the Contract provide for an adjustment to the Contract Sum (considered further in section 4 (Payment)).

The Contract Sum is expressed exclusive of VAT, so it is therefore sensible for the Parties to adopt the same approach for any related sums or adjustments discussed over the course of the project.

It is also expressed in both words and figures, to reduce uncertainty.

Article 3: Collaborative working

Article 3 ***Collaborative working***

The Parties shall work with each other and with other project team members in a co-operative and collaborative manner, in good faith and in a spirit of trust and respect. To that end, each shall support collaborative behaviour and address behaviour which is not collaborative.

Article 3 is a new Article, and was formerly included within (the optional) Supplemental Provision 5 of the 2016 edition of the Contract.

The JCT have now made this provision mandatory, and have given it increased prominence, to reflect the aims of the UK Government's Construction Playbook, which encourages the development of '*long-term, strategic collaborative relationships*'.[3]

Whilst the majority of the Article is largely self-explanatory, the Court of Appeal has provided some helpful guidance on the concept of '*good faith*'.[4]

Very briefly, and as one might expect, there is a core duty on the Parties to act honestly.

However, depending on the context, the requirement to act in good faith may go further, and may prohibit conduct that '*reasonable and honest*' people would regard as '*commercially unacceptable*', albeit not necessarily dishonest.

In simple terms, therefore, the Parties should adopt a sense of 'fair play' in all their dealings with one another.

Article 4: Employer's Agent

Article 4 ***Employer's Agent***

For the purposes of this Contract the Employer's Agent is ____________ of ____________ or such other person as the Employer nominates in its place. Save to the extent that the Employer may otherwise specify by notice to the Contractor, the Employer's Agent shall have full authority to receive and issue applications, consents, instructions, notices, requests or statements and otherwise to act for the Employer under any of the Conditions.

The Contract assumes that the Employer will appoint an agent (defined as the '*Employer's Agent*'), to help them administer the Contract.

Article 4 introduces the Employer's Agent, who has very considerable authority to act on the Employer's behalf in this regard: unless the Employer notifies

[3] Available at: https://www.gov.uk/government/publications/the-construction-playbook

[4] *Re Compound Photonics Group Ltd [2022] EWCA Civ 1371*

otherwise, the default position is that the Employer's Agent may issue and receive applications, consents, instructions, notices, requests, or statements, and more generally, can act for the Employer under any of the Conditions.

In practical terms, therefore, the relationship between the Contractor and the Employer's Agent is often of equal importance to the relationship between the Contractor and the Employer.

In the interests of both collaboration and certainty, it is recommended to copy all three parties into all correspondence, throughout the course of the project.

If it wishes, the Employer may also nominate a replacement Employer's Agent.

Resource 1 page 408	Correspondence: Employer to Contractor	The nomination by the Employer of a replacement Employer's Agent.
Resource 2 page 408	Notice: Employer to Contractor	A revision by the Employer to the Employer's Agent's authority to act on its behalf.

Article 5: Employer's Requirements and Contractor's Proposals

Article 5 ***Employer's Requirements and Contractor's Proposals***

The Employer's Requirements, the Contractor's Proposals and the Contract Sum Analysis are those referred to in the Contract Particulars.

Article 5 simply identifies three of the key Contract Documents considered above, namely the Employer's Requirements, the Contractor's Proposals, and the Contract Sum Analysis.

In each case, they are as referred to against the reference to Article 5 in the Contract Particulars.

Recommendation: Negotiating, revising, and checking the Contract Documents

The Contract Documents (and the Employer's Requirements and Contractor's Proposals in particular) are composed of many other documents, often written by multiple authors.

They are also often subject to extensive negotiation and revision throughout the tender stage, right up until the Contract is executed. Particularly in the case of larger, more complex projects, the terms of the Contract Documents can sometimes be negotiated in a rather disjointed fashion, whereby different representatives of the Parties (lawyers, quantity surveyors, project managers, etc.) will negotiate different documents in isolation.

This approach, whilst often unavoidable, greatly increases the risk of inconsistencies, conflicts, and discrepancies within the Contract Documents.

It is therefore recommended that the Parties leave sufficient time at the end of the negotiation to undertake a comprehensive review of the draft Contract Documents in their entirety, and thereafter to ensure that the final agreed versions of the Employer's Requirements, Contractor's Proposals, and Contract Sum Analysis are accurately identified against the reference to Article 5 in the Contract Particulars.

Article 6: CDM Regulations – Principal Designer and Principal Contractor

Article 6 ***CDM Regulations – Principal Designer and Principal Contractor***

For the purposes of the CDM Regulations:

the Principal Designer is the Contractor

(or) ____________ *of* ____________

or such replacement as the Employer at any time appoints to fulfil that role;

the Principal Contractor is the Contractor

(or) ____________ *of* ____________

or such replacement as the Employer at any time appoints to fulfil that role.

Article 6 introduces the Principal Designer and the Principal Contractor, both for the purposes of the CDM Regulations.

Either role may be carried out by the Contractor, or by another party appointed by the Employer.

The Parties' obligations in respect of the CDM Regulations are considered further below in clause 3.16 (CDM Regulations and Part 2A of the Building Regulations).

Further reading: Principal Designer and Principal Contractor under the CDM Regulations

The Health and Safety Executive has published further guidance on the role of Principal Designer and Principal Contractor for the purposes of the CDM Regulations.[5]

Article 7: Building Regulations – Principal Designer and Principal Contractor

Article 6 ***Building Regulations – Principal Designer and Principal Contractor***

For the purposes of the Building Regulations:

the Principal Designer is the Contractor

(or) __________ of __________

or such replacement as the Employer at any time appoints to fulfil that role;

the Principal Contractor is the Contractor

(or) __________ of __________

or such replacement as the Employer at any time appoints to fulfil that role.

Similar to Article 6 above, Article 7 introduces the Principal Designer and the Principal Contractor, but in this case, for the purposes of the Building Regulations.

This is a new Article, introduced to take account of the 'dutyholder' regime included in Part 2A of the Building Regulations,[6] which sets out a framework of duties and competence requirements for those involved in the procurement, design and carrying out of building work to which the Building Regulations apply.

Again, either role may be carried out by the Contractor, or by another party appointed by the Employer, and regulation 11D(2) of Part 2A of the Building Regulations provides that either role may be carried out by the Principal Designer and Principal Contractor for the purposes of the CDM Regulations.

---

[5] Available at: https://www.hse.gov.uk/construction/cdm/2015/index.htm

[6] Part 2A of the Building Regulations was introduced by the Building Regulations etc. (Amendment) (England) Regulations 2023, pursuant to the Building Safety Act 2022.

The Parties' obligations in respect of Part 2A of the Building Regulations are considered in clause 3.16 (CDM Regulations and Part 2A of the Building Regulations).

Further reading: Principal Designer and Principal Contractor under the Building Regulations

The Health and Safety Executive has published further guidance on the role of Principal Designer and Principal Contractor for the purposes of the Building Regulations.[7]

Article 8: Adjudication

Article 8 ***Adjudication***

If any dispute or difference arises under this Contract, either Party may refer it to adjudication in accordance with clause 9.3.

Article 8 gives the Parties a contractual right to refer any dispute or difference arising under the Contract to adjudication, in addition to the Parties' statutory right under the Housing Grants, Construction and Regeneration Act 1996.

This is considered further in clause 9.3 (Adjudication).

Article 9: Arbitration

Article 9 ***Arbitration***

Where Article 9 applies, then, subject to Article 8 and the exceptions set out below, any dispute or difference between the Parties of any kind whatsoever arising out of or in connection with this Contract shall be referred to arbitration in accordance with clauses 9.4 to 9.9 and the JCT 2024 edition of the Construction Industry Model Arbitration Rules (CIMAR). The exceptions to this Article 9 are:

- *any disputes or differences arising under or in respect of the Construction Industry Scheme or VAT, to the extent that legislation provides another method of resolving such disputes or differences; and*
- *any disputes or differences in connection with the enforcement of any decision of an Adjudicator.*

[7] Available at: https://www.gov.uk/guidance/design-and-building-work-meeting-building-requirements#principal-designers-competence-requirements

Where it applies, Article 9 entitles the Parties to refer any dispute or difference between them arising out of or in connection with the Contract to arbitration, in accordance with both clause 9.4 to clause 9.9 (Arbitration), and the JCT 2024 edition of the Construction Industry Model Arbitration Rules (CIMAR).

This is subject to the right of the Parties to refer any dispute or difference to adjudication, in accordance with Article 8 (Adjudication) above, and the following two exceptions:

- any disputes or differences relating to either the Construction Industry Scheme or to VAT, to the extent that the relevant legislation provides another method of dispute resolution (and in respect of which a tax adviser should be consulted); and
- any disputes or differences in connection with the enforcement of any Adjudicator's decision (relating to Article 8).

Arbitration is considered further below in relation to clause 9.4 to clause 9.9 (Arbitration).

Whether or not Article 9 and clause 9.4 to clause 9.9 apply is identified against the reference to Article 9 in the Contract Particulars. Litigation (described as '*legal proceedings*' under Article 10 below) is the default position, and will apply unless Article 9 and clause 9.4 to clause 9.9 are stated to apply.

Article 10: Legal proceedings

Article 10 ***Legal proceedings***

Subject to Article 8 and (where it applies) to Article 9, the English courts shall have jurisdiction over any dispute or difference between the Parties which arises out of or in connection with this Contract.

Article 10 confirms simply that the English courts have jurisdiction over any dispute or difference between the Parties, which arises out of or in connection with the Contract.

This is subject to the following:

- the right of the Parties to refer any dispute or difference to adjudication, in accordance with Article 8 (Adjudication); and
- where Article 9 (Arbitration) applies, the right of the Parties to refer any dispute or difference to arbitration.

The Contract Particulars

Overview

The Contract Particulars contain many of the project-specific commercial details agreed between the Parties, including key dates and amounts, across a wide range of matters.

Rather than include these throughout the Recitals, Articles, and Conditions, they are contained in one place, for convenience.

Given their commercial importance, each entry in the Contract Particulars should be carefully completed and checked, to ensure that they accurately reflect the agreement made between the Parties. This is particularly the case for those entries where the Contract provides for a default position.

This section therefore provides guidance on the completion of the Contract Particulars (including the default positions), whereas commentary on their related provisions is included elsewhere.

Furthermore, some of the provisions may require specialist advice. In particular:

- The Employer should consult a tax adviser in respect of the entry relating to clause 4.5 (Construction Industry Scheme (CIS)).
- Both Parties should consult an insurance adviser in respect of the entries relating to the following provisions:
  - clause 6.4.1 (Insurance against Personal Injury and Property Damage: Contractor's insurance of its liability);
  - clause 6.5.1 (Insurance against Personal Injury and Property Damage: Contractor's insurance of liability of Employer);
  - clause 6.7 (Insurance of the Works and Existing Structures: Insurance Options and period) and Schedule 3 (Insurance Options);
  - clause 6.10 (Terrorism Cover – policy extensions and premiums) and Schedule 3 (Insurance Options); and
  - clause 6.15 (Professional Indemnity Insurance: Obligation to insure).

As a preliminary point, a number of the entries below require the Parties to identify certain relevant documents. Footnote [8] of the Contract is relevant in respect of these entries, and states: '*If the relevant document or set of documents takes the form of an Annex to this Contract, it is sufficient to refer to that Annex.*'

DOI: 10.1201/9781003485919-7

Provision	Entry	Guidance
Fourth Recital and clause 4.5 (Construction Industry Scheme (CIS))	Construction Industry Scheme (CIS)	The Employer is required to state whether or not it is a '*contractor*', for the purposes of the CIS, at the Base Date. The Employer should consult a tax adviser in this regard.
Fifth Recital	Description of Sections (if any)	If the Works are divided into Sections, each Section must be described here, whether by written description, or by reference to the relevant documents within the Employer's Requirements (or elsewhere).
Sixth Recital	Framework Agreement (if applicable)	If there is a Framework Agreement, its date, title, and the parties to it must be inserted.
Seventh Recital and Part 1 of Schedule 2 (Supplemental Provisions)	Supplemental Provisions – Part 1	This relates to the following Supplemental Provisions: • Supplemental Provision 1 (Named Sub-Contractors); • Supplemental Provision 2 (Valuation of Changes – Contractor's estimates); and • Supplemental Provision 3 (Loss and expense – Contractor's estimates). The Parties are required to state whether or not each of them applies. **Default position** The corresponding note states: '*Where neither entry against one of Supplemental Provisions 1 to 3 below is deleted, that Supplemental Provision does not apply.*' In other words, the default position is that they do not apply. Therefore, if the Parties wish for any of these Supplemental Provisions to apply, they must delete '*does not apply*' against that entry.

(*Continued*)

Provision	Entry	Guidance
Seventh Recital and Part 2 of Schedule 2 (Supplemental Provisions)	Supplemental Provisions – Part 2	This relates to the following Supplemental Provisions: • Supplemental Provision 4 (Acceleration Quotation); • Supplemental Provision 5 (Health and safety); • Supplemental Provision 6 (Cost savings and value improvements); and • Supplemental Provision 7 (Performance Indicators and monitoring). As above, the Parties are required to state whether or not each of them applies. **Default position** The corresponding note states: '*Where neither entry against one of Supplemental Provisions 4 to 7 below is deleted, that Supplemental Provision applies.*' In other words, the default position is that they apply. Therefore, if the Parties wish for any of these Supplemental Provisions not to apply, they must delete '*applies*' against that entry. In respect of the remaining two Supplemental Provisions in Part 2 of Schedule 2, footnote [9] of the Contract states: '*Supplemental Provision 8 (Transparency) applies only where the Employer is a Local or Public Authority or other body to which the Freedom of Information Act 2000 applies; Supplemental Provision 9 (The Public Contracts Regulations 2015) applies only where the Employer is a Local or Public Authority and this Contract is subject to the PC Regulations.*'
Article 5 (Employer's Requirements and Contractor's Proposals)	Employer's Requirements, Contractor's Proposals, Contract Sum Analysis	The Parties are required to identify each of the Employer's Requirements, Contractor's Proposals and Contract Sum Analysis. The corresponding notes for each state: '*State reference numbers and dates or other identifiers of the relevant documents.*'

(*Continued*)

Provision	Entry	Guidance
		As stated above, these Contract Documents are often the subject of extensive negotiation and revision, and so it is important to ensure that the final agreed versions are correctly identified here.
Article 9 (Arbitration)	Arbitration	The Parties are required to state whether or not Article 9 and clause 9.4 to clause 9.9 (Arbitration) apply. **Default position** The corresponding note states: '*If neither entry is deleted, Article 9 and clauses 9.4 to 9.9 do not apply. If disputes and differences are to be determined by arbitration and not by legal proceedings, it must be stated that Article 9 and clauses 9.4 to 9.9 apply.*' In other words, the default position is that Article 9 and clause 9.4 to clause 9.9 do not apply. Therefore, if the Parties wish for those provisions to apply, they must delete '*do not apply*' against that entry.
clause 1.1 (Definitions)	Base Date	The Parties are required to insert the Base Date. The Base Date is referenced in the following provisions: • Fourth Recital, relating to the Construction Industry Scheme (CIS); • clause 1.1 (Definitions), definition of '*Joint Fire Code*'; • clause 2.15.2 (Divergences from Statutory Requirements); • clause 2.26.7 (Relevant Events); • clause 2.26.8 (Relevant Events); • clause 4.4.2 (VAT); • clause 4.21.6 (Relevant Matters); • clause 4.21.7 (Relevant Matters); • clause 5.5 (Daywork); • clause 6.20 (Joint Fire Code – amendments/revisions); and • clause 8.11.1 (Termination by either Party and regulation 73(1)(a) of the PC Regulations).

(*Continued*)

Provision	Entry	Guidance
		It serves as a fixed point in time for each of these provisions, most of which provide for a change in circumstances after that date. In practice, it is usually linked to the date of the Contractor's tender return.
	BIM Protocol (where applicable)	If there is a BIM Protocol, the Parties are required to identify it; the corresponding note states: *'State title, edition, date or other identifiers of the relevant documents.'*
	Date for Completion of the Works	If the Works are not divided into Sections, the Parties are required to insert the Date for Completion of the Works.
	Sections: Dates for Completion of Sections	If the Works are divided into Sections, the Parties are required to insert the Dates for Completion for each Section.
clause 1.7.3 (Notices and other communications)	Addresses for service of notices by the Parties	The Parties are each required to insert both their postal addresses and email addresses, for the purpose of service of notices and other communications. **Default position** The corresponding note states: *'If a Party's address is not stated, it shall, subject to clause 1.7.3, be that shown at the commencement of the Agreement.'*
clause 1.7.4.2 (Notices and other communications)	Service of notices by email	Certain notices must be given in accordance with clause 1.7.4.2 (considered further below). This entry provides for the option to serve these notices by email. If the Parties wish to avail of this option, they are required to insert their email addresses (which may be the same email addresses as those included against the reference to clause 1.7.3 above). **Default position** The corresponding note states: *'If neither entry is deleted or an email address for each Party is not specified, clause 1.7.4.2 shall not apply.'* In other words, if the Parties wish to serve these notices by email, they must delete '*does not apply*' and insert the relevant email addresses.

(*Continued*)

Provision	Entry	Guidance
clause 2.3 (Date of Possession – progress)	Date of Possession of the site	If the Works are not divided into Sections, the Parties are required to insert the Date of Possession of the site.
	Sections: Dates of Possession of Sections	If the Works are divided into Sections, the Parties are required to insert the Dates of Possession for each Section.
clause 2.4 (Deferment of possession)	Deferment of possession of the site	If the Works are not divided into Sections, the Parties are required to state whether or not clause 2.4 applies, which would entitle the Employer to defer the giving of possession of the site for a period not exceeding 6 weeks. If the Parties have agreed a maximum period of deferment which is less than 6 weeks, it must be inserted.
	Sections: deferment of possession of Sections	If the Works are divided into Sections, the Parties are required to state whether or not clause 2.4 (Deferment of possession) applies, which would entitle the Employer to defer the giving of possession of each Section for a period not exceeding 6 weeks. Again, if the Parties have agreed a maximum period of deferment which is less than 6 weeks for any Section, it must be inserted.
clause 2.17.3 (Design Work – liabilities and limitation)	Limit of Contractor's liability for loss of use etc. (if any)	If the Parties have agreed a limit of the Contractor's liability for loss of use etc., in accordance with clause 2.17.3, the monetary amount must be inserted.
clause 2.29.2 (Payment or allowance of liquidated damages)	Liquidated damages	If the Works are not divided into Sections, the rate of liquidated damages must be inserted.
	Sections: rate of liquidated damages for each Section	If the Works are divided into Sections, the rates of liquidated damages for each Section must be inserted.
clause 2.34 (Partial Possession by Employer: Liquidated damages – Relevant Part)	Sections: Section Sums	If the Works are divided into Sections, the values for each of the Section Sums (which, added together, will equal the Contract Sum) must be inserted.

(*Continued*)

Provision	Entry	Guidance
clause 2.35 (Schedules of defects and instructions)	Rectification Period	If the Works are not divided into Sections, the duration of the Rectification Period, starting from the date of practical completion of the Works, must be inserted here. **Default position** The corresponding note states: *'If no other period is stated, the period is 6 months.'*
	Sections: Rectification Periods	If the Works are divided into Sections, the duration of the Rectification Periods, starting from the date of practical completion of each Section, must be inserted here. **Default position** Again, the corresponding note states: *'If no other period is stated, the period is 6 months.'*
clause 4.2 (Contract Sum and Adjustments: Items included in adjustments), clause 4.12 (Gross Valuation – Alternative A), and clause 4.13 (Gross Valuation – Alternative B)	Fluctuations Provision	If the Parties wish for one (or none) of the JCT Fluctuations Options to apply, they are required to select the relevant option. **Default position** The corresponding note states: *'Unless another option or entry is selected, JCT Fluctuations Option A applies.'* In other words, the default position is JCT Fluctuations Option A. Footnote [13] of the Contract further states: *'Unless the Fluctuations Provision is to be JCT Fluctuations Option A, delete all but one of the asterisked choices. The printed text of the JCT Fluctuations Options is no longer included in JCT contract documents but continues to be available on the JCT website www.jctltd.co.uk…'* Footnote [14] of the Contract further states: *'JCT Fluctuations Option C can only operate if a schedule to which rule 11b of the JCT Formula Rules refers is included in the Contract Documents.'*

(*Continued*)

Provision	Entry	Guidance
		JCT Fluctuations Option A (paragraph A.12) or JCT Fluctuations Option B (paragraph B.13) If either JCT Fluctuations Options A or JCT Fluctuations Option B apply, the Parties are required to insert the percentage addition, in relation to paragraph A.12 or paragraph B.13 respectively. JCT Fluctuations Option C (paragraph C.1.2) – JCT Formula Rules If JCT Fluctuations Option C applies, the Parties required to insert the following: • the Base Month, in relation of Rule 3; • the Non-Adjustable Element, also in relation of Rule 3; and • whether Part I or Part II of section 2 of the JCT Formula Rules applies, in relation to Rule 10 and Rule 30(i). **Default position** The corresponding note states: '*Unless Part II is stated to apply, Part I applies.*' In other words, Part I is the default position. Footnote [15] of the Contract further states: '*The Part to be deleted depends upon which method of formula adjustment (Part I – Work Category Method or Part II – Work Group Method) is applicable.*'
clause 4.6 (Advance payment)	Advance payment	If clause 4.6 applies, and the Parties have agreed for an advance payment to be paid to the Contractor (which is not applicable where the Employer is a Local or Public Authority), the following must be inserted: • the amount of the advance payment, whether as a monetary figure, or as a percentage of the Contract Sum; • the date on which it will be paid to the Contractor; and • the manner in which it will be reimbursed to the Employer. Advance Payment Bond If an Advance Payment Bond is required, this must be stated.

(*Continued*)

Provision	Entry	Guidance
		Default position The corresponding note states: *'Where an advance payment is to be made, an advance payment bond is required unless stated that it is not required.'*
clause 4.7.1 (Interim Payments – Contractor's Interim Payment Applications, due dates and Payment Notices)	Method of payment – alternatives	The Parties must identify which of the two alternative methods of payment they have agreed: • Alternative A (Stage Payments); or • Alternative B (Periodic Payments). Footnote [17] of the Contract states: *'Delete whichever Alternative is not applicable. Where Interim Payments are to be made by stages (including by quantity of units and sub-units completed) make the appropriate entries or prepare and insert a separate schedule of cumulative stage values.'* **Default position** The corresponding note states: *'If no Alternative is selected, Alternative B applies.'* In other words, Alternative B is the default. If Alternative A (Stage Payments) is selected, the Parties are required to either: • identify the document in which the stages are set out; or • insert a brief description of the stages, along with their cumulative value (with the cumulative value of the final stage equalling the Contract Sum).
clause 4.7.2 (Interim Payments – Contractor's Interim Payment Applications, due dates and Payment Notices)	Interim Payments – Interim Valuation Dates	The Parties are required to insert the date of the first Interim Valuation Date, which Footnote [19] of the Contract states should not be more than one month after the Date of Possession. This applies for both Alternative A (Stage Payments) and Alternative B (Periodic Payments). Subsequent Interim Valuation Dates will take place on the same date in each month, or the *nearest* (not 'next') Business Day in that month.

(*Continued*)

Provision	Entry	Guidance
		Default position The corresponding note states: '*...if no date is stated, the first Interim Valuation Date is one month after the Date of Possession.*'
clause 4.15.4 (Listed Items)	Listed Items – uniquely identified	If the Parties have agreed that the Contractor must provide a bond for payment in respect of uniquely identified Listed Items, the amount of that bond must be inserted here.
clause 4.15.5 (Listed Items)	Listed Items – not uniquely identified	If clause 4.15.5 applies, to the effect that the Contractor must provide a bond for payment in respect of Listed Items which are not uniquely identified, the amount of that bond must be inserted here.
clause 4.17 (Retention Bond)	Contractor's Retention Bond	If clause 4.17 applies, to the effect that the Contractor will provide a Retention Bond, the following particulars must be inserted: • the maximum aggregate sum (for the purposes of clause 2 of the bond); and • the expiry date (for the purposes of clause 6.3 of the bond). The clause references relate to the form of Retention Bond included in Part 3 of Schedule 6 (Forms of Bonds). **Default position** The corresponding note states: '*Not applicable where the Employer is a Local or Public Authority and, in other cases, not applicable unless stated to apply, with relevant particulars given below*'.
clause 4.18.1 (Retention – amounts and periods)	Retention Percentage	The Parties are required to insert the Retention Percentage. **Default position** The corresponding note states: '*The percentage is 3 per cent unless a different rate is stated; if no retention is required, insert 'Nil' or '0'.*'

(*Continued*)

Provision	Entry	Guidance
clause 4.21.6 and clause 4.21.7 (Relevant Matters)	Relevant Matters	The Parties are required to state whether or not the following Relevant Matters apply: • clause 4.21.6 (relating to the effects of an epidemic on the execution of the Works); and • clause 4.21.7 (relating to legislation, the exercise of statutory powers, and the publication of guidance) **Default position** The corresponding note states: *'In the case of each clause referred to below, where neither entry against the clause is deleted, that clause does not apply.'* In other words, the default position is that neither clause applies.
clause 5.5 (The Valuation Rules: Daywork)	Daywork	The Parties are required to identify the document which sets out the Percentage Additions to each section of the prime cost or, if they apply in respect of labour, the All-Inclusive Rates.
clause 6.4.1 (Insurance against Personal Injury and Property Damage: Contractor's insurance of its liability)	Contractor's Public Liability insurance: injury to persons or property	The Parties must insert the *minimum* level of cover required. (The Contractor may also insure for a greater sum, if it wishes.) The type of cover is stated as being '... *for any one occurrence or series of occurrences arising out of one event*'. The Parties should consult their respective insurance advisers when completing this entry.
clause 6.5.1 (Insurance against Personal Injury and Property Damage: Contractor's insurance of liability of Employer)	Insurance – liability of Employer	If the Employer's Requirements states that this insurance is required, the Parties must insert the minimum amount of indemnity required. The type of cover is stated as being '... *for any one occurrence or series of occurrences arising out of one event*'. Footnote [20] of the Contract states: '... *If the indemnity is to be for an aggregate amount and not for any one occurrence or series of occurrences, the entry should be amended to make this clear.'* The Parties should consult their respective insurance advisers when completing this entry.

(*Continued*)

Provision	Entry	Guidance
clause 6.7 (Insurance of the Works and Existing Structures: Insurance Options and period) and Schedule 3 (Insurance Options)	Works insurance – Insurance Option applicable	The Parties are required to identify which Insurance Option is applicable: • Insurance Option A (New Buildings – All Risks Insurance of the Works by the Contractor); • Insurance Option B (New Buildings – All Risks Insurance of the Works by the Employer); or • Insurance Option C (Joint Names Insurance by the Employer of Existing Structures and Works in or Extensions to them). The Parties must also state the percentage to cover professional fees. **Default position** The corresponding note states: '*If no other percentage is stated, it shall be 15 per cent.*' <u>Insurance Option A</u> If Insurance Option A applies, and cover is to be provided under the Contractor's annual policy (in accordance with paragraph A.2 of Schedule 3), the Contractor is required to insert the annual renewal date of that policy. <u>Insurance Option C</u> If Insurance Option C applies, the Parties must state either: • that paragraph C.1 of Schedule 3 applies; or • that paragraph C.1 of Schedule 3 is replaced by the C.1 Replacement Schedule, and identify that document. **Default position** The corresponding note states: '*Unless otherwise stated, paragraph C.1 applies.*'

(*Continued*)

Provision	Entry	Guidance
		Footnote [21] of the Contract states: *'Insurance Option C is for use in the case of alterations of or extensions to Existing Structures. Under that option, the Employer is required to take out a Joint Names Policy for All Risks Insurance for the Works and also, if paragraph C.1 applies, a Joint Names Policy to insure the Existing Structures and their contents owned by it or for which it is responsible against loss or damage by the Specified Perils…'* The Parties should consult their respective insurance advisers when completing this entry.
clause 6.10 (Terrorism Cover – policy extensions and premiums) and Schedule 3 (Insurance Options)	Terrorism Cover – details of the required cover	The Parties must identify the document containing details of the required Terrorism Cover. **Default position** The corresponding note states: *'Unless otherwise stated, Pool Re Cover is required.'* Footnote [22] of the Contract further states: *'Obtaining Terrorism Cover for the Works, which unless otherwise agreed is necessary in order to comply with the requirements of Insurance Option A, B or C, will involve an additional premium and in certain situations has been difficult to effect...'* The Parties should consult their respective insurance advisers when completing this entry.
clause 6.15 (Professional Indemnity Insurance: Obligation to insure)	Professional Indemnity insurance	The Parties are required to state whether the amount of indemnity required is either: • for any one claim or series of claims arising out of one event; or • the aggregate amount for any one period of insurance. **Default position** The corresponding note states: *'If an alternative is not selected the amount shall be the aggregate amount for any one period of insurance. A period of insurance for these purposes shall be one year unless otherwise stated.'*

(*Continued*)

Provision	Entry	Guidance
		The Parties are then required to state the monetary amount of indemnity required. **Default position** The corresponding note states: *'If no amount is stated, insurance under clause 6.15 shall not be required.'* The Parties must then state: • any sub-limits within the overall level of cover; • any specific exclusions listed in the relevant schedules (or other policy documents) to the relevant policy; and • the expiry of the required period. **Default position** The corresponding note states: *'If no period is selected, the expiry date shall be 6 years from the date of practical completion of the Works.'* The Parties should consult their respective insurance advisers when completing this entry.
clause 6.17 (Joint Fire Code – compliance: Application of clauses)	The Joint Fire Code	The Parties are required to state whether or not the Joint Fire Code applies. If it applies, the Parties are then required to also state whether the insurer under either Insurance Option A, Insurance Option B, or Insurance Option C (paragraph C.2) has specified that the Works are a 'Large Project'. Footnote [24] of the Contract further states: *'Where Insurance Option A applies these entries are made on information supplied by the Contractor.'*
clause 6.20 (Joint Fire Code – amendments/ revisions)	Joint Fire Code – amendments/ revisions	The Parties are required to state whether the cost, if any, of compliance with any amendments or revisions to the Joint Fire Code will be borne by either the Employer or the Contractor. **Default position** The corresponding note states: *'The cost shall be borne by the Contractor unless otherwise stated.'*

(*Continued*)

Provision	Entry	Guidance
clause 7.2 (Assignment: Rights of enforcement)	Assignment/ grant by Employer of rights under clause 7.2	Whether or not the Works are divided into Sections, the Parties are first required to state whether clause 7.2 applies. **Default position** The corresponding note states: *'If neither entry is deleted, clause 7.2 applies.'* In other words, the default position is that clause 7.2 applies.
	Sections: rights under clause 7.2	If the Works are divided into Sections, the Parties are then required to state whether rights under clause 7.2 apply to each Section. **Default position** The corresponding note states: *'If clause 7.2 applies, amend the entry if rights under that clause are to apply to certain Sections only.'* In other words, the default position is that, if clause 7.2 applies, it will apply to all Sections.
clause 7.3.1 (Performance Bonds and Guarantees)	Performance bond or guarantee from bank or other approved surety	The Parties must state whether or not a performance bond or guarantee from a bank or other approved surety is required. **Default position** The corresponding note states: *'If this entry is not completed or the required form is not specified, a performance bond or guarantee is not required.'* If a performance bond or guarantee is required, then the Parties are also required to state: • the identity of the document setting out the required form of the bond or guarantee; • the initial value of the bond or guarantee, expressed as a percentage of the Contract Sum; • the period of validity (if it is not already specified in the required form); and

(Continued)

Provision	Entry	Guidance
		Default position The corresponding note states: '*If no entry is selected, the date shall be the date of practical completion of the Works.*' • any reduction in value (if it is not already specified in the required form), if the bond or guarantee expires later than the date of practical completion of the Works, expressed as a percentage reduction in the initial value. **Default position** The corresponding note states: '*If no other percentage is stated, it shall be 50 per cent.*' Footnote [25] of the Contract further states: '*If a performance bond is required, the identity of the issuer as well as the operative terms of the bond should be agreed prior to execution of the Contract.*'
clause 7.3.2 (Performance Bonds and Guarantees)	Guarantee from the Contractor's parent company	The Parties are required to state whether or not a guarantee from the Contractor's parent company is required. If a guarantee from the Contractor's parent company is required, then the Parties are also required to state: • the parent company's name and company registration number; and • the identity of the document setting out the required form of the guarantee.
clause 7.4 (Clauses 7.7 to 7.11 – Preliminary: Rights Particulars)	Third Party Rights and Collateral Warranties	The Parties are required to identify the Rights Particulars document, which provides details of the requirements for the grant by the Contractor and sub-contractors of P&T Rights, Funder Rights, and/or (in the case of sub-contractors) Employer Rights in respect of the Works, either as third party rights or by collateral warranties.

(*Continued*)

Provision	Entry	Guidance
		Footnote [26] of the Contract further states: *'The relevant Rights Particulars should identify the beneficiaries (by name, class or description) and the sub-contractors which are also required to grant rights, specify whether rights are to be granted at each level as Third Party Rights or by way of Collateral Warranties, state in those cases where the default provision is not to apply which alternative provision is to apply in its place and give any other details required to complete the terms of the rights or warranties that are to be given...'*
clause 8.9.2 (Termination by Contractor: Default by Employer)	Period of suspension	The Parties are required to insert the period of suspension, in respect of clause 8.9.2. **Default position** The corresponding note states: *'If none is stated, the period is 2 months.'*
clause 8.11.1.1 to clause 8.11.1.7 (Termination by either Party and regulation 73(1)(a) of the PC Regulations)	Period of suspension	The Parties are required insert the period of suspension, in respect of clause 8.11.1.1 to clause 8.11.1.7. **Default position** The corresponding note states: *'If none is stated, the period is 2 months.'*
clause 9.1 (Notification and negotiation of disputes)	Notification and negotiation of disputes	The Parties are required to insert the names of their respective nominees.
clause 9.3.1 (Adjudication)	Adjudication	The Parties should either: • insert the name of the Adjudicator, and select the nominating body in circumstances where the named Adjudicator is unwilling or unable to act; or • select the nominating body only. In either case, the nominating body may be selected from the list provided, but the Parties may agree a different nominating body, if they wish.

(*Continued*)

Provision	Entry	Guidance
		Default position The corresponding note states: *'If a body has not been selected from those listed below or another body chosen and inserted, the nominating body shall be one of the bodies listed below selected by the Party requiring the reference to adjudication.'*
clause 9.5.1 (Notice of reference to arbitration)	Arbitration: appointor of Arbitrator (any of any replacement)	The Parties are required to insert the name of the appointor of the Arbitrator. Footnote [29] of the Contract states: *'This only applies where the Contract Particulars state (against the reference to Article 9) that Article 9 and clauses 9.4 to 9.9 (Arbitration) apply.'* The appointor may be selected from the list provided, but the Parties may agree a different appointor, if they wish. **Default position** The corresponding note states: *'If no body is selected from those listed below or another body chosen and inserted, the appointor shall be the President or a Vice-President of the body listed below selected by the Party serving the first notice or arbitration under clause 9.5. For any subsequently served notice of arbitration from any Party under clause 9.5, the appointor shall be the President or a Vice-President of the same body that was selected for the first notice of arbitration.'*

Resource 3 page 409	Notice: Contractor or Employer to the other Party	The replacement of the Employer's nominee or Contractor's nominee, relating to clause 9.1.

Attestation

Overview

The Attestation section is included at the end of the Agreement, after the Contract Particulars.

It is where the Parties execute the Agreement, either 'under hand' (sometimes described as a 'simple' contract) or as a deed.

Notes on execution

The Contract provides a section entitled '*Notes on Execution*' (contained on page 22).

In summary, the Parties may execute the Agreement either:

- under hand, as a simple contract (using the form provided on page 23);
- as a deed (using one of the forms (A) to (D) provided on page 25 for the Employer, and on page 27 for the Contractor); or
- using another form, for example, if a Party wishes an attorney to execute the Agreement on its behalf.

Very briefly, a deed requires a number of additional formalities, and has a longer limitation period for bringing a claim in contract (generally 12 years) than a simple contract (generally 6 years).

Electronic execution

The Parties may wish to execute the Agreement electronically; something increasingly common following a report by the Law Commission in 2019 entitled '*Electronic execution of documents*', which reached the following general conclusion:

> *'An electronic signature is capable in law of being used to validly execute a document (including a deed) provided that (i) the person signing the document intends to authenticate the document and (ii) any formalities relating to execution of that document are satisfied'.*[8]

There are now several electronic signature platforms available to facilitate electronic execution in this regard.

[8] Law Commission (2019) *Electronic execution of documents*, page 2. Available at: https://lawcom.gov.uk/project/electronic-execution-of-documents/

DOI: 10.1201/9781003485919-8

Notes on execution as a deed

The Contract also provides a further section entitled '*Notes on Execution as a Deed*' (contained on page 24).

In summary, there are four options, (A) to (D).

Options (A) to (C) are applicable to private or public companies:

(A) execution by either (i) a Director and the Company Secretary, or (ii) two Directors of the company.
(B) execution by affixing the common seal of the company (or other body corporate), in the presence of either (i) a Director and the Company Secretary, or (ii) two Directors of the company.
(C) execution by the attested signature of a single Director of the company, in the presence of a witness.

Option (D) is applicable to individuals:

(D) execution by the attested signature of the individual, in the presence of a witness.

The Conditions

Section 1: Definitions and Interpretation

Section overview

Section 1 includes a number of preliminary provisions, to assist the Parties with both the interpretation and administration of the Contract.

It will therefore need to be cross-referenced extensively, when considering other provisions.

The following clauses assist with the interpretation of the Contract:

- clause 1.1 (Definitions)
- clause 1.2 (Reference to clauses etc.)
- clause 1.3 (Agreement etc. to be read as a whole)
- clause 1.4 (Headings, references to persons, legislation etc.)
- clause 1.6 (Contracts (Rights of Third Parties) Act 1999)
- clause 1.8 (Effect of Final Statement)
- clause 1.9 (Effect of payments other than payment of Final Statement)
- clause 1.11 (Applicable law)

The following clauses assist with the administration of the Contract:

- clause 1.5 (Reckoning periods of days)
- clause 1.7 (Notices and other communications)
- clause 1.10 (Consents and approvals)

Sub-sections

- Definitions (clause 1.1)
- Interpretation (clause 1.2 – clause 1.11)

DOI: 10.1201/9781003485919-11

Definitions

Overview

This sub-section contains only clause 1.1 (Definitions), which provides a list of defined terms (identifiable by their capitalised form), which are used throughout the Agreement and the Conditions.

It is therefore an important reference for the Parties, to assist with their interpretation of the Contract.

Unless the context otherwise requires, or the Agreement or the Conditions specifically provide otherwise, the defined terms within the list are to have the meanings included within clause 1.1.

This helps create consistency and establish certainty throughout the Contract.

Key provisions

- clause 1.1 (Definitions)

Clause 1.1 (Definitions)

Clause 1.1 sets out the defined terms in a number of different ways.

In some instances, it sets out the definition in its entirety. For example:

clause 1.1 ***Site Materials:*** *all unfixed materials and goods delivered to and placed on or adjacent to the Works which are intended for incorporation therein.*

In other instances, it directs the Parties to another provision within the Contract Documents. For example:

clause 1.1 ***Relevant Event:*** *see clause 2.26.*

DOI: 10.1201/9781003485919-12

In other instances, it references several other defined terms and provisions, which must therefore also be considered. For example:

> *clause 1.1* ***Employer's Persons:*** *all persons employed, engaged or authorised by the Employer, excluding the Contractor, Contractor's Persons, and any Statutory Provider but including any such third party as is referred to in clause 3.15.2 or 3.15.4.*

Where relevant and appropriate, reference to these defined terms has been made extensively throughout Part 1 of this book, rather than discussing them further here, in isolation.

Recommendation: Cross-referencing defined terms

When considering any provision in the Contract containing a defined term (which is most of them), the Parties are encouraged to cross-reference any corresponding definitions included within clause 1.1.

For example, when considering the Relevant Event contained within clause 2.26.11 (Relevant Events), namely '*loss or damage occasioned by any Specified Peril*', it will be necessary to refer to the definition of '*Specified Perils*'.

In turn, the definition of Specified Perils included within clause 1.1 cross-references clause 6.8 (Insurance of the Works and Existing Structures: Related definitions).

Extensive and multiple cross-referencing of this nature is an unavoidable part of interpreting and administering the Contract, but the process becomes quicker with familiarity, and it is ultimately worthwhile.

Recommendation: Communicating using defined terms

The Parties are further encouraged to use any relevant defined terms accurately in all verbal and written communications, to reduce uncertainty and misunderstanding.

This is particularly the case in the following circumstances:

- when there are alternative terms derived from other standard forms of contract. For example, the JCT Standard Building Contract refers to '*Variations*', rather than '*Changes*'; or
- when there are alternative terms derived from colloquial or conventional use. For example, the Parties should refer to the '*Rectification Period*', rather than, say, a '*defects liability period*', and to '*practical completion*', rather than, say, '*handover*'.

In summary, interpreting the Contract can often prove challenging enough, without further complicating the matter by using terms which are not defined within it. Put another way, it is best to view the Contract and its defined terms as a 'path', from which the Parties are advised not to stray.

The resources for contract administration included within Part 2 adopt this approach.

Terms defined elsewhere

As stated above, clause 1.1 does not contain an exhaustive list of all the defined terms in the Contract.

Furthermore, not all defined terms use the capitalised form (for example, a '*person*' or '*legislation*'), and in some instances, the same defined term can have different meanings under different provisions (for example, '*Termination Payment*' has three different meanings).

The table below therefore identifies these terms, along with their corresponding provisions (in the order in which they appear in the Contract).

Defined term	Provision
the Works	First Recital
the Employer's Requirements	First Recital
the Contractor's Proposals	Second Recital
the Contract Sum Analysis	Second Recital
the Contract Sum	Article 2 (Contract Sum)
the C.1 Replacement Schedule	the entry against the reference to clause 6.7 (Insurance Options and period) and Schedule 3 (Insurance Options) in the Contract Particulars
a Large Project	the entry against the reference to clause 6.17 (Joint Fire Code – compliance: Application of clauses) in the Contract Particulars
the Rights Particulars	the entry against the reference clause 7.4 (Clauses 7.7 to 7.11 – Preliminary: Rights Particulars) in the Contract Particulars
a person	clause 1.4.4 (Headings, references to persons, legislation etc.)
legislation	clause 1.4.5 (Headings, references to persons, legislation etc.)
the relevant statement	clause 1.8.1 (Effect of Final Statement)
a Pre-Agreed Adjustment	clause 2.23.2 (Adjustment of Completion Date: Related definitions and interpretation)
a Relevant Omission	clause 2.23.3 (Adjustment of Completion Date: Related definitions and interpretation)
the Practical Completion Statement	clause 2.27.1 (Practical completion)
a Section Completion Statement	clause 2.27.2 (Practical completion)
a Non-Completion Notice	clause 2.28 (Non-Completion Notice)
the Relevant Part	clause 2.30 (Partial Possession by Employer: Contractor's consent)
the Relevant Date	clause 2.30 (Partial Possession by Employer: Contractor's consent)

(*Continued*)

Defined term	Provision
a Notice of Completion of Making Good	clause 2.36 (Notice of Completion of Making Good)
an Interim Payment Application	clause 4.7.3 (Interim Payments – Contractor's Interim Payment Applications, due dates and Payment Notices)
a Payment Notice	clause 4.7.5 (Interim Payments – Contractor's Interim Payment Applications, due dates and Payment Notices)
the paying Party	clause 4.8 (Relevant statement and Final Payment Notice)
a Final Payment Notice	clause 4.8 (Relevant statement and Final Payment Notice)
a Pay Less Notice	clause 4.9.5 (Interim and final payments – final date and amount)
the Retention	clause 4.14.1 (Sums due as Interim Payments)
the required bond	clause 4.15.4 (Listed Items)
the Retention Bond	clause 4.17.2 (Retention Bond)
the Surety	clause 4.17.2 (Retention Bond)
the Employer's Final Statement	clause 4.24.4 (Final Statement and Final Payment)
the relevant statement	clause 4.24.5.3 (Final Statement and Final Payment)
a Change	clause 5.1 (Definition of Changes)
a Valuation	clause 5.2 (Valuation of Changes and provisional sum work)
the Valuation Rules	clause 5.2 (Valuation of Changes and provisional sum work)
the Works Insurance Policy	clause 6.7.2 (Insurance of the Works and Existing Structures: Insurance Options and period)
All Risks Insurance	clause 6.8 (Insurance of the Works and Existing Structures: Related definitions)
Excepted Risks	clause 6.8 (Insurance of the Works and Existing Structures: Related definitions)
Joint Named Policy	clause 6.8 (Insurance of the Works and Existing Structures: Related definitions)
Pool Re Cover	clause 6.8 (Insurance of the Works and Existing Structures: Related definitions)
Specified Perils	clause 6.8 (Insurance of the Works and Existing Structures: Related definitions)
Terrorism Cover	clause 6.8 (Insurance of the Works and Existing Structures: Related definitions)

(*Continued*)

Defined term	Provision
the cessation date	clause 6.11.1 (Terrorism Cover – non-availability – Employer's options)
reinstatement work	clause 6.13.4 (Loss or damage – insurance claims and reinstatement)
the Remedial Measures	clause 6.19.1 (Breach of Joint Fire Code – Remedial Measures)
the Consultants	clause 7.4.3 (Clauses 7.7 to 7.11 – Preliminary: Rights Particulars)
a specified default	clause 8.4.1 (Termination by Employer: Default by Contractor), or clause 8.9.1 (Termination by Contractor: Default by Employer)
Termination Payment	clause 8.7.6 (Termination by Employer: Consequences of termination under clauses 8.4 to 8.6), clause 8.8.3 (Termination by Employer: Employer's decision not to complete the Works), or clause 8.12.5 (Consequences of Termination under clauses 8.9 to 8.11, etc.)
a specified suspension event	clause 8.9.2 (Termination by Contractor: Default by Employer)
Named Sub-Contract Work	paragraph 1.1, Supplemental Provision 1 (Named Sub-Contractors)
a Named Sub-Contractor	paragraph 1.1, Supplemental Provision 1 (Named Sub-Contractors)
an Acceleration Quotation	paragraph 4.1.1, Supplemental Provision 4 (Acceleration Quotation)
a Confirmed Acceptance	paragraph 4.3, Supplemental Provision 4 (Acceleration Quotation)
FOIA (Freedom of Information Act 2000)	Supplemental Provision 8 (Transparency)
P&T Rights	Part 1 of Schedule 5 (Third Party Rights for Purchasers and Tenants)
Funder Rights	Part 2 of Schedule 5 (Third Party Rights for a Funder)

Interpretation

Overview

This sub-section includes several provisions which assist the Parties in both the interpretation and administration of the Contract.

In particular, the Parties should ensure that they have:

- calculated any periods in accordance with clause 1.5 (Reckoning periods of days);
- sent any notices or other communications in accordance with clause 1.7 (Notices and other communications);
- undertaken a comprehensive review of the Final Statement (in the case of the Employer) or the Employer's Final Statement (in the case of the Contractor), given the effects of clause 1.8 (Effect of Final Statement); and
- provided reasons, in circumstances where they have delayed or withheld any consent or approval, in accordance with clause 1.10 (Consents and approvals).

Key provisions

- clause 1.2 (Reference to clauses etc.)
- clause 1.3 (Agreement etc. to be read as a whole)
- clause 1.4 (Headings, references to persons, legislation etc.)
- clause 1.5 (Reckoning periods of days)
- clause 1.6 (Contracts (Rights of Third Parties) Act 1999)
- clause 1.7 (Notices and other communications)
- clause 1.8 (Effect of Final Statement)
- clause 1.9 (Effect of payments other than payment of Final Statement)
- clause 1.10 (Consents and approvals)
- clause 1.11 (Applicable law)

Clause 1.2: Reference to clauses etc.

1.2 *Unless otherwise stated, a reference in the Agreement or in these Conditions to a clause or Schedule is to that clause in or Schedule to these Conditions and, unless the context otherwise requires, a reference in a Schedule to a paragraph is to that paragraph of that Schedule.*

DOI: 10.1201/9781003485919-13

Clause 1.2 assists with the interpretation of the (very frequent) references to the words '*clauses*', '*Schedules*', or '*paragraphs*' within the Agreement and the Conditions.

In summary:

- unless otherwise stated, any reference in the Agreement or in the Conditions to a '*clause*' is a reference to a clause in the Conditions;
- unless otherwise stated, any reference in the Agreement or in the Conditions to a '*Schedule*' is a reference to one of the six Schedules to the Conditions; and
- unless the context otherwise requires, a reference in a Schedule to a '*paragraph*' is a reference to a paragraph of that Schedule.

Clause 1.3: Agreement etc. to be read as a whole

> *1.3 The Agreement and these Conditions are to be read as a whole. Nothing contained in any other Contract Document or any Framework Agreement, irrespective of their terms, shall override or modify the Agreement or these Conditions.*

Clause 1.3 assists with the interpretation of the Contract Documents as a whole, by effectively establishing a hierarchy.

The Agreement and the Conditions are to be '*read as a whole*'; in other words, construed or understood together.

In the event of a conflict between:

(i) the Agreement or the Conditions; and
(ii) any other Contract Document (including the Employer's Requirements, Contractor's Proposals, Contract Sum Analysis, or (where applicable) the BIM Protocol), or any Framework Agreement,

then the Agreement or Conditions will take precedence.

This is '*irrespective of their terms*'; in other words, no matter what those other Contract Documents may state.

Recommendation: Negotiating, revising, and checking the Contract Documents

Similar to the recommendation made in respect of Article 5 (Employer's Requirements and Contractor's Proposals), in light of the effect of clause 1.3, the Parties are encouraged to leave sufficient time at the end of the contract negotiations to undertake a comprehensive review of the draft Contract Documents in their entirety, to remove any inconsistencies, conflicts, and discrepancies in or between the Agreement, the Conditions, and the remaining Contract Documents.

Resource 4 page 410	Notice: Contractor or Employer to the other Party	A notice of a conflict within the Contract Documents.

Market practice: Hierarchy of Contract Documents

Clause 1.3 is often amended to provide for a more prescriptive hierarchy of the Contract Documents, in the event of a conflict. For example, it may provide that the Contract Documents are to be read and construed in the following order:

1. the Agreement
2. the Conditions
3. the Employer's Requirements
4. the Contractor's Proposals
5. the Contract Sum Analysis
6. the BIM Protocol (where applicable)

Clause 1.4: Headings, references to persons, legislation etc.

1.4 *In the Agreement and these Conditions, unless the context otherwise requires:*

1.4.1 *the headings, notes and footnotes are included for convenience only and shall not affect the interpretation of this Contract;*

1.4.2 *the singular includes the plural and vice versa;*

1.4.3 *a gender includes any other gender;*

1.4.4 *a reference to a 'person' includes any individual, firm, partnership, company and any other body corporate;*

1.4.5 *a reference to a statute, statutory instrument or other subordinate legislation ('legislation') is to such legislation as amended and in force from time to time, including any legislation which re-enacts or consolidates it, with or without modification, and including corresponding legislation in any other relevant part of the United Kingdom; and*

1.4.6 *references to documents shall, where there is a BIM Protocol or other protocol relating to the supply of documents or information, be deemed to include information in a form or medium conforming to that protocol.*

Similar to clause 1.2 (Reference to clauses etc.), clause 1.4 assists with the interpretation of particular words and phrases within the Agreement and the Conditions (unless the context otherwise requires), including:

- headings, notes, and footnotes, which do not affect the interpretation of the Contract;
- the singular and plural;
- gender;
- persons, to include individuals, firms, partnerships, companies, or any other body corporate;
- legislation, to mean such legislation as is amended and in force from time to time (rather than the original legislation as enacted), including any re-enactment or consolidation of that legislation, and including any corresponding legislation in Scotland, Wales, or Northern Ireland; and
- the BIM Protocol.

Clause 1.5: Reckoning periods of days

> *1.5 Where under this Contract an act is required to be done within a specified period of days after or from a specified date, the period shall begin immediately after that date. Where the period would include a day which is a Public Holiday that day shall be excluded.*

Clause 1.5 assists with the calculation of periods, which is required extensively throughout the Conditions.

Where the Contract requires an act to be done within a specified period of days after or from a specified date, the period begins immediately *after* that date.

Where the period includes a day which is a Public Holiday, that day will be excluded.

'*Public Holidays*' are defined in clause 1.1 (Definitions), and include Christmas Day, Good Friday, and bank holidays (under the Banking and Financial Dealings Act 1971).

The Parties are therefore encouraged to mark all Public Holidays on their project calendar, so that they are properly accounted for.

In summary, when calculating any such relevant period, the Parties must:

- ensure that the calculation begins immediately after the specified date (i.e. starting from the next day); and
- if there are any Public Holidays during that period, ensure that they been accounted for.

As an example, clause 2.35.1 (Schedules of defects and instructions) requires that any relevant defects, shrinkages, and other faults must be specified by the Employer in a schedule of defects, which the Employer is required to deliver to the Contractor as an instruction *'not later than 14 days after the expiry of that Rectification Period'*.

Let us assume that the relevant Rectification Period expires on 21 December.

In accordance with clause 1.5, therefore, the Employer must deliver the schedule of defects as an instruction not later than 07 January the following year:

- the 14-day period commences immediately after 21 December, namely 22 December; and
- Christmas Day (25 December), Boxing Day (26 December), and New Year's Day (01 January) are excluded, because they are all Public Holidays.

Clause 1.6: Contracts (Rights of Third Parties) Act 1999

> 1.6 *Other than such rights of any Purchasers, Tenants and/or Funder as take effect pursuant to clauses 7.7 and/or 7.8, nothing in this Contract confers or is intended to confer any right to enforce any of its terms on any person which is not a party to it.*

Clause 1.6 states that nothing in the Contract confers (or is intended to confer) any right to enforce any of its terms, on any person other than the Contractor or the Employer.

The exception is in relation to the rights of either of the following (considered below):

- any Purchasers or Tenants, in accordance with clause 7.7 (Third Party Rights from Contractor: Rights for Purchasers and Tenants); or
- any Funder, in accordance with clause 7.8 (Third Party Rights from Contractor: Rights for a Funder).

Very briefly, the clause is required because, unless expressly excluded in this way, the Contracts (Rights of Third Parties) Act 1999 might otherwise entitle a third party to enforce the terms of the Contract, if those terms confer a benefit upon them.

Clause 1.7: Notices and other communications

> 1.7.1 *Any notice or other communication between the Employer (or Employer's Agent) and the Contractor that is expressly referred to in the Agreement or these Conditions (including, without limitation, each application, approval, consent, confirmation, counter-notice, decision, instruction or other notification) shall be in writing.*

Clause 1.7 assists the Parties with the administration of the Contract, by setting out the rules relating to the service of notices and other communications between them.

The Contract refers to several different types of correspondence. Clause 1.7.1 identifies some of these, namely notices, applications, approvals, consents, confirmations, counter-notices, decisions, and instructions.

It states that any such correspondence expressly referred to in the Agreement or the Conditions must be in writing, and thereby helps avoid any uncertainty between the Parties as to the nature and content of any correspondence between them.

(However, the Parties are still encouraged to maintain a sustained dialogue throughout the course of the project, which can promote collaboration, co-operation, and shared understanding in a way that written correspondence alone likely cannot.)

By contrast, a number of provisions deal with communications not in writing, in certain limited circumstances:

- clause 1.7.5, in cases of emergency; and
- clause 3.7 (Instructions other than in writing).

1.7.2 *Subject to clause 1.7.4, each such notice or other communication and any documents to be supplied may or (where so required) shall be sent or transmitted by the means (electronic or otherwise) and in such format as the Parties have agreed or may from time to time agree in writing for the purposes of this Contract.*

Clause 1.7.2 gives the Parties broad discretion (subject to clause 1.7.4 below) to agree between them how notices, communications, and any documents to be supplied are to be sent.

This agreement should be made in writing, and may be revised throughout the course of the project.

Resource 5 page 410	Communication Protocol: Employer and Contractor	A communication protocol template to govern the exchange of notices, communications, and documents between the Parties.

1.7.3 *Subject to clauses 1.7.2 and 1.7.4, any notice, communication or document may be given or served by any effective means and shall be duly given or served if:*

1.7.3.1 *delivered by hand or sent by pre-paid post to the recipient's address stated in the Contract Particulars against clause 1.7.3, or to such other address as the recipient may from time to time notify to the sender, or if no such address is then current, the recipient's last known principal business address or (where a body corporate) its registered or principal office; or*

1.7.3.2 *sent by email to the recipient's email address stated in the Contract Particulars against clause 1.7.3, or to such other email address as the recipient may from time to time notify to the sender.*

Clause 1.7.3 states that any notice, communication or document may be given or served by '*any effective means*', which again gives the Parties a broad discretion (subject to clause 1.7.2 and clause 1.7.4) to use email, pre-paid post, or even delivery by hand.

Notwithstanding this discretion, email is submitted to have significant advantages over the other methods (including speed, cost, evidence of delivery, and the ability to copy multiple parties), and is therefore recommended.

The clause then clarifies what constitutes effective service:

- Delivery by hand or by pre-paid post: if delivered to (i) the recipient's address stated in the Contract Particulars, (ii) such other address as the recipient may notify to the sender, or (iii) if no such address is then current, the recipient's last known principal business address or its registered or principal office.
- Email: if sent to (i) the recipient's email address stated in the Contract Particulars, or (ii) such other email address as the recipient may notify to the sender.

> *1.7.4 Any notice expressly required by this Contract to be given in accordance with this clause 1.7.4:*
>
> *1.7.4.1 shall be delivered by hand or sent by Signed For 1st class or Special Delivery Guaranteed post (or any method of posting as has replaced either method and is then current) and, where sent by post in that manner, shall be deemed to have been received on the second Business Day after the date of posting; or*
>
> *1.7.4.2 (where clause 1.7.4.2 is stated in the Contract Particulars to apply) shall be sent by email to the recipient's email address stated in the Contract Particulars against clause 1.7.4.2, or to such other email address as the recipient may from time to time notify to the sender (provided that such notification states that it is a notice of change under this clause 1.7.4.2), and shall be deemed to have been received on the next Business Day after the day on which it was sent.*

Several provisions within the Contract require notices to be given in accordance with clause 1.7.4, which provides two alternative methods of service:

- Delivery by hand or sent by Signed For 1st class or Special Delivery Guaranteed post (or any replacement[9]), where it will be deemed to have been received on the second Business Day after the date of posting; or
- Email (where clause 1.7.4.2 is stated in the Contract Particulars to apply), if sent to (i) the recipient's email address stated in the Contract Particulars against the reference to clause 1.7.4.2, or (ii) such other email address as the recipient may notify to the sender. However, if sent in this manner, it should be noted that it will be deemed to have been received on the *next Business Day* after the

[9] A list of the Royal Mail's services in this regard is available at: https://www.royalmail.com/sending/uk

day on which it was sent, even though in reality it will likely have been received moments after being sent. The Parties should therefore take care to give the relevant notices as soon as possible.

Notwithstanding this, and similar to clause 1.7.3 above, email is still submitted to have significant advantages over the other methods, and so the Parties are encouraged to state that clause 1.7.4.2 in the Contract Particulars applies.

Recommendation: Email correspondence

The Parties may also wish to consider the following further measures in respect of email correspondence under both clause 1.7.3 and clause 1.7.4.2, to promote certainty:

- Each Party should use the same email address for both clause 1.7.3 and clause 1.7.4, to avoid confusion.
- In addition, the Parties could each include multiple email addresses.
- Alternatively, the Parties could achieve the same effect by ensuring that multiple representatives for each Party have access to the designated email address account.
- The Parties could also consider setting up an automatic email reply, to notify the sender that their email has been received.

The full list of clauses which require notices to be given in accordance with clause 1.7.4 is as follows:

- clause 6.14 (Loss or damage to Existing Structures – right of termination);
- clause 7.7 (Rights for Purchasers and Tenants);
- clause 7.8 (Rights for a Funder);
- clause 7.9 (Contractor's Warranties – Purchasers and Tenants);
- clause 7.10 (Contractor's Warranty – Funder);
- clause 7.11 (Third Party Rights and Collateral Warranties from Sub-Contractors); and
- all notices required by section 8 (Termination), except for the notices referred to in clause 8.13 (Termination Payment – final date, notices and amount). This includes the following clauses:
    - clause 8.4.1 (Termination by Employer: Default by Contractor);
    - clause 8.4.2 (Termination by Employer: Default by Contractor);
    - clause 8.4.3 (Termination by Employer: Default by Contractor);
    - clause 8.5.1 (Termination by Employer: Insolvency of Contractor);
    - clause 8.5.2 (Termination by Employer: Insolvency of Contractor);
    - clause 8.6 (Termination by Employer: Corruption and regulation 73(1)(b) of the PC Regulations);
    - clause 8.7.4 (Termination by Employer: Consequences of termination under clauses 8.4 to 8.6);
    - clause 8.8.1 (Termination by Employer: Employer's decision not to complete the Works);

- clause 8.9.1 (Termination by Contractor: Default by Employer);
- clause 8.9.2 (Termination by Contractor: Default by Employer);
- clause 8.9.3 (Termination by Contractor: Default by Employer);
- clause 8.9.4 (Termination by Contractor: Default by Employer);
- clause 8.10.1 (Termination by Contractor: Insolvency of Employer);
- clause 8.10.2 (Termination by Contractor: Insolvency of Employer);
- clause 8.11.1 (Termination by either Party and regulation 73(1)(a) of the PC Regulations) (note that clause 8.11.1 refers to two separate notices); and
- clause 8.11.3 (Termination by either Party and regulation 73(1)(a) of the PC Regulations).

Resource 6 page 412	Notice: Contractor or Employer to the other Party	A notice to the other Party confirming a change of email address.

1.7.5 If in an emergency any communication is made orally with respect to health and safety, risk of damage to property or insurance matters, written confirmation of it shall be sent as soon thereafter as is reasonably practicable.

Clause 1.7.5 provides a limited exception to the general requirement under clause 1.7.1 for all communications between the Parties to be in writing, which is in the event of an emergency relating to one of the following:

- health and safety;
- risk of damage to property; or
- insurance matters.

However, as soon as is reasonably practicable after the emergency, written confirmation of the oral communication must be sent.

Resource 7 page 412	Confirmation: Contractor or Employer to the other Party	The confirmation of an oral communication, in the case of an emergency.

Clause 1.8: Effect of Final Statement

1.8.1 *As from the due date for the final payment specified in clause 4.24.5 and in addition to the effects referred to in clause 4.24.6, the Final Statement or, as the case may be, the Employer's Final Statement ('the relevant statement') shall, except as provided in clauses 1.8.2 and 4.24.6 (and save in respect of fraud), have effect in any proceedings under or arising out of or in connection with this Contract (whether by adjudication, arbitration or legal proceedings) as conclusive evidence that:*

1.8.1.1 *where and to the extent that any particular quality of any materials or goods or any particular standard of an item of workmanship was expressly described in the Employer's Requirements, or in any instruction issued by the Employer under these Conditions, to be for the Employer's approval, the particular quality or standard was to the Employer's reasonable satisfaction, but the relevant statement shall not be conclusive evidence that they or any other materials, goods or workmanship comply with any other requirement or term of this Contract;*

1.8.1.2 *all and only such extensions of time, if any, as are due under clause 2.25 have been given; and*

1.8.1.3 *the reimbursement of direct loss and/or expense, if any, due to the Contractor as agreed, ascertained or valued in accordance with these Conditions is in final settlement of all and any claims which the Contractor has or may have arising out of the occurrence of any Relevant Matters, whether such claim be for breach of contract, duty of care, statutory duty or otherwise.*

In the interest of creating certainty, clause 1.8.1 provides that the Final Statement or the Employer's Final Statement (considered further in clause 4.24 (Final Statement and Final Payment)) constitutes conclusive evidence of the following three matters, in the context of any adjudication, arbitration, or litigation (considered further in section 9 (Settlement of Disputes)):

- **Materials, goods, or workmanship:** where (i) any particular quality of any materials or goods, or (ii) any standard of workmanship, was expressly described in either the Employer's Requirements, or in an instruction issued by the Employer, to be for the Employer's approval, the particular quality or standard *was* to the Employer's reasonable satisfaction. However, this is not conclusive evidence that they or any other materials, goods or workmanship comply with any *other* requirement under the Contract.
- **Extensions of time:** any extensions of time which were due, were given (considered further in clause 2.25 (Fixing Completion Date)).
- **Reimbursement of loss and expense:** the reimbursement of any direct loss and/or expense due to the Contractor as either agreed between the Parties, or otherwise ascertained or valued, is in final settlement of all and any claims which

the Contractor has arising out of the occurrence of any of the Relevant Matters, whether for breach of contract, duty of care, statutory duty, or otherwise.

This takes effect from the due date for the final payment specified in clause 4.24.5.

In summary, it effectively settles the Contractor's entitlement to extensions of time and reimbursement of direct loss and/or expense, and establishes (to a limited extent) that the Contractor has complied with certain requirements under the Contract to provide materials, goods or workmanship to the Employer's satisfaction.

Clause 1.8.1 is in addition to the effects referred to in clause 4.24.6 (considered further below), namely that the Final Statement or the Employer's Final Statement become conclusive as to the amount of the sum due under clause 4.24.2.

The Contractor should therefore have careful regard to these significant effects when preparing the Final Statement (as should the Employer, if preparing the Employer's Final Statement).

There are a number of exceptions:

- the exception provided in clause 1.8.2 (considered below);
- the exception provided in clause 4.24.6, where either (i) the Employer disputes anything in the Final Statement, or (ii) the Contractor disputes anything in the Employer's Final Statement; and
- in respect of any fraud.

Market practice: Deletion of clause 1.8.1.1

It is common in practice for clause 1.8.1.1 to be deleted, leaving only clause 1.8.1.2 (settling the Contractor's claims for extensions of time) and clause 1.8.1.3 (settling the Contractor's claims for reimbursement of direct loss and/or expense).

1.8.2 The effects of the relevant statement specified in clauses 1.8.1 and 4.24.6 shall be suspended pending the conclusion of any adjudication, arbitration or other proceedings, and shall thereafter be subject to the terms of any decision, award or judgment in and any settlement of those proceedings:

1.8.2.1 where those proceedings are commenced before or within 28 days after the due date for the final payment; or

1.8.2.2 in the case of an adjudication commenced within the period referred to in clause 1.8.2.1 in which the Adjudicator gives their decision after the due date for the final payment, where arbitration or legal proceedings to determine the dispute or difference in question are commenced within 28 days of the date of that decision but not otherwise.

Clause 1.8.2 sets out one of the three exceptions to clause 1.8.1 above, in circumstances where either (i) the Employer wishes to challenge the Final Statement, or (ii) the Contractor wishes to challenge the Employer's Final Statement.

In such instances, the effects of the Final Statement or the Employer's Final Statement (as set out in both clause 1.8.1 and clause 4.24.6) will be suspended, pending the outcome of any subsequent:

- adjudication (considered further in clause 9.3 (Adjudication));
- arbitration (considered further in clause 9.4 – clause 9.9 (Arbitration)); or
- any other legal proceedings (such as litigation).

Following the conclusion of these proceedings, the Final Statement or the Employer's Final Statement will be subject to the terms of:

- the Adjudicator's decision, in the case of an adjudication;
- the award, in the case of an arbitration;
- the judgment, in the case of legal proceedings; or
- any settlement, in respect of any of the above.

This is provided that either:

- any of these proceedings are commenced within 28 days after the due date for the final payment, as set out in clause 4.24.5; or
- in the case of an adjudication commenced within the 28-day period above, in which the Adjudicator gives their decision after the due date for the final payment, where a subsequent arbitration or legal proceedings to determine the dispute or difference are commenced within 28 days of the date of the Adjudicator's decision. This is because an Adjudicator's decision is binding on the Parties unless and until the dispute is finally determined by arbitration, legal proceedings, or settlement.

Recommendation: Review of the Final Statement or Employer's Final Statement

Given the 28-day deadlines specified in clause 1.8.2, and the considerable preparation involved in any subsequent adjudication, arbitration, or legal proceedings, the Employer is encouraged to undertake a careful review of the Final Statement (as is the Contractor, in the case of the Employer's Final Statement), as soon as it is received.

1.8.3 For the purposes of clause 1.8.2 any proceedings shall be treated as concluded if during any period of 12 months commencing on or after the issue of the relevant statement neither Party takes a further step in them.

Clause 1.8.3 broadens the definition of the '*conclusion*' of any adjudication, arbitration, or legal proceedings, as referred to in clause 1.8.2, to include inactivity by the Parties.

This is designed to prevent the matter remaining unresolved indefinitely.

If any such adjudication, arbitration or legal proceedings is commenced after the issue of the Final Statement or Employer's Final Statement, and for a period of 12 months neither Party takes a further step in them, those proceedings will be considered concluded, even though a decision, award or judgment has not been given.

Clause 1.9: Effect of payments other than payment of Final Statement

1.9 Save as stated in clause 1.8, no payment by the Employer shall of itself be conclusive evidence that any works, any materials or goods or any design to which it relates are in accordance with this Contract.

Clause 1.9 clarifies that a payment made by the Employer to the Contractor will not, by itself, constitute conclusive evidence that (i) any works, (ii) any materials or goods, or (iii) any design, to which that payment relates, are in accordance with the Contract.

It addresses the possibility that the Employer may make a payment in respect of certain works, materials or goods, or design, which the Employer subsequently considers not to be in accordance with the Contract.

However, rather than seek to rely on clause 1.9, it is evidently preferable for the Employer to verify, as far as possible, that any works, any materials or goods, or any design, are in accordance with the Contract, before making any related payment.

The only exception to clause 1.9 is set out in clause 1.8 (Effect of Final Statement).

Clause 1.10: Consents and approvals

1.10 Where consent or approval of either Party is expressly required under these Conditions and is requested, such consent or approval shall not be unreasonably delayed or withheld, except in the case of either Party's consent under clause 7.1 where the giving of such consent shall be at the sole discretion of the Party from which it is sought.

Several provisions in the Conditions require one Party to obtain the consent or approval from the other Party.

Clause 1.10 states that, if such a request is made, the other Party's consent or approval must not be unreasonably delayed or withheld.

In any related correspondence, therefore, the Parties are encouraged to provide their reasons for delaying or withholding their consent or approval, in order to comply with clause 1.10.

The only exception is in relation to clause 7.1 (Assignment: General), where the Parties have complete discretion as to whether to give or withhold their consent to the other Party assigning the Contract, or any of the other Party's rights under the Contract.

Resource 8 page 413	Reminder: Employer and Contractor	A reminder that consents or approvals must not be unreasonably delayed or withheld.

Market practice: Meaning of 'approval'

Clause 1.10 is often amended to the effect that the meaning of '*approval*' is narrowed, in the context of the Employer giving its approval to the Contractor (but not the other way around).

In particular, the term '*approval*' is often defined as meaning '*an acceptance of general principles only*', and a clarification is usually added to the effect that no such approval will relieve the Contractor from any of its obligations under the Contract.

This is to avoid any suggestion from the Contractor that an '*approval*' from the Employer constitutes an acknowledgement that the Contractor has complied with the Contract.

In such instances, the Contractor should therefore focus on the terms of the Contract, rather than any statement of approval from the Employer (or Employer's Agent).

Clause 1.11: Applicable law

1.11 This Contract shall be governed by and construed in accordance with the law of England.

Clause 1.11 simply states that the Contract will be governed by, and construed in accordance with, English law.

If both Parties are based in another jurisdiction, they may wish to amend it.

Section 2: Carrying out the Works

Section overview

Section 2 contains several of the Parties' key obligations under the Contract relating to the main stages of the construction phase.

In summary, the Contractor must:

- complete the design for the Works; and
- carry out and complete the construction of the Works.

The Employer must:

- give possession of the site to the Contractor;
- where appropriate, give extensions of time by adjusting the Completion Date; and
- issue the Practical Completion Statement (or each Section Completion Statement), when practical completion has been achieved.

Section 2 also includes the procedures relating to the following:

- the taking of partial possession of the Works (or Section) by the Employer, before the issue of the Practical Completion Statement (or Section Completion Statement);
- the Employer's right to withhold or deduct liquidated damages, for late completion; and
- the making good of defects during the Rectification Period.

Sub-sections

- Contractor's Obligations (clause 2.1 – clause 2.2)
- Possession (clause 2.3 – clause 2.6)
- Supply of Documents, Setting Out etc. (clause 2.7 – clause 2.9)
- Discrepancies and Divergences (clause 2.10 – clause 2.16)
- Design Work – liabilities and limitation (clause 2.17)
- Fees, Royalties and Patent Rights (clause 2.18 – clause 2.20)
- Unfixed Materials and Goods – property, risk etc. (clause 2.21 – clause 2.22)
- Adjustment of Completion Date (clause 2.23 – clause 2.26)
- Practical Completion, Lateness and Liquidated Damages (clause 2.27 – clause 2.29)
- Partial Possession by Employer (clause 2.30 – clause 2.34)
- Defects (clause 2.35 – clause 2.36)
- Contractor's Design Documents (clause 2.37 – clause 2.38)

DOI: 10.1201/9781003485919-15

Contractor's Obligations

Overview

This sub-section contains some of the Contractor's key obligations under the Contract, relating to:

- completing the design for the Works;
- carrying out and completing the Works;
- complying with Employer's instructions;
- ensuring that materials, goods, and workmanship are in accordance with the Contract; and
- suggesting amendments to the Works which may result in an improvement in environmental performance and sustainability.

Key provisions

- clause 2.1 (General obligations)
- clause 2.2 (Materials, goods and workmanship)

Clause 2.1: General obligations

2.1.1 *The Contractor shall carry out and complete the Works in a proper and workmanlike manner and in compliance with the Contract Documents, the Construction Phase Plan and Statutory Requirements and for that purpose shall complete the design for the Works including the selection of any specifications for the kinds and standards of the materials, goods and workmanship to be used in the construction of the Works so far as not described or stated in the Employer's Requirements or Contractor's Proposals, and shall give all notices required by the Statutory Requirements.*

Clause 2.1.1 requires the Contractor to both carry out and complete the Works:

- in a '*proper and workmanlike*' manner; and
- in compliance with:
  - the Contract Documents;
  - the Construction Phase Plan; and
  - the Statutory Requirements.

DOI: 10.1201/9781003485919-16

It further requires the Contractor to complete the design for the Works.

This includes choosing the specifications for the kinds and standards of materials, goods, and workmanship to be used in the construction of the Works, if they have not already been described in either the Employer's Requirements or the Contractor's Proposals.

Finally, clause 2.1.1 requires the Contractor to give all notices required by the Statutory Requirements. For example, relating to the Development Control Requirements (thought to relate primarily to planning), or the Building Regulations.

> 2.1.2 *The Contractor's obligation to the Employer to comply with the Statutory Requirements under clause 2.1.1 shall not apply to the extent that the relevant part or parts of the Employer's Requirements state specifically that the Employer's Requirements comply with the Statutory Requirements.*

Clause 2.1.2 applies where a part of the Employer's Requirements specifically states that the Employer's Requirements comply with the Statutory Requirements.

In these circumstances, the Contractor's obligation to the Employer under clause 2.1.1 to comply with the Statutory Requirements does *not* apply, on the basis that the Employer has expressly stated that the relevant part of the Employer's Requirements *already* complies with the Statutory Requirements.

Clause 2.15.2.3 (Divergences from Statutory Requirements) below applies where any such statement in the Employer's Requirements proves to be incorrect.

> 2.1.3 *The Contractor shall pass to the Employer all approvals received by the Contractor in connection with the Statutory Requirements.*

Clause 2.1.3 relates to any approvals received by the Contractor in connection with the Statutory Requirements. For example, relating to the Development Control Requirements or the Building Regulations.

The Contractor must provide all such approvals to the Employer.

Resource 9 page 414	Correspondence: Contractor to Employer	The provision by the Contractor to the Employer of approvals received by the Contractor, in connection with the Statutory Requirements.
Resource 10 page 414	Tracker: Contractor and Employer	A tracker template for recording approvals in connection with the Statutory Requirements.

Market practice: Responsibility for obtaining permissions, approvals, consents, etc.

Clause 2.1.3 is often expanded to include an allocation of responsibility between the Parties for obtaining any necessary permissions, approvals, and consents, in connection with the Works.

2.1.4 *The Contractor shall comply with any instruction and be bound by any decision of the Employer issued or made under or pursuant to these Conditions and any such instruction or decision shall have effect except to the extent that it is varied by the Employer or under the dispute resolution procedures of this Contract.*

Clause 2.1.4 relates to the Contractor's obligations in respect of Employer's instructions and decisions.

It provides that the Contractor:

(i) is required to comply with any Employer's instructions, provided they are issued in accordance with the Conditions; and
(ii) will be bound by any decision by the Employer, again, provided they are made in accordance with the Conditions.

The only exception is where any such instruction or decision is varied by either:

- the Employer itself; or
- the dispute resolution procedures of the Contract (whether by negotiation, mediation, adjudication, arbitration, or litigation), considered further in section 9 (Settlement of Disputes).

Resource 11 page 415	Reminder: Employer to Contractor	A reminder from the Employer to the Contractor of the Contractor's obligations to comply with Employer's instructions, and to be bound by the Employer's decisions.

2.1.5 *The Contractor is encouraged to suggest economically viable amendments to the Works which, if instructed as a Change, may result in an improvement in environmental performance and sustainability in the carrying out of the Works or of the completed Works and a reduction in environmental impact, provided that no such instruction shall extend the Contractor's obligations in relation to design under this Contract.*

Clause 2.1.5 relates to potential amendments to the Works which, if they are instructed as a Change, may help:

- improve the environmental performance and sustainability; or
- reduce the environmental impact,

of either (i) the carrying out of the Works, or (ii) the completed Works.

The Contractor is encouraged to suggest any such amendments, provided they are '*economically viable*'.

However, clause 2.1.5 provides that any such instructions will not extend (or increase) the Contractor's obligations in respect of design.

Resource 12 page 415	Proposed change: Contractor to Employer	A proposed change to the Works by the Contractor, to improve environmental performance and sustainability.

Clause 2.2: Materials, goods and workmanship

> 2.2.1 *All materials and goods for the Works shall, so far as procurable, be of the kinds and standards described in the Employer's Requirements or, if not there specifically described, as described in the Contractor's Proposals or other Contractor's Design Documents. The Contractor shall not substitute any materials or goods so described without the Employer's consent which, if given, shall not relieve the Contractor of its other obligations.*

Clause 2.2 sets out the Contractor's key obligations in respect of both (i) materials and goods, and (ii) workmanship.

Its primary aim is to ensure that those materials, goods, and workmanship are in compliance with the Contract Documents.

Clause 2.2.1 relates to all goods and materials for the Works, which must be of the kinds and standards as described in either:

(i) the Employer's Requirements; or, if they have not been specifically described in the Employer's Requirements, either
(ii) the Contractor's Proposals; or
(iii) other Contractor's Design Documents, in accordance with Schedule 1 (Design Submission Procedure), considered below.

This obligation is subject to those materials and goods being '*procurable*' (or obtainable). For example, certain goods specified in the Employer's Requirements may be discontinued, and are no longer available.

Furthermore, clause 2.2.1 prohibits the Contractor from substituting any specified materials or goods, unless the Employer has given its consent.

The Contractor should consider any such proposed substitution carefully, because even if the Employer does provide its consent, the Contractor will still be required to ensure that the substituted materials or goods enable the Contractor to comply with all its other obligations under the Contract (for example, a particular performance specification in the Employer's Requirements).

Resource 13 page 416	Request for consent: Contractor to Employer	A request by the Contractor for the Employer's consent to substitute materials or goods.
Resource 14 page 416	Request for consent given: Employer to Contractor	Consent given by the Employer for the Contractor to substitute materials or goods.
Resource 15 page 417	Consent withheld: Employer to Contractor	Consent withheld by the Employer for the Contractor to substitute materials or goods.

2.2.2 *The Contractor shall provide to the Employer all information that the Employer reasonably requests regarding the environmental impact of the supply and use of materials and goods which the Contractor selects.*

Clause 2.2.2 relates to the environmental impact of materials and goods used for the Works.

It requires the Contractor to provide the Employer information relating to the environmental impact of both (i) the supply (for example, including manufacture and transportation), and (ii) the use (for example, the extent of any wastage), of any materials and goods selected by the Contractor.

The Contractor is therefore encouraged to obtain this information, as part of the procurement of all materials and goods.

Resource 16 page 417	Request: Employer to Contractor	A request by the Employer for information on the environmental impact of materials and goods selected by the Contractor.

> 2.2.3 *Workmanship for the Works shall be of the standards described in the Employer's Requirements or, if not there specifically described, as described in the Contractor's Proposals or other Contractor's Design Documents.*

Clause 2.2.3 is similar to clause 2.2.1, but relates to workmanship, rather than materials and goods.

'*Workmanship*' is not defined, but broadly relates to *how* the Contractor carries out its work, so that it complies with the Contract Documents.

For example, the publication *'Materials and workmanship: Approved Document 7'*, relating to the Building Regulations, states the following in relation to workmanship:

'...you will meet the requirements of regulation 7 [Materials and workmanship] if...Workmanship is such that, where relevant, materials are adequately mixed or prepared and applied, used or fixed so as to perform adequately the functions for which they are intended.'[10]

Clause 2.2.3 requires that workmanship for the Works be of the kinds and standards as described in either:

- the Employer's Requirements; or, if they have not been specifically described in the Employer's Requirements, either
- the Contractor's Proposals; or
- other Contractor's Design Documents, in accordance with Schedule 1 (Design Submission Procedure), considered below.

> 2.2.4 *The Contractor shall before carrying out the relevant work and/or ordering the relevant goods or materials provide the Employer with such samples of the standard of workmanship or the quality of the goods or materials which the Contractor intends to provide as are specifically referred to in the Employer's Requirements or in the Contractor's Proposals.*

Clause 2.2.4 relates to the provision of samples, to reassure the Employer that materials, goods, and workmanship are in accordance with the Contract.

If the Employer's Requirements or Contractor's Proposals specifically refer to samples in respect of materials, goods or workmanship, then the Contractor must first provide the Employer with samples of (i) the quality of those materials or goods, or (ii) the standard of workmanship, before the materials or goods are ordered, or the relevant work is carried out.

[10] Ministry of Housing, Communities & Local Government (2018 to 2021) *Materials and workmanship: Approved Document 7*, at page 3. Available at: https://www.gov.uk/government/publications/material-and-workmanship-approved-document-7

Resource 17 page 418	Correspondence: Contractor to Employer	The provision of samples by the Contractor to the Employer.
Resource 18 page 418	Tracker: Contractor and Employer	A tracker template for recording the provision of samples.

2.2.5 *The Contractor shall at the Employer's request provide the Employer with reasonable proof that the materials and goods used comply with this clause 2.2.*

Clause 2.2.5 also aims to provide the Employer with reassurance that materials and goods comply with clause 2.2.

At the Employer's request, the Contractor must provide the Employer with '*reasonable proof*' that materials and goods used comply with clause 2.2 (and clause 2.2.1 and clause 2.2.4 in particular). This proof could take several forms, for example, product data sheets, associated certificates, and other manufacturers' literature.

Resource 19 page 419	Request: Employer to Contractor	A request by the Employer for reasonable proof from the Contractor that materials and goods are compliant.
Resource 20 page 419	Reply: Contractor to Employer	The provision by the Contractor to the Employer of reasonable proof that materials and goods are compliant.

2.2.6 *The Contractor shall take all reasonable steps to encourage Contractor's Persons to be registered cardholders under the Construction Skills Certification Scheme (CSCS) or qualified under an equivalent recognised qualification scheme.*

Clause 2.2.6 relates to workmanship, and aims to ensure that the Contractor's Persons are suitably trained, qualified, and experienced.

It requires the Contractor to take '*all reasonable steps*' to encourage all Contractor's Persons to be either:

(i) registered cardholders under the Construction Skills Certification Scheme ('CSCS'); or
(ii) qualified under an equivalent recognised qualification scheme.

Resource 21 page 420	Confirmation: Contractor to Employer	Confirmation by the Contractor to the Employer of steps taken to encourage Contractor's Persons to be CSCS registered cardholders.

Possession

Overview

This sub-section relates to the possession, control, and use of the site. It includes:

- the Employer's obligation to give possession of the site to the Contractor, to enable the Contractor to carry out and complete the Works;
- the option for the Employer to defer giving possession of the site, for a limited period;
- the potential for the Employer to occupy or use the site prior to practical completion; and
- the potential for works *not* forming part of the Contract to be carried out by the Employer or Employer's Persons.

Key provisions

- clause 2.3 (Date of Possession – progress)
- clause 2.4 (Deferment of possession)
- clause 2.5 (Early use by Employer)
- clause 2.6 (Work not forming part of the Contract)

Clause 2.3: Date of Possession – progress

2.3 *On the Date of Possession of the site or, in the case of a Section, possession of the relevant part of the site shall be given to the Contractor and the Contractor shall thereupon begin the construction of the Works or Section and regularly and diligently proceed with and complete the same on or before the relevant Completion Date. For Works insurance purposes the Contractor shall retain possession:*

2.3.1 *of the site and the Works up to and including the date of issue of the Practical Completion Statement; or*

2.3.2 *of each Section and the relevant part of the site up to and including the date of issue of the Section Completion Statement for that Section and, in respect of any balance of the site, up to and including the date of issue of the Practical Completion Statement*

and, subject to clause 2.30 and section 8, the Employer shall not be entitled to take possession of any part or parts of the Works or Section until such date.

DOI: 10.1201/9781003485919-17

Clause 2.3 governs the giving of possession (and in effect, control) of the site (or if the Works are divided into Sections, the part of the site relating to each Section) by the Employer, to allow the Contractor to carry out the Works.

The Employer is required to give possession on the relevant Date of Possession, identified against the entry to clause 2.3 in the Contract Particulars. (This is subject to a potential ability to defer the giving of possession, considered in clause 2.4 below.)

Once the Employer has given possession, the Contractor must then '*regularly and diligently*' carry out and complete the Works on (or before) the relevant Completion Date.

The Court of Appeal has provided helpful guidance as to the meaning of '*regularly and diligently*':

'*Taken together the obligation upon the contractor is essentially to proceed continuously, industriously and efficiently with appropriate physical resources so as to progress the works steadily towards completion substantially in accordance with the contractual requirements as to time, sequence and quality of work.*'[11]

For the purposes of the Works insurance (considered further in clause 6.7 (Insurance of the Works and Existing Structures: Insurance Options and period)), the Contractor will then retain possession of both the site and the Works, up to the date the Employer issues the Practical Completion Statement.

If the Works are divided into Sections, the Contractor will then retain possession of:

(i) each Section and the relevant part of the site, up to the date the Employer issues the relevant Section Completion Statement; and
(ii) any remaining areas of the site, up to the date the Employer issues the Practical Completion Statement.

During this period, the Employer will be not entitled to take possession of any part of the Works, subject to the following:

- the taking of partial possession by the Employer, in accordance with clause 2.30 (Partial Possession by Employer: Contractor's consent); or
- the termination of the Contractor's employment under the Contract, in accordance with section 8 (Termination).

Resource 22 page 420	Confirmation: Employer to Contractor	Confirmation by the Employer that it has given possession of the site (or part of it) to the Contractor.
Resource 23 page 421	Confirmation: Contractor to Employer	Confirmation by the Contractor that it has taken possession of the site (or part of it) from the Employer.

[11] *West Faulkner Associates v London Borough of Newham* [1994] 71 BLR 1.

Clause 2.4: Deferment of possession

> 2.4 *If the Contract Particulars state that clause 2.4 applies in respect of the Works or a Section, the Employer may defer the giving of possession of the site or relevant part of it for a period not exceeding 6 weeks or lesser period stated in the Contract Particulars, calculated from the relevant Date of Possession.*

Clause 2.4 provides the Employer with greater flexibility over when it is required to give possession of the site to the Contractor.

If it applies (as stated against the reference to clause 2.4 in the Contract Particulars), it entitles the Employer to defer giving possession of (i) the site, in respect of the Works, or (ii) the relevant part of the site, in respect of a Section, for up to either:

- 6 weeks; or
- any lesser period stated in the Contract Particulars,

in both cases, calculated from the relevant Date of Possession (included against the reference to clause 2.3 in the Contract Particulars).

Resource 24 page 422	Notice: Employer to Contractor	A notice from the Employer that it is deferring giving possession of the site (or part of it) to the Contractor.

Clause 2.5: Early use by Employer

> 2.5.1 *Notwithstanding clause 2.3, the Employer may, with the Contractor's consent, use or occupy the site or the Works or part of them, whether for storage or otherwise, before the date of issue of the Practical Completion Statement or relevant Section Completion Statement. Before the Contractor gives its consent to such use or occupation, the Party responsible for the Works Insurance Policy and/or, where there are Existing Structures, for any insurance cover relating to them shall notify the insurers and obtain confirmation that such use or occupation will not prejudice the insurance.*

Clause 2.5 entitles the Employer to use or occupy the site or the Works *before* the Practical Completion Statement or relevant Section Completion Statement has been provided (for example, for storage).

This is to be distinguished from the Employer taking partial *possession* of the site, considered further in clause 2.30 (Partial Possession by Employer: Contractor's consent). The Court has provided helpful guidance on this distinction.[12]

However, given that this use or occupation might adversely affect or impede the Contractor from carrying out the Works, the Contractor must first give its consent to such early use.

Before it does, the Party responsible for either:

- the Works Insurance Policy (namely the Contractor, if Insurance Option A applies, or the Employer, if Insurance Option B or Insurance Option C applies); and/or
- insurance cover relating to the Existing Structures,

must notify the insurers of the proposed early use, and obtain the insurers' confirmation that this will not prejudice (or adversely affect) the insurance.

Resource 25 page 422	Request: Employer to Contractor	A request by the Employer for the Contractor's consent for early use.
Resource 26 page 423	Request for consent given: Contractor to Employer	Consent given by the Contractor to the Employer for early use.
Resource 27 page 423	Request for consent withheld: Contractor to Employer	Consent withheld by the Contractor to the Employer for early use.
Resource 28 page 424	Request: Contractor or Employer to insurer	A request by the Contractor or Employer for confirmation from the insurers under the Works Insurance Policy (and in relation to the Existing Structures) that early use will not prejudice the insurance.

2.5.2 *Where Insurance Option A applies and/or the Contractor is to cover its own or any other risks in relation to any Existing Structures and any insurers' confirmation is conditional on an additional premium being paid, the Contractor shall notify the Employer of the amount of such additional premium. If the Employer continues to require such use or occupation, any additional premium shall be added to the Contract Sum and the Contractor shall if requested produce the receipt for it to the Employer.*

[12] *Impresa Castelli SpA v Cola Holdings Ltd* [2002] EWHC 1363 (TCC)

Clause 2.5.2 is relevant where:

- Insurance Option A (New Buildings – All Risks Insurance of the Works by the Contractor) applies; and/or
- the Contractor is required to cover its own or any other risks relating to the Existing Structures; and
- any insurers' confirmation (under clause 2.5.1) is conditional upon an additional premium being paid.

In these circumstances, the Contractor must notify the Employer of that additional premium.

If the Employer still wishes to use or occupy the site or the Works, then the amount of that additional premium will be added to the Contract Sum, and, if requested, the Contractor will provide the receipt to the Employer.

Resource 29 page 424	Notice: Contractor to Employer	A notice from the Contractor to the Employer of an additional insurance premium for early use.
Resource 30 page 425	Confirmation: Employer to Contractor	An acceptance by the Employer of an additional insurance premium for early use.

Clause 2.6: Work not forming part of the Contract

2.6 *In regard to any work not forming part of this Contract which the Employer requires to be carried out by the Employer itself or by any Employer's Person:*

2.6.1 *where the Contract Documents provide the information necessary to enable the Contractor to carry out and complete the Works or each relevant Section in accordance with this Contract, the Contractor shall permit the execution of such work;*

2.6.2 *where the Contract Documents do not provide the information referred to in clause 2.6.1, the Employer may with the Contractor's consent arrange for the execution of that work.*

Clause 2.6 relates to work which does *not* form part of the Contract between the Contractor and the Employer, but which the Employer requires to be carried out on site (whether by the Employer itself, or by Employer's Persons).

(For example, the Works under the Contract might relate to shell and core works only, but the Employer also requires fit out works to be carried out.)

Such work might adversely affect or impede the Contractor from carrying out the Works under the Contract, and could cause considerable disruption and interference.

Clause 2.6 therefore distinguishes between the circumstance where (i) the Contract Documents *do* provide the information necessary (in respect of such work not forming part of the Contract) to enable the Contractor to carry out and complete the Works in accordance with the Contract, and (ii) where the Contract Documents do *not*:

- where they do, the Contractor must permit the execution of such work not forming part of the Contract, in accordance with clause 2.6.1.
- where they do not, then the Contractor must first give its consent, before the Employer may arrange for that work to be carried out, in accordance with clause 2.6.2.

Recommendation: Contract Documents

If the Employer *does* require work not forming part of the Contract to be carried out (whether by the Employer or an Employer's Person), then, as far as possible, it should ensure that the Contract Documents provide all of the information necessary to enable the Contractor to carry out and complete the Works in accordance with the Contract (including completing the Works on or before the Completion Date).

This information typically includes the scope of the work not forming part of the Contract, the parties involved, and details relating to access, facilities, coordination, programmes and sequencing, as well as any applicable obligations under the CDM Regulations and Part 2A of the Building Regulations.

Clause 2.6.1

Resource 31 page 425	Reminder: Employer to Contractor	A reminder from the Employer to the Contractor of work not forming part of the Contract.
Resource 32 page 426	Confirmation: Contractor to Employer	Permission given by the Contractor to the Employer for the execution of work not forming part of the Contract.

Clause 2.6.2

Resource 33 page 427	Request: Employer to Contractor	A request by the Employer for the Contractor's consent for the Employer to arrange for the execution of work not forming part of the Contract.
Resource 34 page 427	Request for consent given: Contractor to Employer	Consent given by the Contractor for the Employer to arrange for the execution of work not forming part of the Contract.
Resource 35 page 428	Request for consent withheld: Contractor to Employer	Consent withheld by the Contractor for the Employer to arrange for the execution of work not forming part of the Contract.

Market practice: Contractor's consent

Clause 2.6 is often expanded to introduce an 'access regime' (to be included in the Employer's Requirements), which imposes obligations on the Parties in respect of the coordination of the Works with the work not forming part of the Contract.

Supply of Documents, Setting Out etc.

Overview

This brief sub-section includes the following provisions, which broadly relate to documents:

- the custody of the Contract Documents, and requirements relating to confidentiality;
- the Contractor's obligation to submit Contractor's Design Documents in accordance with the Design Submission Procedure; and
- the Employer's obligation to define the boundaries of the site.

Key provisions

- clause 2.7 (Contract Documents)
- clause 2.8 (Construction information)
- clause 2.9 (Site boundaries)

Clause 2.7: Contract Documents

2.7.1 *The Contract Documents shall remain in the custody of the Employer and shall be available at all reasonable times for inspection by the Contractor.*

Clause 2.7 is drafted on the basis that there is only one original (paper) copy of the Contract Documents.

It provides that the Employer will keep the original copy of the Contract Documents, but requires the Employer to make them available for inspection by the Contractor at '*all reasonable times*' (likely to relate to the site working hours or standard business hours).

Increasingly in practice, however, the Contract is often executed electronically or in counterpart, such that the Employer and the Contractor each has an original copy.

DOI: 10.1201/9781003485919-18

Recommendation: Record keeping for the Contract Documents

The Parties are encouraged to maintain an electronic copy of the Contract Documents, separated into sub-folders as follows:

- the Agreement and the Conditions (including any schedule of amendments);
- the Employer's Requirements;
- the Contractor's Proposals;
- the Contract Sum Analysis;
- the Construction Phase Plan; and
- any other Contract Documents.

These should be in a 'searchable' PDF format, to allow the Parties to locate specific provisions more easily.

Resource 36 page 428	Request: Contractor to Employer	A request by the Contractor to the Employer to inspect the Contract Documents.

2.7.2 *Immediately after the execution of this Contract the Employer, without charge to the Contractor, shall (unless previously provided or unless the BIM Protocol or other communications protocol requires otherwise) provide the Contractor with one copy, certified on behalf of the Employer, of the Contract Documents, together with any pre-construction information required to be provided to the Contractor under regulation 4(4) of the CDM Regulations and any building information required to be provided to the Contractor under regulation 11A(4) of the Building Regulations.*

Again, clause 2.7.2 assumes that there is only one original copy of the Contract Documents, and requires the Employer to provide a certified copy to the Contractor immediately after the execution of the Contract, along with:

(i) any pre-construction information required to be provided to the Contractor under regulation 4(4) of the CDM Regulations; and
(ii) any building information required to be provided to the Contractor under regulation 11A(4) of the Building Regulations.

Regulation 4(4) of the CDM Regulations

Regulation 4(4) of the CDM Regulations states:

> *'(4) A client [the Employer] must provide pre-construction information as soon as is practicable to every designer and contractor appointed, or being considered for appointment, to the project.'*
>
> '*Pre-construction information*' is defined in regulation 2(1) as follows:
>
> *"pre-construction information' means information in the client's possession or which is reasonably obtainable by or on behalf of the client, which is relevant to the construction work and is of an appropriate level of detail and proportionate to the risks involved, including—*
>
> *(a) information about—*
> *(i) the project;*
> *(ii) planning and management of the project;*
> *(iii) health and safety hazards, including design and construction hazards and how they will be addressed; and*
> *(b) information in any existing health and safety file;"*

Regulation 11A(4) of the Building Regulations

Similarly, Regulation 11A(4) of the Building Regulations states:

> *'(4) A client [the Employer] must provide building information as soon as is practicable to every designer and contractor on the project.'*
>
> '*Building information*' is defined in regulation 11Q(1) as follows:
>
> *"building information' means information in the client's possession or which is reasonably obtainable by or on behalf of the client, which is relevant to the building work or the design work, including information about—*
>
> *(a) the work;*
> *(b) planning and management of the project;*
> *(c) issues relating to compliance with any relevant requirement and how they were addressed;'*

There is considerable overlap between the two regulations, both of which are broadly drafted.

The Employer's obligations under clause 2.7.2 apply unless:

- the Employer has previously provided the Contractor with a copy of this information; or
- the BIM Protocol (or another communications protocol) states that the Employer does not have to provide a copy to the Contractor.

Resource 37 page 428	Correspondence: Employer to Contractor	The provision by the Employer to the Contractor of a certified copy of the Contract Documents, pre-construction information under the CDM Regulations, and building information under the Building Regulations.

2.7.3 The Contractor shall keep on the site or accessible there to the Employer's Agent at all reasonable times copies of the Contract Documents and the Contractor's Design Documents.

Clause 2.7.3 requires the Contractor to keep a copy of both the Contract Documents and the Contractor's Design Documents on site, and make them available to the Employer's Agent, at '*all reasonable times*' (again, likely to relate to the site working hours or standard business hours).

Increasingly in practice, there will likely be an electronic copy of the Contract Documents and the Contractor's Design Documents, typically stored on a document management platform, for which the Contractor can provide the necessary access to the Employer's Agent.

Resource 38 page 429	Correspondence: Contractor to Employer's Agent	A reminder by the Contractor to the Employer's Agent of its entitlement to access copies of the Contract Documents and the Contractor's Design Documents.

2.7.4 Neither Party shall divulge or use except for the purposes of this Contract:

2.7.4.1 the Contract Documents and the Contractor's Design Documents; or

2.7.4.2 any confidential information of the other Party,

save that the Employer may use in connection with the maintenance, use, repair, advertisement, letting or sale of the Works any of the documents supplied by the Contractor.

Apart from the exceptions listed below, and except for the purposes of the Contract, clause 2.7.4 prohibits both Parties from disclosing or using any of the following:

- the Contract Documents;
- the Contractor's Design Documents; or
- any confidential information of the other Party.

In other words, these documents and information must be kept confidential.

There are a number of exceptions in the case of the Employer, which may use any documents provided by the Contractor in connection with the maintenance, use, repair, advertisement, letting, or sale of the Works. For example, to assist with the preparation of marketing literature.

> 2.7.5 *Where the Employer is a Local or Public Authority or other body of the type mentioned in Supplemental Provision 8, its obligations of confidentiality shall be subject to that Supplemental Provision.*

Clause 2.7.5 applies where the Employer is a Local or Public Authority, or another body to which the provisions of the Freedom of Information Act 2000 apply.

In these circumstances, the Employer's obligations in respect of confidentiality are subject to Supplemental Provision 8 (Transparency), considered below, which relates to potential disclosures of the Contract Documents under the Freedom of Information Act 2000.

Clause 2.8: Construction information

> 2.8 *Save for any Contractor's Design Documents contained in the Contractor's Proposals, the Contractor shall without charge provide to the Employer copies of the Contractor's Design Documents as and when necessary from time to time in accordance with the Design Submission Procedure and the Contractor shall not commence any work to which such a document relates before that procedure has been complied with.*

Clause 2.8 relates to the provision by the Contractor of Contractor's Design Documents, in accordance with the Design Submission Procedure, which is included in Schedule 1 (Design Submission Procedure), considered below.

'*Contractor's Design Documents*' are defined in clause 1.1 (Definitions) as follows:

> 1.1 **Contractor's Design Documents:** *the drawings, details and specifications of materials, goods and workmanship and other related documents and information prepared by or for the Contractor in relation to the design of the Works (including such as are contained in the Contractor's Proposals), together, where applicable, with any other design documents or information to be provided by the Contractor under the BIM Protocol.*

Clause 2.8 requires the Contractor to provide the Employer with copies of all Contractor's Design Documents (unless they are already contained in the Contractor's Proposals), in accordance with the Design Submission Procedure.

Furthermore, the Contractor must not begin carrying out any work relating to a Contractor's Design Document until the Design Submission Procedure has been complied with. This is to help ensure that the Contractor's Design Document, and therefore any related work, is in accordance with the Contract.

The Contractor should therefore take care not to undertake work relating to a Contractor's Design Document until the Design Submission Procedure has been complied with, because it otherwise risks not being paid for that work (as provided in paragraph 6 of the Design Submission Procedure).

Resource 39 page 429	Correspondence: Contractor to Employer	Confirmation from the Contractor to the Employer of the provision of Contractor's Design Documents.
Resource 40 page 430	Reminder: Employer to Contractor	A reminder from the Employer to the Contractor of its obligation to comply with the Design Submission Procedure.

Clause 2.9: Site boundaries

2.9 The Employer shall define the boundaries of the site.

Clause 2.9 simply provides that it is the Employer's responsibility to define the boundaries of the site.

This is typically achieved by reference to an appropriately annotated site plan drawing.

Resource 41 page 431	Confirmation: Employer to Contractor	Confirmation by the Employer to the Contractor of the boundaries of the site.

Market practice: Marking out

Clause 2.9 is often expanded to make the Contractor responsible for marking out the boundaries of the site, in accordance with the definition provided by the Employer.

Discrepancies and Divergences

Overview

This sub-section primarily relates to the resolution of various errors or difficulties, described as '*inadequacies*', '*discrepancies*', and '*divergences*', within or between the following:

- the Employer's Requirements (including any Changes to the Employer's Requirements);
- the definition of the site boundary;
- Employer's instructions;
- the Contractor's Proposals;
- Contractor's Design Documents; and
- the Statutory Requirements.

Whilst there are a number of exceptions (in respect of divergences from the Statutory Requirements), the general principle is that the cost of resolving any such matters will be borne by the Party responsible for the relevant document.

Key provisions

- clause 2.10 (Divergence in Employer's Requirements and definition of site boundary)
- clause 2.11 (Preparation of Employer's Requirements)
- clause 2.12 (Employer's Requirements – inadequacy)
- clause 2.13 (Notice of discrepancies etc.)
- clause 2.14 (Discrepancies in documents)
- clause 2.15 (Divergences from Statutory Requirements)
- clause 2.16 (Emergency compliance with Statutory Requirements)

Clause 2.10: Divergence in Employer's Requirements and definition of site boundary

2.10.1 *Any divergence between the Employer's Requirements and the definition of the site boundary as provided under clause 2.9 shall be corrected by an instruction issued by the Employer and such instruction shall be treated as a Change.*

DOI: 10.1201/9781003485919-19

Clause 2.10 provides for where there is a divergence between (i) the Employer's Requirements, and (ii) the definition of the site boundary, for which the Employer is responsible under clause 2.9 above.

In these circumstances, the Employer must issue an instruction to correct the divergence, which will be treated as a Change.

> 2.10.2 *If the Employer or the Contractor becomes aware of any such divergence it shall immediately give the other notice with details.*

Clause 2.10.2 requires both Parties to immediately notify the other if they become aware of any such divergence between the Employer's Requirements and the definition of the site boundary.

Resource 42 page 431	Notice: Contractor to Employer	A notice from the Contractor to the Employer of a divergence between the Employer's Requirements and the definition of the site boundary, and a request for an instruction.
Resource 43 page 432	Notice: Employer to Contractor	A notice from the Employer to the Contractor of a divergence between the Employer's Requirements and the definition of the site boundary, with an accompanying instruction.

Clause 2.11: Preparation of Employer's Requirements

> 2.11 *Subject to clause 2.15, the Contractor shall not be responsible for the contents of the Employer's Requirements or for verifying the adequacy of any design contained within them.*

Clause 2.11 provides that the Contractor is not responsible for either:

- the contents of the Employer's Requirements; or
- verifying the adequacy of any design contained within the Employer's Requirements.

The only exception is in the case of divergences between the Employer's Requirements and the Statutory Requirements, considered further in clause 2.15 (Divergences from Statutory Requirements) below.

Market practice: Responsibility for the design in the Employer's Requirements

Clause 2.11 (along with clause 2.12 to clause 2.15) is often amended to the effect that the Contractor accepts responsibility for any design contained within the Employer's Requirements. This includes responsibility for resolving any inadequacy, discrepancy, or divergence within or between the Employer's Requirements and the Contractor's Proposals or Contractor's Design Documents, and the Statutory Requirements.

This creates a 'single point of responsibility' for the Employer, but it potentially greatly increases the risk for the Contractor, which will then be responsible for design for which it may not have had any involvement in preparing.

In such circumstances, the Contractor will need to ensure that it has had sufficient opportunity during the tender period to undertake a comprehensive review of the Employer's Requirements.

Clause 2.12: Employer's Requirements – inadequacy

2.12.1 *If an inadequacy is found in any design in the Employer's Requirements and the Contractor under clause 2.11 is not responsible for verifying its adequacy, then, if or to the extent that the inadequacy is not dealt with in the Contractor's Proposals, the Employer's Requirements shall be corrected, altered or modified accordingly.*

Clause 2.12 relates to inadequacies in any design included in the Employer's Requirements. '*Inadequacy*' is not a defined term, but it could include any error, mistake, omission, problem, shortcoming, or other defectiveness in the design.

If any such inadequacy is discovered in any design in the Employer's Requirements, and the Contractor is *not* responsible under clause 2.11 for verifying the adequacy of that design, then that inadequacy:

- will be dealt with as provided for in the Contractor's Proposals; or, failing that,
- will be corrected, altered, or modified accordingly.

2.12.2 *Subject to clause 2.15, any correction, alteration or modification under clause 2.12.1 shall be treated as a Change.*

Clause 2.12.2 provides that any such correction, alteration, or modification described under clause 2.12.1 will be treated as a Change.

This is on the basis that the Contractor is not responsible for verifying the adequacy of any design contained within the Employer's Requirements, in accordance with clause 2.11.

Again, the only exception is in the case of divergences between the Employer's Requirements and the Statutory Requirements, considered further in clause 2.15 (Divergences from Statutory Requirements) below.

Resource 44 page 432	Notice: Contractor to Employer	A notice from the Contractor to the Employer of an inadequacy in the Employer's Requirements, which is dealt with in the Contractor's Proposals.
Resource 45 page 433	Notice: Contractor to Employer	A notice from the Contractor to the Employer of an inadequacy in the Employer's Requirements (which is not dealt with in the Contractor's Proposals), and a request for an instruction.

Clause 2.13: Notice of discrepancies etc.

2.13 *If the Contractor becomes aware of any inadequacy as is referred to in clause 2.12 or any other discrepancy or divergence in or between any of the following, namely:*

2.13.1 *the Employer's Requirements;*

2.13.2 *the Contractor's Proposals and other Contractor's Design Documents; and*

2.13.3 *any instruction issued by the Employer under these Conditions;*

it shall immediately give notice with appropriate details to the Employer, and the Employer shall issue instructions in that regard.

Clause 2.13 applies where the Contractor discovers either:

- any inadequacy in the design in the Employer's Requirements; or
- any discrepancy or divergence in or between the following:
  - the Employer's Requirements;
  - the Contractor's Proposals and other Contractor's Design Documents; and
  - any Employer's instruction.

In these circumstances, the Contractor must immediately notify the Employer with appropriate details, following which the Employer must issue instructions to address the matter.

Resource 46 page 433	Instruction: Employer to Contractor	An instruction from the Employer following a notice of inadequacy in the design in the Employer's Requirements, under clause 2.12.

Clause 2.14: Discrepancies in documents

2.14.1 *Where the discrepancy or divergence to be notified under clause 2.13 is within or between the Contractor's Proposals and/or other Contractor's Design Documents, the Contractor shall notify the Employer of its proposed amendment to remove the discrepancy or divergence and (subject to compliance with Statutory Requirements) the Employer shall decide between the discrepant items or otherwise may accept the Contractor's proposed amendment; the Contractor shall be obliged to comply with the decision or acceptance by the Employer without cost to the Employer.*

Clause 2.14 provides how any such discrepancies or divergences notified by the Contractor under clause 2.13 are to be resolved.

The resolution will depend on where those discrepancies or divergences are located.

Clause 2.14.1 provides for where the discrepancy or divergence is within or between documents prepared by the Contractor:

- the Contractor's Proposals; and/or
- other Contractor's Design Documents.

In these circumstances, the Contractor must notify the Employer of the Contractor's proposed amendment to remove the discrepancy or divergence.

The Employer must then either:

- decide between the discrepant items; or otherwise,
- accept the Contractor's proposed amendment.

(In both cases, this is subject to the obligation to comply with the Statutory Requirements.)

The Contractor is then required to comply with (i) the Employer's decision, or (ii) the Employer's acceptance of the Contractor's proposed amendment, at the Contractor's cost.

Resource 47 page 434	Notice: Contractor to Employer	A notice from the Contractor to the Employer of a discrepancy or divergence in or between the Contractor's Proposals and/or other Contractor's Design Documents.
Resource 48 page 434	Instruction: Employer to Contractor	An instruction from the Employer following a notice of discrepancy or divergence in or between the Contractor's Proposals and/or other Contractor's Design Documents.

2.14.2 *Where the discrepancy is within the Employer's Requirements (including any Change to them) the Contractor's Proposals shall prevail (subject to compliance with Statutory Requirements), without any adjustment of the Contract Sum. Where the Contractor's Proposals do not deal with the discrepancy, the Contractor shall notify the Employer of its proposed amendment to deal with the discrepancy and the Employer shall either agree the proposed amendment or decide how the discrepancy is to be dealt with; that agreement or decision shall be notified to the Contractor and treated as a Change.*

Clause 2.14.2 provides for where the discrepancy is within the Employer's Requirements (including any Changes to the Employer's Requirements, for example, following an Employer's instruction under clause 3.9 (Instructions requiring Changes)).

If the Contractor's Proposals deal with the discrepancy, then they will prevail (or take precedence), and there will be no adjustment (whether increase or decrease) of the Contract Sum. (Again, this is subject to the obligation to comply with the Statutory Requirements.)

If the Contractor's Proposals do *not* deal with the discrepancy, the Contractor must notify the Employer of the Contractor's proposed amendment to deal with the discrepancy.

The Employer must then either:

- agree to the Contractor's proposed amendment to deal with the discrepancy; or
- otherwise decide how the discrepancy is to be dealt with.

In both cases, the Employer must notify the Contractor of their agreement or decision, which will be treated as a Change.

Resource 49 page 435	Notice: Contractor to Employer	A notice from the Contractor to the Employer of a discrepancy in or between the Employer's Requirements and/or instructions.
Resource 50 page 436	Instruction: Employer to Contractor	An instruction from the Employer following a notice of a discrepancy in or between the Employer's Requirements and/or instructions.

Clause 2.15: Divergences from Statutory Requirements

2.15.1 If the Contractor or Employer becomes aware of a divergence between the Statutory Requirements and

2.15.1 the Employer's Requirements (including any Change); or

2.15.2 the Contractor's Proposals or other Contractor's Design Documents,

it shall immediately give the other notice specifying the divergence and the Contractor shall notify the Employer of its proposed amendment for removing the divergence. With the Employer's consent, the Contractor shall entirely at its own cost, save as provided in clause 2.15.2, complete the design and construction of the Works in accordance with the amendment and the Employer shall note the amendment on the Contract Documents.

Clause 2.15 relates to divergences from the Statutory Requirements.

If the Contractor or the Employer discovers any such divergence between the Statutory Requirements and either:

- the Employer's Requirements (including any Changes to the Employer's Requirements, for example, following an Employer's instruction under clause 3.9 (Instructions requiring Changes)); or
- the Contractor's Proposals, or other Contractor's Design Documents,

they must immediately notify the other Party, stating the divergence.

The Contractor is then obliged to notify the Employer of the Contractor's proposed amendment for removing the divergence, to ensure compliance with the Statutory Requirements.

Provided the Contractor has obtained the Employer's consent, it must then complete the design and construction of the Works in accordance with its proposed amendment.

This will be at the Contractor's cost, unless one of the circumstances in clause 2.15.2 applies.

The Employer is then also required to note that amendment on the Contract Documents.

Resource 51 page 436	Notice: Contractor to Employer	A notice from the Contractor to the Employer of a divergence between the Statutory Requirements and the Employer's Requirements (including any Change), the Contractor's Proposals, or other Contractor's Design Documents, and a request for the Employer's consent to proceed with the Contractor's proposed amendment to remove the divergence.
Resource 52 page 437	Notice: Employer to Contractor	A notice from the Employer to the Contractor of a divergence between the Statutory Requirements and the Employer's Requirements (including any Change), the Contractor's Proposals, or other Contractor's Design Documents, and a request for a proposed amendment to remove the divergence.
Resource 53 page 438	Consent given: Employer to Contractor	Consent given by the Employer to the Contractor, for the Contractor's proposed amendment, following a notice of a divergence between the Statutory Requirements and the Employer's Requirements (including any Change), the Contractor's Proposals, or other Contractor's Design Documents.
Resource 54 page 438	Consent withheld: Employer to Contractor	Consent withheld by the Employer to the Contractor, for the Contractor's proposed amendment, following a notice of a divergence between the Statutory Requirements and the Employer's Requirements (including any Change), the Contractor's Proposals, or other Contractor's Design Documents.

2.15.2.1 *If after the Base Date there is a change in the Statutory Requirements which necessitates an alteration or modification to the Works, such alteration or modification shall be treated as a Change.*

Clause 2.15.2.1 applies where:

- there is a change in the Statutory Requirements *after* the Base Date (as identified against the reference to clause 1.1 in the Contract Particulars); and
- that change necessitates an alteration or modification to the Works, in order to ensure compliance with the Statutory Requirements.

In these circumstances, the necessary alteration or modification will be treated as a Change.

Resource 55 page 439	Confirmation and request: Contractor to Employer	Confirmation by the Contractor to the Employer of a change in the Statutory Requirements after the Base Date which necessitates an alteration or modification to the Works, and a request for an instruction.
Resource 56 page 439	Confirmation and instruction: Employer to Contractor	Confirmation by the Employer to the Contractor of a change in the Statutory Requirements after the Base Date which necessitates an alteration or modification to the Works, and an instruction.

Market practice: Change in the Statutory Requirements after the Base Date

Clause 2.15.2.1 is often amended to make the Contractor liable for any changes to the Statutory Requirements after the Base Date, provided those changes were 'reasonably foreseeable'.

2.15.2.2 *If any amendment to the Contractor's Proposals becomes necessary for conformity with the terms of any permission or approval made by a decision of the relevant authority after the Base Date for the purposes of Development Control Requirements, such amendment shall be treated as a Change provided that such treatment is not precluded in the Employer's Requirements.*

Clause 2.15.2.2 applies where:

- a permission is given, or an approval is made, by a decision of the relevant authority for the purposes of Development Control Requirements (thought to relate primarily to planning), *after* the base date; and
- that decision necessitates an amendment to the Contractor's Proposals, so that the Contractor's Proposals conform with the terms of the permission given or approval made.

In these circumstances, the amendment to the Contractor's Proposals will be treated as a Change, unless this is prohibited in the Employer's Requirements.

Resource 57 page 440	Confirmation: Contractor to Employer	Confirmation from the Contractor to the Employer of an amendment to the Contractor's Proposals for conformity with the Development Control Requirements, and a request for an instruction.
Resource 58 page 441	Confirmation and instruction: Employer to Contractor	Confirmation from the Employer to the Contractor of an amendment to the Contractor's Proposals for conformity with the Development Control Requirements, and an instruction.

2.15.2.3 *If any amendment to the part or parts of the Employer's Requirements to which clause 2.1.2 refers becomes necessary for conformity with Statutory Requirements the Employer shall issue an instruction requiring a Change.*

Clause 2.15.2.3 applies in circumstances where any part of the Employer's Requirements specifically states that the Employer's Requirements comply with the Statutory Requirements.

If this statement proves to be incorrect, and an amendment to that part of the Employer's Requirements therefore becomes necessary in order to comply with the Statutory Requirements, then the Employer must issue an instruction requiring a Change.

Resource 59 page 442	Confirmation and request for instruction: Contractor to Employer	Confirmation from the Contractor to the Employer of an amendment to the Employer's Requirements to conform with the Statutory Requirements, and a request for an instruction.
Resource 60 page 442	Confirmation and instruction: Employer to Contractor	Confirmation from the Employer to the Contractor of an amendment to the Employer's Requirements to conform with the Statutory Requirements, and an instruction.
Resource 61 page 443	Tracker: Contractor and Employer	A tracker template for recording inadequacies, divergences, and discrepancies under clause 2.11 to clause 2.15.

Clause 2.16: Emergency compliance with Statutory Requirements

2.16.1 *If in an emergency compliance with the Statutory Requirements necessitates the Contractor supplying materials and/or executing work before receiving the Employer's consent under clause 2.15.1, the Contractor shall supply such limited materials and execute such limited work as are reasonably necessary to secure immediate compliance.*

Clause 2.16 applies in the (hopefully rare) circumstances where the Contractor is required to comply with the Statutory Requirements in an emergency. For example, relating to compliance with the Building Regulations.

It effectively takes precedence over clause 2.15.1, which requires the Contractor to obtain the Employer's consent before removing any divergence between the Statutory Requirements and either (i) the Employer's Requirements (including any Change), or (ii) the Contractor's Proposals, or other Contractor's Design Documents.

It applies in the case of an emergency, where it is necessary for the Contractor to either (i) supply certain materials, or (ii) carry out certain work, in order to comply with the Statutory Requirements, *before* the Contractor has obtained the Employer's consent under clause 2.15.1.

In these circumstances, the Contractor is required to (i) supply such *limited* materials, and (ii) carry out such *limited* work, (in other words, the minimum amount) as are reasonably necessary to ensure immediate compliance with the Statutory Requirements.

2.16.2 The Contractor shall forthwith notify the Employer of the emergency and the steps that it is taking under clause 2.16.1.

Clause 2.16.2 then requires the Contractor to immediately notify the Employer of both (i) the emergency, and (ii) the steps the Contractor is taking under clause 2.16.1 above.

Resource 62 page 443	Notice: Contractor to Employer	A notice from the Contractor to the Employer of an emergency compliance with the Statutory Requirements.

Design Work – liabilities and limitation

Overview

This sub-section contains only one provision, relating to the Contractor's standard of care in respect of design work.

In other words, the degree of care that the Contractor must take in carrying out such design work.

Key provisions

- Clause 2.17 (Design Work – liabilities and limitation)

Clause 2.17: Design Work – liabilities and limitation

2.17.1.1 Insofar as its design of the Works is comprised in the Contractor's Proposals and in what it is to complete in accordance with the Employer's Requirements and these Conditions (including any further design that it is required to carry out as a result of a Change), the Contractor warrants, subject to clause 2.17.1.2, that it shall use the reasonable skill and care in such design to be expected of a qualified and experienced architect (or other appropriate professional designer) undertaking the design of works similar in scope and character to such design of the Works.

Clause 2.17.1.1 sets out the Contractor's standard of care in respect of design work.

It applies to the extent that the Contractor's design of the Works is comprised in:

- the Contractor's Proposals; and
- what the Contractor is required to complete in accordance with the Employer's Requirements and the Conditions (including any *further* design that the Contractor is required to carry out, as a result of a Change).

Subject to the limitation in clause 2.17.1.2 below, the Contractor warrants (or promises) to the Employer that it will use the reasonable skill and care in respect of such design work as would be expected of both a qualified and experienced

DOI: 10.1201/9781003485919-20

architect (or other appropriate professional designer, such as a structural engineer, or a mechanical and electrical consultant) carrying out the design of works similar in scope and character to the design of the Works under the Contract.

The Contractor will therefore only be liable to the Employer under clause 2.17.1.1 for problems with its design in circumstances where the Contractor did *not* exercise this standard of care. In other words, where the Contractor was negligent.

> 2.17.1.2 *To the extent permitted by the Statutory Requirements, the Contractor shall have no greater duty, obligation or liability than to exercise reasonable skill and care as provided in clause 2.17.1.1 in respect of such design and under no circumstances shall the Contractor be subject to any duty, obligation or liability which requires that any such design shall be fit for its purpose.*

Clause 2.17.1.2 relates to 'fitness for purpose', which is a higher standard of care than that contained in clause 2.17.1, as it would effectively require the Contractor to promise to the Employer that its design would be fit for the Employer's intended purpose, no matter what.

It is therefore an 'absolute' obligation, with 'strict' liability: the Employer would not be required to demonstrate that the Contractor had been negligent, only that the design was not fit for its intended purpose.

The Contractor's Professional Indemnity insurance policy typically would not cover this standard of care.

Clause 2.17.1.2 therefore includes an express 'fitness for purpose' *exclusion* in respect of design.

As far as it is allowed by the Statutory Requirements, it provides that, in respect of design:

- the Contractor will be under no greater duty, obligation, or liability than that included in clause 2.17.1.1 above; and
- the Contractor will *not* be subject to any duty, obligation, or liability which requires that the design must be fit for its purpose.

> 2.17.2 *Where and to the extent that this Contract involves the Contractor in taking on work for or in connection with the provision of a dwelling or dwellings or work in relation to a dwelling or dwellings, clause 2.17.1 includes the Contractor's liability under the Defective Premises Act 1972.*

Clause 2.17.2 relates to the Defective Premises Act 1972.

It applies where the Contract involves the Contractor carrying out work for or in connection with (i) the provision of a dwelling (or dwellings), or (ii) in relation to a dwelling (or dwellings). In other words, homes.

In these circumstances, the Contractor's liability under clause 2.17.1 will also include its liability under the Defective Premises Act 1972.

Section 1(1) (*Duty to build dwellings properly*) of the Defective Premises Act 1972 states:

> *'A person taking on work for or in connection with the provision of a dwelling (whether the dwelling is provided by the erection or by the conversion or enlargement of a building) owes a duty—*
>
> *(a) if the dwelling is provided to the order of any person, to that person; and*
> *(b) without prejudice to paragraph (a) above, to every person who acquires an interest (whether legal or equitable) in the dwelling;*
>
> *to see that the work which he takes on is done in a workmanlike or, as the case may be, professional manner, with proper materials and so that as regards that work the dwelling will be fit for habitation when completed.'*

2.17.3 Where or to the extent that this Contract does not involve the Contractor in taking on work for or in connection with the provision of a dwelling or dwellings or work in relation to a dwelling or dwellings to which that Act applies, the Contractor's liability for loss of use, loss of profit or other consequential loss arising in respect of the Contractor's liability under clause 2.17.1 shall be limited to the amount, if any, stated in the Contract Particulars; but such a limitation shall not apply to or be affected by any liability for liquidated damages under clause 2.29.

Clause 2.17.3 provides for an optional cap on the Contractor's liability under clause 2.17.1, in certain limited circumstances.

In contrast with clause 2.17.2, clause 2.17.3 applies where the Contract does *not* involve the Contractor carrying out work for or in connection with (i) the provision of a dwelling (or dwellings), or (ii) in relation to a dwelling (or dwellings), to which the Defective Premises Act 1972 applies.

In these circumstances, the Contractor's liability under clause 2.17.1 will be limited to the amount (if any) stated against the reference to clause 2.17.3 in the Contract Particulars, in respect of the following:

- loss of use;
- loss of profit; or
- other consequential (or 'indirect') loss.

However, this cap on liability does *not* extend to the Contractor's liability to pay liquidated damages under clause 2.29 (Payment or allowance of liquidated damages) for late completion.

It also does not limit or affect the Contractor's liability for either:

- *direct* loss, including the cost of remedial work; or
- in circumstances where the problem cannot be remedied, any resultant diminution (or reduction) in value of the Works.

Fees, Royalties and Patent Rights

Overview

This sub-section relates to fees or charges which may be payable in accordance with the Statutory Requirements, along with the potential infringement of any patent rights during the course of carrying out of the Works.

Key provisions

- clause 2.18 (Fees or charges legally demandable)
- clause 2.19 (Patent rights and royalties – Contractor's indemnity)
- clause 2.20 (Patent rights – Instructions)

Clause 2.18: Fees or charges legally demandable

2.18 *The Contractor shall pay all fees or charges (including any rates or taxes) legally demandable under any Statutory Requirement and indemnify the Employer against any liability resulting from any failure to do so. No adjustment shall be made to the Contract Sum in respect of the amount of any such fees or charges (including any rates or taxes other than VAT) unless they are stated by way of a Provisional Sum in the Employer's Requirements, in which case clause 4.2 shall apply.*

Clause 2.18 requires the Contractor to pay all fees, charges, rates, or taxes which may be demanded under any of the Statutory Requirements. For example, relating to Development Control Requirements (thought to relate primarily to planning), or the Building Regulations.

Furthermore, the Contractor is also obliged to indemnify the Employer against any liability the Employer may incur, as a result of the Contractor's failure to make payment in this regard.

The Contract Sum will not be adjusted in respect of any such fees, charges, rates, or taxes (with the exception of VAT), unless they are included as a Provisional Sum in the Employer's Requirements. In these circumstances, clause 4.2 (Contract Sum and Adjustments: Items included in adjustments) will apply, which provides for the corresponding adjustment of the Contract Sum.

DOI: 10.1201/9781003485919-21

Clause 2.19: Patent rights and royalties – Contractor's indemnity

> 2.19 *Where the carrying out of the Works involves the supply or use of any patented article, process or invention or drawings or models of buildings that are the subject of copyright, other than drawings or models provided by the Employer, all royalties or other sums payable in respect of such supply and use shall be deemed to have been included in the Contract Sum or, where appropriate, the quoted adjustment to that sum, and the Contractor shall indemnify the Employer from and against all claims and proceedings which may be brought or made against the Employer, and all damages, costs and expense to which it may be put, by reason of the Contractor infringing or being held to have infringed any patent rights in relation to any such articles, processes and inventions or infringing or being held to have infringed copyright.*

Clause 2.19 applies where the carrying out of the Works involves either:

- any patented article, process, or invention; or
- drawings or models of buildings which are the subject of copyright (except for drawings or models provided by the Employer).

In these circumstances, any associated royalties or other sums in respect of the supply or use of such items are deemed included within the Contract Sum (or the quoted adjustment to the Contract Sum, where appropriate).

Clause 2.19 also includes an indemnity from the Contractor, relating to any claims which may be brought against the Employer, and all related damages, costs, and expense which it may incur, if caused by an infringement of the relevant patent rights or copyright by the Contractor.

Clause 2.20: Patent rights – Instructions

> 2.20.1 *Where in order to comply with the Employer's instructions, it is necessary for the Contractor in carrying out the Works to supply and/or use any patented article, process or invention, the Contractor shall not be liable in respect of any infringement or alleged infringement of any patent rights relating to such article, process or invention and all royalties, damages or other sums which the Contractor may be liable to pay to the persons entitled to such rights shall be added to the Contract Sum.*

Clause 2.20.1 applies where the Contractor is required to supply or use any patented article, process, or invention in carrying out the Works, in order to comply with an Employer's instruction.

In these circumstances:

- the Contractor will *not* be liable in respect of any infringement (or alleged infringement) of any related patent rights; and
- any royalties, damages, or other sums for which the Contractor is liable will be added to the Contract Sum.

> 2.20.2 *If prior to the instructions being carried out the Employer or the Contractor is or becomes aware that such supply or use may infringe any patent rights, it shall promptly notify and consult the other and the instructions shall not take effect unless confirmed by the Employer.*

Clause 2.20.2 applies where the Employer or the Contractor discover that the supply or use of any such patented article, process, or invention may infringe any patent rights, if the Contractor were to comply with the Employer's instruction.

In these circumstances:

- it must promptly notify and consult the other Party; and
- the Employer must confirm the instruction, if it wishes for the instruction to take effect.

This gives the Employer the opportunity to avoid incurring any liability relating to the infringement of those patent rights.

Resource 63 page 444	Notice: Contractor to Employer	A notice from the Contractor to the Employer of a potential infringement of patent rights.
Resource 64 page 444	Notice: Employer to Contractor	A notice from the Employer to the Contractor of a potential infringement of patent rights.
Resource 65 page 445	Confirmation: Employer to Contractor	Confirmation or withdrawal of an instruction by the Employer, relating to a potential infringement of patent rights.

Unfixed Materials and Goods – property, risk etc.

Overview

This sub-section relates primarily to the treatment of 'unfixed' materials and goods, and distinguishes between:

- Site Materials, which are on site; and
- Listed Items, which are off site.

Key provisions

- clause 2.21 (Materials and goods – on site)
- clause 2.22 (Materials and goods – off site)

Clause 2.21: Materials and goods – on site

> 2.21 *Site Materials shall not be removed from storage on or adjacent to the Works except for use on the Works without the Employer's consent. Where their value has been included in any Interim Payment, they shall upon such payment become the Employer's property, but, subject to Insurance Option B or C (if applicable), the Contractor shall remain responsible for loss or damage to them.*

Clause 2.21 relates to the '*Site Materials*', which are defined in clause 1.1 (Definitions) as follows:

> 1.1 ***Site Materials:*** *all unfixed materials and goods delivered to and placed on or adjacent to the Works which are intended for incorporation therein.*

Clause 2.21 prohibits the Contractor from removing the Site Materials from (or adjacent to) the Works unless:

- the Contractor has done so in order to use those Site Materials on the Works; or
- the Employer has given its consent.

DOI: 10.1201/9781003485919-22

(The Contractor will hopefully be able to manage its logistics sufficiently so that any such removal would not be required in any event.)

Clause 2.21 further provides that if the value of any Site Materials has been included in any Interim Payment, then once that payment has been made:

- those Site Materials will become the Employer's property; but
- subject to Insurance Option B (New Buildings – All Risks Insurance of the Works by the Employer) or Insurance Option C (Joint Names Insurance by the Employer of Existing Structures and Works in or Extensions to them), if either of them applies, the Contractor will still remain responsible for any loss or damage to those Site Materials.

Resource 66 page 445	Request: Contractor to Employer	A request from the Contractor to the Employer for consent to remove Site Materials from storage on or adjacent to the Works.
Resource 67 page 446	Consent given: Employer to Contractor	Consent given by the Employer for the Contractor to remove Site Materials from storage on or adjacent to the Works.
Resource 68 page 446	Consent withheld: Employer to Contractor	Consent withheld by the Employer for the Contractor to remove Site Materials from storage on or adjacent to the Works.

Clause 2.22: Materials and goods – off site

2.22 *Where the value of any Listed Items has in accordance with clause 4.15 been included in any Interim Payment, those items shall become the Employer's property and thereafter the Contractor shall not, except for use upon the Works, remove or cause or permit them to be moved or removed from the premises where they are. The Contractor shall be responsible for any loss of or damage to them and for the cost of their storage, handling and insurance until they are delivered to and placed on or adjacent to the Works. As from such delivery the provisions of clause 2.21 (except the words "Where their value" to "Employer's property, but,") shall apply to such items.*

Clause 2.22 relates to materials and goods off site, and in particular, '*Listed Items*', which are defined in clause 1.1 (Definitions) as follows:

> 1.1 ***Listed Items:*** *materials, goods and/or items prefabricated for inclusion in the Works which are listed as such items by the Employer in a list supplied to the Contractor and annexed to the Employer's Requirements.*

Clause 2.22 provides that, once the value of any Listed Items has been included in any Interim Payment, in accordance with clause 4.15 (Listed Items):

- those Listed Items will become the Employer's property;
- the Contractor must not then move them from their current premises, except for use on the Works;
- the Contractor will be responsible for any loss or damage to them; and
- the Contractor will also be liable for the cost of their storage, handling, and insurance, until they are delivered and placed on (or adjacent to) the Works.

Once they have been delivered to the site, they become Site Materials, and therefore (in accordance with clause 2.21) they must not be removed from (or adjacent to) the Works unless:

- the Contractor has done so in order to use those Site Materials on the Works; or
- the Employer has given its consent.

Again, subject to Insurance Option B or Insurance Option C, if either of them applies, the Contractor will still remain responsible for any loss or damage to those Site Materials.

Resource 69 page 447	Tracker: Contractor and Employer	A tracker template for recording Listed Items.

Adjustment of Completion Date

Overview

This sub-section relates to delays in the progress of the Works, which are a very common occurrence.

In some circumstances, the Contractor will be entitled to an extension of time as a result, by adjusting the relevant Completion Date. In others, it will not, and may therefore be required to pay liquidated damages to the Employer, for late completion.

Furthermore, it is for the Employer to decide whether to give an extension of time.

In that commercial context, it is unsurprising that disputes over the Contractor's entitlement to extensions of time are very common.

This sub-section sets out the following:

- the circumstances where the Contractor is entitled to an extension of time;
- the steps the Contractor must take to realise that entitlement; and
- the Employer's corresponding obligations in assessing the Contractor's entitlement to an extension of time.

Key provisions

- clause 2.23 (Related definitions and interpretation)
- clause 2.24 (Notice by Contractor of delay to progress)
- clause 2.25 (Fixing Completion Date)
- clause 2.26 (Relevant Events)

Clause 2.23: Related definitions and interpretation

2.23 In clauses 2.24 to 2.26 and, so far as relevant, in the other clauses of these Conditions:

2.23.1 any reference to delay or extension of time includes any further delay or further extension of time;

2.23.2 'Pre-agreed Adjustment' means the fixing of a revised Completion Date for the Works or a Section under Supplemental Provision 2 or by the Confirmed Acceptance of an Acceleration Quotation;

2.23.3 'Relevant Omission' means the omission of any work or obligation through an instruction for a Change under clause 3.9.

DOI: 10.1201/9781003485919-23

Clause 2.23 assists with the interpretation of the three key clauses relating to extensions of time, namely clause 2.24 to clause 2.26.

Clause 2.23.1 provides that any reference to either delay or an extension of time includes any *further* delay or extension of time, to help clarify that the extension of time mechanism set out in clause 2.24 to clause 2.26 may be used more than once, as there may be multiple delays and extensions of time.

Clause 2.23 also includes two further definitions:

- A '*Pre-agreed Adjustment*' relates to the following two provisions which allow the Parties to agree an adjustment to the Completion Date:
  - Supplemental Provision 2 (Valuation of Changes – Contractor's estimates), in relation to compliance with Employer's instructions under clause 3.9 (Instructions requiring Changes); and
  - Supplemental Provision 4 (Acceleration Quotation), in circumstances where the Employer wishes to achieve practical completion before the relevant Completion Date.

  It is relevant for the purposes of clause 2.25.4 and the definition of '*Completion Date*' included in clause 1.1 (Definitions).
- A '*Relevant Omission*' is where the Employer issues an instruction to the Contractor under clause 3.9 (Instructions requiring Changes), omitting any work or obligation. It is relevant for the purposes of clause 2.25 below.

Clause 2.24: Notice by Contractor of delay to progress

2.24.1 *If and whenever it becomes reasonably apparent that the progress of the Works or any Section is being or is likely to be delayed the Contractor shall forthwith give notice to the Employer of the material circumstances, including the cause or causes of the delay, and shall identify in the notice any event which in its opinion is a Relevant Event.*

Clause 2.24.1 sets out the Contractor's obligations if it considers that the progress of the Works (or any Section) is either (i) being delayed, or (ii) is *likely* to be delayed.

In these circumstances, clause 2.24.1 requires the Contractor to give a notice to the Employer immediately, upon the delay becoming reasonably apparent.

This notice must contain the following:

- **the '*material circumstances*' of the delay**: in other words, the Contractor must describe what has happened. This could include:
  - the part of the Works (or Section) affected;
  - the activities affected (for example, particular design, procurement, or construction activities);
  - the parties involved; and
  - the relevant dates and timeline of events (for example, relating to when certain activities were planned to take place, and when they actually took place);

- **the '*cause or causes of the delay*'**: the Contractor must explain what has caused the delay. There may be more than one cause, and the circumstances may well be very complicated. For example, it may be necessary to distinguish between superficial 'immediate' causes and the 'root' causes; and
- **Relevant Events**: if the Contractor considers that any of the events identified in its notice is a '*Relevant Event*' (as defined in clause 2.26 below), it must state so.

Recommendation: Notification of all delays

Clause 2.24.1 requires the Contractor to notify the Employer of *any* delay to the progress of the Works, irrespective of whether that delay might entitle the Contractor to an extension of time.

It is recognised that, on one level, there is evident pressure on the Parties to emphasise those delays which could best serve their respective commercial interests. A Contractor has a commercial interest in notifying the Employer of a Relevant Event, but may be reluctant in notifying the Employer of a delay for which it is responsible. Equally (and notwithstanding its obligation to act impartially in respect of these matters), the Employer has a commercial interest in demonstrating that an event is not a Relevant Event.

However, it is ultimately considered to be in the best commercial interests of both Parties (and the progress of the Works) to promote transparency and accountability, as this should help resolve any disputes more quickly. This also aligns with the Parties' shared obligation under Article 3 (Collaborative working), to work with each other in a collaborative manner.

The Contractor is therefore encouraged to see the benefit in notifying the Employer of all delays to the progress of the Works, as clause 2.24.1 requires.

Market practice: 'Time bars' to extensions of time

Clause 2.24 is often amended to the effect that the Contractor is obliged to notify the Employer of any delays within a specific timeframe, failing which it will *not* be entitled to any extension of time. These are often described as 'time bar' or 'condition precedent' provisions.

They evidently pose a greater risk to the Contractor, and therefore require more careful administration.

2.24.2 *In respect of each event identified in the notice the Contractor shall, if practicable in such notice or otherwise in writing as soon as possible thereafter, give particulars of its expected effects, including an estimate of any expected delay in the completion of the Works or any Section beyond the relevant Completion Date.*

Clause 2.24.2 imposes further requirements relating to the events included in the Contractor's notice under clause 2.24.1.

It requires the Contractor to notify the Employer of the following, either (i) within the notice given under clause 2.24.1, or (ii) as soon as possible afterwards, in respect of each delay:

- the effects the Contractor expects the delay will have on the progress of the Works. This could include:
  - the activities affected (for example, particular design, procurement, or construction activities);
  - the manner in which those activities will be affected (for example, whether design work can be completed, or an order for materials placed); and
  - how this will ultimately delay the progress of the Works; and
- the estimated duration of the period between (i) the Completion Date (by when the Works should have completed), and (ii) when the Works are now expected to achieve practical completion, as a result of the delay.

This information is often provided in the form of a programme.

Recommendation: Estimates of the delay

Clause 2.24.2 requires the Contractor to give an '*estimate*' of the expected delay, which may not be entirely accurate. For complex construction projects spanning a number of years, these estimates will very likely not be entirely accurate.

However, the Contractor should avoid waiting until it has a more precise understanding of the extent of the delay. It is encouraged to provide its initial estimate (supported by any relevant underlying assumptions) as soon as possible, and then update that estimate in accordance with clause 2.24.3 below, as often as is necessary.

Resource 70 page 447	Notice: Contractor to Employer	A notice from the Contractor to the Employer of a delay to the progress of the Works, and particulars of its expected effects.

2.24.3 *The Contractor shall forthwith notify the Employer of any material change in the estimated delay or in any other particulars.*

Once the initial notices under clause 2.24.1 and clause 2.24.2 have been given, clause 2.24.3 requires the Contractor to keep the Employer updated.

It requires the Contractor to notify the Employer immediately, if the Contractor considers that there is any '*material*' (or significant) change in either:

- the Contractor's estimate of the delay; or
- any other particulars.

It may be necessary for the Contractor to update the Employer in accordance with clause 2.24.3 on multiple occasions.

Resource 71 page 448	Notice: Contractor to Employer	A notice from the Contractor to the Employer of a material change in the estimated delay, or in other particulars.

2.24.4 *If, on receiving the particulars under clause 2.24.2, or any notification under clause 2.24.3, the Employer requires such further information as is reasonably necessary to enable it to reach a decision under clause 2.25.1, then, not later than 14 days from receipt of the particulars or notification, it shall notify the Contractor stating the further information required, and the Contractor shall supply such further information.*

Clause 2.24.4 entitles the Employer to request further information from the Contractor, which is reasonably necessary to enable the Employer to make its decision under clause 2.25.1 (Fixing Completion Date), regarding the Contractor's entitlement to an extension of time.

The Employer must notify the Contractor of any further information it requires, not later than 14 days from receipt of either:

- the initial particulars of the delay, in accordance with clause 2.24.2; or
- any notification from the Contractor of any material change in the estimated delay or in any other particulars, in accordance with clause 2.24.3.

The Contractor is then required to provide that further information to the Employer.

Resource 72 page 449	Notice: Employer to Contractor	A notice from the Employer to the Contractor requesting further information to enable the Employer to reach a decision under clause 2.25.1.
Resource 73 page 450	Correspondence: Contractor to Employer	The provision by the Contractor to the Employer of further information, further to the Employer's notice under clause 2.24.4.
Resource 74 page 450	Reminder: Contractor to Employer	A reminder from the Contractor to the Employer of its obligation to notify the Contractor of its decision under clause 2.25.1 as soon as reasonably practicable.

Clause 2.25: Fixing Completion Date

2.25.1 *If on receiving a notice and particulars under clauses 2.24.1 and 2.24.2 or (where applicable) a notification under clause 2.24.3:*

2.25.1.1 *any of the events which are stated to be a cause of delay is a Relevant Event; and*

2.25.1.2 *completion of the Works or of any Section is likely to be delayed thereby beyond the relevant Completion Date,*

then, save where these Conditions expressly provide otherwise, the Employer shall give an extension of time by fixing such later date as the Completion Date for the Works or Section as it then estimates to be fair and reasonable.

Clause 2.25.1 sets out the Employer's obligations, once the Contractor has given:

- the notice under clause 2.24.1;
- the particulars under clause 2.24.2; and, where applicable,
- a notification under clause 2.24.3.

If both:

- any of the events which the Contractor has stated to be a cause of delay is a Relevant Event (as defined in clause 2.26 below); and
- completion of the Works (or Section) is likely to be delayed because of that Relevant Event, beyond the relevant Completion Date,

then the Employer must give an extension of time.

This is achieved by fixing a later Completion Date for the Works (or Section) as the Employer then estimates to be '*fair and reasonable*'.

The exception is where the Conditions expressly state that there will be *no* adjustment of the Completion Date, or that the Contractor will *not* be entitled to an extension of time (for example, in clause 3.14 (Workmanship not in accordance with the Contract)).

Resource 75 page 451	Notice: Employer to Contractor	A notice from the Employer to the Contractor, giving an extension of time.
Resource 76 page 453	Notice: Employer to Contractor	A notice from the Employer to the Contractor, not giving an extension of time.

2.25.2 *The Employer shall notify the Contractor of its decision in respect of each notice and particulars under clauses 2.24.1 and 2.24.2 and each notification (if any) under clause 2.24.3, whether or not an extension is given, as soon as is reasonably practicable and in any event not later than 8 weeks from the later of:*

2.25.2.1 *the date of receipt of the particulars or notification; or*

2.25.2.2 *the date of receipt of the further information required under clause 2.24.4.*

Where the period from receipt to the Completion Date is less than 8 weeks, it shall endeavour to do so prior to the Completion Date.

Whether or not the Employer gives an extension of time, clause 2.25.2 requires the Employer to notify the Contractor of its decision in respect of each of the following:

- the notice under clause 2.24.1, and the particulars under clause 2.24.2; and, where applicable,
- a notification under clause 2.24.3.

The notice must be given as soon as '*reasonably practicable*', and in any event not later than 8 weeks from the later of the following:

- the date the Employer received (i) the particulars under clause 2.24.2, or (ii) the notification under clause 2.24.3; or
- the date the Employer received the further information required by the Employer, under clause 2.24.4.

If the period between (i) the receipt of the particulars, notification, or further information required, and (ii) the Completion Date, is less than 8 weeks, then the Employer must try to notify the Contractor of its decision prior to that Completion Date.

2.25.3 *The Employer shall in its decision state:*

2.25.3.1 *the extension of time that it has attributed to each Relevant Event; and*

2.25.3.2 *(in the case of a decision under clause 2.25.4 or 2.25.5) the reduction in time that it has attributed to each Relevant Omission.*

Clause 2.25.3 provides that, in the notice of its decision to the Contractor under clause 2.25.2, the Employer must state:

- the extension of time that the Employer has attributed to each Relevant Event; and
- in the case of a decision under clause 2.25.4 or clause 2.25.5 below, the *reduction* in time that the Employer has attributed to each '*Relevant Omission*' (defined in clause 2.23.3 above).

Recommendation: Record keeping of delays

Assessing the Contractor's entitlement to extensions of time is often a very difficult exercise for the Employer, particularly in the context of multiple delays on a complex project.

For both Parties, it requires consistent, disciplined, active management, along with regular communication.

It is amongst the most frequently cited advice on this subject, but worth repeating: the quality and availability of contemporaneous records is critical to ensuring that the Employer can perform its obligations under clause 2.25.1, and the Contractor obtains the extensions of time to which it is entitled in principle.

2.25.4 After the first fixing of a later Completion Date in respect of the Works or a Section, either under clause 2.25.1 or by a Pre-agreed Adjustment, but subject to clauses 2.25.6.3 and 2.25.6.4, the Employer may by notice to the Contractor, giving the details referred to in clause 2.25.3, fix a Completion Date for the Works or that Section earlier than that previously so fixed if the fixing of such earlier Completion Date is fair and reasonable, having regard to any Relevant Omissions for which instructions have been issued after the last occasion on which a new Completion Date was fixed for the Works or for that Section.

Clause 2.25.4 applies where a later Completion Date has already been fixed in respect of the Works (or Section), either:

- under clause 2.25.1; or
- by a '*Pre-agreed Adjustment*' (defined in clause 2.23.2 above).

In these circumstances, the Employer may give a notice to the Contractor, fixing a Completion Date for the Works (or Section) *earlier* than that previously fixed, if it is '*fair and reasonable*', having regard to any Relevant Omissions for which the Employer has issued instructions *after* the last time a new Completion Date was fixed for the Works (or Section).

This notice must include the reduction in time that the Employer has attributed to each Relevant Omission, in accordance with clause 2.25.3.

The fixing of an earlier Completion Date under clause 2.25.4 is subject to both:

- clause 2.25.6.3, which prohibits the Employer from fixing a Completion Date earlier than the (original) Date for Completion (as stated against the reference to clause 1.1 in the Contract Particulars); and
- clause 2.25.6.4, which prohibits the Employer from altering the length of any Pre-Agreed Adjustment, unless both (i) that Pre-Agreed Adjustment relates to a Change, and (ii) that Change is itself the subject of a Relevant Omission.

2.25.5 *After the Completion Date for the Works or for a Section, if this occurs before the date of practical completion, the Employer may, and not later than the expiry of 12 weeks after the date of practical completion shall, by notice to the Contractor, giving the details referred to in clause 2.25.3:*

2.25.5.1 *fix a Completion Date for the Works or for the Section later than that previously fixed if that is fair and reasonable having regard to any Relevant Events, whether on reviewing a previous decision or otherwise and whether or not the Relevant Event has been specifically notified by the Contractor under clause 2.24; or*

2.25.5.2 *subject to clauses 2.25.6.3 and 2.25.6.4, fix a Completion Date earlier than that previously fixed if that is fair and reasonable having regard to any instructions for Relevant Omissions issued after the last occasion on which a new Completion Date was fixed for the Works or Section; or*

2.25.5.3 *confirm the Completion Date previously fixed.*

Clause 2.25.5 requires the Employer to undertake a final review of the Contractor's entitlement to extensions of time.

The Employer must notify the Contractor of the outcome of this review by no later than 12 weeks after the date of practical completion of the Works or relevant Section, and this notice must include the details set out in clause 2.25.3 above.

The Employer must decide between the following three options:

- **Award a further extension of time, by fixing a later Completion Date.** Again, similar to clause 2.25.1, it must be '*fair and reasonable*', having regard to any Relevant Events, and (i) whether or not the Employer has reviewed a previous decision it made, and (ii) whether or not the Contractor has notified the Employer of the Relevant Event under clause 2.24.
- **Fix an *earlier* Completion Date.** Again, similar to clause 2.25.1, it must be '*fair and reasonable*', having regard to any instructions for Relevant Omissions which were issued by the Employer *after* the last time a new Completion Date was fixed.

- **Maintain the Completion Date as currently stated.** This is on the basis that no further extensions (or reductions) of time are due.

Resource 77 page 454	Notice: Employer to Contractor	A notice from the Employer to the Contractor of the final review of the Completion Dates.

2.25.6 Provided always that:

Clause 2.25.6 imposes a number of further qualifications on both:

- the Contractor's entitlement to extensions of time, and
- the Employer's decisions under clause 2.25.4 and clause 2.25.5.2 above.

2.25.6.1 the Contractor shall constantly use its best endeavours to prevent delay in the progress of the Works or any Section, however caused, and to prevent the completion of the Works or Section being delayed or further delayed beyond the relevant Completion Date;

Clause 2.25.6.1 requires that the Contractor must constantly use its '*best endeavours*' (or best efforts) to:

- prevent any delay in the progress of the Works (or any Section), however that delay is caused; and
- prevent practical completion of the Works (or any Section) being delayed (or further delayed) beyond the Completion Date.

Resource 78 page 455	Reminder: Employer to Contractor	A reminder from the Employer to the Contractor of its obligation to use best endeavours to prevent delay.

2.25.6.2 in the event of any delay the Contractor shall do all that may reasonably be required to the satisfaction of the Employer to proceed with the Works or Section;

Clause 2.25.6.2 requires that, if there is a delay, the Contractor must do everything to proceed with the Works (or Section), which may reasonably be required of it, to the Employer's satisfaction.

For example, if the delay in question will not prevent the Contractor from proceeding with other work (for example, design work, or procurement), then the Contractor must continue with that work. If the Employer has any specific requirements in this regard, it should notify the Contractor as soon as possible.

Resource 79 page 455	Reminder: Employer to Contractor	A reminder from the Employer to the Contractor to do all that may reasonably be required to the satisfaction of the Employer to proceed with the Works.

2.25.6.3 no decision of the Employer under clause 2.25.4 or 2.25.5.2 shall fix a Completion Date for the Works or any Section earlier than the relevant Date for Completion; and

In making a decision under clause 2.25.4 or clause 2.25.5.2, clause 2.25.6.3 prohibits the Employer from fixing a Completion Date for the Works (or any Section) earlier than the (original) Date for Completion (as stated against the reference to clause 1.1 in the Contract Particulars).

2.25.6.4 no decision under clause 2.25.4 or 2.25.5.2 shall alter the length of any Pre-agreed Adjustment except where that adjustment relates to a Change and that Change is itself the subject of a Relevant Omission.

In making a decision under clause 2.25.4 or clause 2.25.5.2, clause 2.25.6.4 prohibits the Employer from altering the length of any Pre-Agreed Adjustment, unless both (i) that Pre-Agreed Adjustment relates to a Change, and (ii) that Change is itself the subject of a Relevant Omission.

In other words, the Employer may not unilaterally undo an agreement previously made by the Parties.

Market practice: Culpable delay

Clause 2.25.6 will often be amended to expressly state that, save in respect of '*Specified Perils*' (defined in clause 6.8 (Insurance of the Works and Existing Structures: Related definitions)), the Contractor will not be entitled to an extension of time to the extent that it has (or any Contractor's Persons have) caused or contributed to that delay. This is sometimes described as 'culpable' delay.

Clause 2.26: Relevant Events

2.26 The following are the Relevant Events referred to in clauses 2.24 and 2.25:

Clause 2.26 includes Relevant Events which will, in principle, entitle the Contractor to an extension of time, in accordance with clause 2.25.

The Relevant Events partly overlap with the Relevant Matters considered in clause 4.21 (Relevant Matters) below, which relate to the Contractor's entitlement to reimbursement of loss and expense.

2.26.1 Changes and any other matters or instructions which under these Conditions are to be treated as, or as requiring, a Change;

Changes are one of the most common Relevant Events in practice, and are defined in clause 5.1 (Definition of Changes).

This Relevant Event applies:

- where the Employer instructs a Change which amends the Employer's Requirements, in accordance with clause 3.9 (Instructions requiring Changes); and
- to matters or Employer's instructions which are to be *treated* as a Change.

2.26.2.1 Employer's instructions:

under clause 2.13, except for any instructions relating to a discrepancy or divergence in or between the Contractor's Proposals and/or other Contractor's Design Documents;

Clause 2.26.2.1 relates to the Employer's instructions under clause 2.13 (Notice of discrepancies etc.), in respect of inadequacies, discrepancies, and divergences.

However, it does *not* apply in respect of instructions relating to any discrepancy or divergence in or between the Contractor's Proposals and/or other Contractor's Design Documents, on the basis that these documents were prepared by the Contractor.

Market practice: Responsibility for the design in the Employer's Requirements

Clause 2.26.2.1 is often removed, on the same basis that clause 2.11 to clause 2.15 are often amended, to the effect that the Contractor accepts responsibility for any design contained within the Employer's Requirements, including any inadequacy, discrepancy, or divergence in them.

> 2.26.2.2 *under clause 3.10 or 3.11; or*

Clause 2.26.2.2 relates to the Employer's instructions under:

- clause 3.10 (Postponement of work), where the Employer instructs the Contractor to postpone any work to be carried out under the Contract; and
- clause 3.11 (Instructions on Provisional Sums), where the Employer issues instructions regarding the expenditure of any Provisional Sums included in the Employer's Requirements.

> 2.26.2.3 *for the opening up for inspection or testing of any work, materials or goods under clause 3.12 or 3.13.3 (including making good), unless the inspection or test shows that the work, materials or goods are not in accordance with this Contract;*

Clause 2.26.2.3 relates to the Employer's instructions for the Contractor to either (i) open up for inspection, or (ii) test, any work, materials, or goods, in accordance with either:

- clause 3.12 (Inspection – tests); or
- clause 3.13.3 (Work not in accordance with the Contract), in respect of any potential further non-compliance,

along with any associated making good.

However, it will *not* apply in circumstances where the relevant work, materials, or goods are found not to be in accordance with the Contract.

> 2.26.3 *deferment of the giving of possession of the site or any Section under clause 2.4;*

Clause 2.26.3 applies where the Employer elects to defer the giving of possession of the site (or any Section) to the Contractor, in accordance with clause 2.4 (Deferment of possession), where it is stated against the reference to clause 2.4 in the Contract Particulars to apply.

> 2.26.4 *compliance with clause 3.15.1 or with the Employer's instructions under clause 3.15.2 or (except to the extent that the presence of asbestos or contaminated material has been identified in the Contract Documents and/or any such material has been brought on to the site by the Contractor or any Contractor's Person) compliance with clause 3.15.3 or with Employer's instructions under clause 3.15.4;*

Clause 2.26.4 relates to compliance with:

- clause 3.15.1 (Antiquities, asbestos, contaminated material and unexploded ordnance), and the Employer's instructions under clause 3.15.2, both in relation to fossils, antiquities, and other objects of interest or value found on the site; or
- clause 3.15.3, and the Employer's instructions under clause 3.15.4, both in relation to asbestos, contaminated material, and unexploded ordnance. However, this does not apply to the extent that (i) the presence of asbestos or contaminated material was identified in the Contract Documents, or (ii) any asbestos or contaminated material was brought on to the site by either the Contractor or Contractor's Persons.

2.26.5 *suspension by the Contractor under clause 4.11 of the performance of any or all of its obligations under this Contract;*

Clause 2.26.5 applies where the Contractor suspends any or all of its obligations under the Contract, in accordance with clause 4.11 (Contractor's right of suspension), in circumstances where the Employer fails to pay a sum payable to the Contractor in accordance with clause 4.9 (Interim and final payments – final date and amount), along with any VAT properly chargeable, by the final date for payment.

2.26.6 *any impediment, prevention or default, whether by act or omission, by the Employer or any Employer's Person, except to the extent caused or contributed to by any default, whether by act or omission, of the Contractor or any Contractor's Person;*

Along with Changes (in clause 2.26.1 above), acts of prevention by the Employer (or Employer's Persons) are a common Relevant Event in practice.

Clause 2.26.6 applies to any '*impediment, prevention or default*' by either the Employer or any Employer's Persons, and whether this was caused by an act or an omission (in other words, a failure to do something).

This is very broadly drafted, to reflect the wide range of potentially applicable acts or omissions by or on behalf of the Employer.

However, it does not apply to the extent that such impediment, prevention, or default was caused, or contributed to, by any default of the Contractor (or any Contractor's Person), again, whether by act or omission.

In other words, the Contractor will not be entitled to an extension of time for this Relevant Event to the extent that it has caused or contributed to the underlying problem.

2.26.7 *an epidemic:*

2.26.7.1 *first occurring after the Base Date which affects the execution of the Works; or*

2.26.7.2 *first occurring before the Base Date whose effects change after the Base Date and any such change affects the execution of the Works*

by limiting the availability or use of labour or the availability of persons engaged in providing services for the Works where such labour and/or persons are necessary for the proper carrying out of the Works, or preventing the Contractor from, or delaying the Contractor in, securing such goods or materials or such services as are necessary for the proper carrying out of the Works;

Clause 2.26.7 relates to epidemics either (i) first occurring *after* the Base Date, which affects the execution of the Works, or (ii) first occurring *before* the Base Date, but whose effects change *after* the Base Date, and any such change affects the execution of the Works, by:

- limiting (i) the availability or use of labour, or (ii) the availability of persons engaged in providing services for the Works, where such labour or persons are necessary to properly carry out the Works; or
- either (i) preventing the Contractor from, or (ii) or delaying the Contractor in, securing goods, materials, or services as are necessary to properly carry out the Works.

2.26.8 *the occurrence after the Base Date of any of the following which affects the execution of the Works:*

2.26.8.1 *the passing into law of any statute, statutory instrument or other subordinate legislation, regulation or bye-law, whether to make a new law or change or repeal an existing law;*

2.26.8.2 *the exercise of any statutory power, except to the extent caused or contributed to by any default, whether by act or omission, of the Contractor or any Contractor's Person; or*

2.26.8.3 *the publication of any guidance,*

by the United Kingdom Government or any of the devolved administrations, or any Local or Public Authority or any equivalent authority governed by public law in any of the devolved administrations, or in the case of guidance by the Construction Leadership Council or its successor;

Clause 2.26.8 relates to the occurrence after the Base Date of any of the following, which affects the execution of the Works:

- the passing into law of legislation, regulations, or bye-laws (whether to make a new law, or to change or repeal an existing law);
- the exercise of any statutory power, except to the extent that the exercise of the power has been caused or contributed to by any default (whether by act or omission) of the Contractor, or any Contractor's Person; or
- the publication of guidance,

by (i) the United Kingdom Government (or any devolved administration), or (ii) any Local or Public Authority (or any equivalent authority governed by public law in any devolved administration), or (iii) in the case of guidance, by the Construction Leadership Council, or its successor.

> 2.26.9 *the carrying out by a Statutory Provider of work in pursuance of its statutory obligations in relation to the Works, or the failure to carry out such work;*

Clause 2.26.9 relates to either:

- the carrying out of work by a Statutory Provider, in pursuance of its statutory obligations; or
- the failure by a Statutory Provider to carry out this work.

'*Statutory Providers*' are defined in clause 1.1 as follows:

> *1.1* ***Statutory Providers:*** *any person executing work solely in pursuance of its statutory obligations, including any persons employed, engaged or authorised by such person upon or in connection with that work.*

This means that it does not necessarily apply to *any* work carried out by *any* utility company, including those relating to gas, water, electricity, and telecommunications.

It is therefore necessary to first establish whether the Statutory Provider is carrying out works relating to its '*statutory obligations*'. In some circumstances, a Statutory Provider may be carrying out *both* works which are in pursuance of its statutory obligations, and those which are not.

The Parties should also consider the relevant underlying legislation (for example, the Gas Act 1986, or the Electricity Act 1989).

Market practice: Additional requirements for Statutory Providers

Clause 2.26.7 is often amended to impose additional requirements on the Contractor, to help ensure that Statutory Providers carry out their work in accordance with the overall programme for the Works. These include:

- providing Statutory Providers with sufficient notice (and sending reminders, if necessary);
- agreeing programmes and terms with Statutory Providers; and
- providing Statutory Providers with all the information and access they require to carry out their works.

The Contractor is encouraged to copy the Employer into all such correspondence with Statutory Providers.

2.26.10 exceptionally adverse weather conditions;

This Relevant Event relates to '*exceptionally*' adverse weather conditions. This is potentially a high threshold. For example, the Contractor may be prevented from pouring concrete as a result of very low temperatures, thereby causing a delay, but these weather conditions may not be considered '*exceptionally*' adverse, if they occur in January.

Market practice: Met Office Records

In light of the increasing impact of climate change and global warming, it is likely to become increasingly difficult to establish what are '*exceptionally*' adverse weather conditions.

Clause 2.26.10 is often amended to clarify the meaning of '*exceptionally*' adverse, by having regard to the records from the Met Office weather station nearest to the site, over a given period (say, 10 years).

The JCT offers both '*Weather Planning Reports*' and '*Downtime Reports*' to assist with this, in collaboration with the Met Office.[13]

2.26.11 loss or damage occasioned by any Specified Peril;

Clause 2.26.11 relates to loss or damage caused by any Specified Peril.

[13] Available at: https://www.jctltd.co.uk/jct-met-office-weather-reports

The '*Specified Perils*' are defined in clause 6.8 as follows:

> 6.8 *Specified Perils: fire, lightning, explosion, storm, flood, escape of water from any water tank, apparatus or pipe, earthquake, aircraft and other aerial devices or articles dropped therefrom, riot and civil commotion, but excluding Excepted Risks.*

> 2.26.12 *civil commotion or the use or threat of terrorism and/or the activities of the relevant authorities in dealing with such event or threat;*

Clause 2.26.12 relates to delays caused by either:

- '*civil commotion*' (or disturbance);
- terrorism;
- the *threat* of terrorism; and/or
- the activities of the relevant authorities (for example, the Government, the armed forces, or the emergency services) in dealing with the above.

> 2.26.13 *strike, lock-out or local combination of any trades employed upon the Works or engaged in the preparation, manufacture or transportation of any of the goods or materials required for them or any persons engaged in design work for the Works;*

Clause 2.26.13 relates to delays caused by industrial action affecting:

- any trade involved in the Works;
- any trade engaged in the preparation, manufacture, or transportation of goods or materials required for the Works; or
- anyone undertaking design for the works.

> 2.26.14 *delay in receipt of any necessary permission or approval of any statutory body which the Contractor has taken all practicable steps to avoid or reduce;*

Clause 2.26.14 relates to delay in receiving any permission or approval from any statutory body (for example, in respect of Development Control Requirements, or the Building Regulations), which are necessary for the Works to be carried out or completed.

This is provided that the Contractor has taken '*all practicable steps*' to avoid or reduce this delay. For example, by ensuring that any associated applications are made accurately and in good time, and supported by all relevant information.

2.26.15 force majeure.

Clause 2.26.15 relates to delays caused by '*force majeure*', which is not a defined term, but is thought to cover matters beyond the control of either Party, such as war, or a natural disaster.

Resource 80 page 456	Checklist: Contractor	A checklist for the Contractor for drafting delay notices and providing further particulars.
Resource 81 page 457	Checklist: Employer	A checklist for the Employer for giving extensions of time.
Resource 82 page 458	Tracker: Contractor and Employer	A tracker template for delay notices, for both the Contractor and the Employer.

Practical Completion, Lateness, and Liquidated Damages

Overview

This sub-section relates to practical completion of the Works, and includes both:

- the conditions for practical completion to be achieved; and
- the consequences if practical completion is not achieved by the relevant Completion Date, namely the potential withholding or deduction of liquidated damages by the Employer.

Key provisions

- clause 2.27 (Practical completion)
- clause 2.28 (Non-Completion Notice)
- clause 2.29 (Payment or allowance of liquidated damages)

Clause 2.27: Practical completion

2.27 *When practical completion of the Works or a Section is achieved and the Contractor has complied sufficiently with clauses 2.37 and 3.16 in respect of the supply of documents and information, then:*

2.27.1 *in the case of the Works, the Employer shall forthwith issue a statement to that effect ('the Practical Completion Statement');*

2.27.2 *in the case of a Section, it shall forthwith issue a statement of practical completion of that Section (a 'Section Completion Statement');*

and practical completion of the Works or the Section shall be deemed for all the purposes of this Contract to have taken place on the date stated in that statement.

Clause 2.27 relates to practical completion of the Works (or a Section).

DOI: 10.1201/9781003485919-24

It applies where the following conditions have been met:

- '*practical completion*' of the Works (or a Section) has been achieved. '*Practical completion*' is not defined in the Contract, but in general terms, it is when all of the Works have been carried out and completed, and the Employer is able to take possession;
- the Contractor has complied sufficiently with clause 2.37 (As-built Drawings), in respect of the provision of Contractor's Design Documents and related information that (i) show or describe the Works as built, or (ii) relate to the maintenance and operation of them, or their installations; and
- the Contractor has complied sufficiently with clause 3.16 (CDM Regulations and Part 2A of the Building Regulations), in relation to the supply of documents and information (for example, the provision of the health and safety file to the Employer, where the Contractor is the Principal Designer for the purposes of the CDM Regulations).

When these conditions have been met, the Employer must then issue:

- the Practical Completion Statement, in the case of the Works; or
- a Section Completion Statement, in the case of a Section.

(In the case of the last Section Completion Statement, the Employer should also issue the Practical Completion Statement for the Works at the same time.)

Practical completion of the Works (or relevant Section) will be considered to have taken place on the date stated in the Practical Completion Statement (or Section Completion Statement).

The achievement of practical completion of the Works has a number of significant implications:

- the Contractor will no longer retain possession of the site, in accordance with clause 2.3 (Date of Possession – progress);
- the Employer will have no entitlement to withhold or deduct liquidated damages for late completion, in accordance with clause 2.29 (Payment or allowance of liquidated damages), for the period after the date of practical completion;
- the Rectification Period will commence, in relation to clause 2.35 (Schedules of defects and instructions);
- the Employer may only deduct half the Retention (and will therefore be required to release the other half to the Contractor), in accordance with clause 4.18 (Retention – amounts and periods); and
- the Works Insurance Policy will no longer need to be maintained, in accordance with clause 6.7 (Insurance of the Works and Existing Structures: Insurance Options and period).

Resource 83 page 459	Reminder: Contractor to Employer	A reminder from the Contractor to the Employer of the Contractor's anticipated date of practical completion.
Resource 84 page 460	Employer to Contractor: Practical Completion Statement	A Practical Completion Statement template, from the Employer to the Contractor.
Resource 85 page 460	Employer to Contractor: Section Completion Statement	A Section Completion Statement template, from the Employer to the Contractor.
Resource 86 page 461	Checklist: Contractor and Employer	A checklist for the requirements of practical completion.

Market practice: Pre-conditions to practical completion

Clause 2.27 is often amended to include additional preconditions to practical completion, and to expressly permit the Employer to issue the Practical Completion Statement, notwithstanding the existence of minor defects and outstanding works (sometimes described as 'snags').

Clause 2.28: Non-Completion Notice

2.28 *If the Contractor fails to complete the Works or a Section by the relevant Completion Date, the Employer shall issue a notice to that effect (a 'Non-Completion Notice'). If a new Completion Date is fixed after the issue of such a notice, such fixing shall cancel that notice and the Employer shall where necessary issue a further notice.*

Clause 2.28 applies where the Contractor has not completed the Works or Section by the relevant Completion Date.

In these circumstances, the Employer is required to issue a notice confirming this, defined as a '*Non-Completion Notice*'.

Clause 2.28 further provides that, if the relevant Completion Date is adjusted *after* the issue of the Non-Completion Notice, this will have the effect of cancelling that Non-Completion Notice.

Where the Contractor then fails to complete the Works or Section by that adjusted Completion Date, the Employer is required to issue a further Non-Completion Notice.

The issue of a Non-Completion Notice is a precondition to the withholding or deduction of liquidated damages by the Employer for late completion, in accordance with clause 2.29 below.

Resource 87 page 462	Notice: Employer to Contractor	A Non-Completion Notice template, from the Employer to the Contractor.
Resource 88 page 463	Notice: Employer to Contractor	A further Non-Completion Notice template from the Employer to the Contractor, if a new Completion Date was fixed after the issue of the initial Non-Completion Notice.

Clause 2.29: Payment or allowance of liquidated damages

2.29.1 Provided:

2.29.1.1 the Employer has issued a Non-Completion Notice for the Works or a Section; and

2.29.1.2 the Employer has notified the Contractor before the due date for the final payment under clause 4.24.5 or (if applicable) the Termination Payment under clause 8.7.5, 8.8.2 or 8.12.4 that it may require payment of, or may withhold or deduct, liquidated damages,

the Employer may, not later than 5 days before the final date for payment of the amount payable under clause 4.24 or (if applicable) 8.7.6, 8.8.3 or 8.12.5, give notice to the Contractor in the terms set out in clause 2.29.2.

Clause 2.29 relates to liquidated damages, which may be withheld or deducted by the Employer in circumstances where the Contractor fails to complete the Works or a Section by the relevant Completion Date.

It aims to compensate the Employer for the loss it has suffered as a result, in a way that can be easily calculated, by reference to a rate of liquidated damages (typically calculated on a weekly basis).

Clause 2.29.1 sets out the steps the Employer must take before it may withhold or deduct liquidated damages.

Provided both that:

- the Employer has issued a Non-Completion Notice for the Works (or relevant Section), in accordance with clause 2.28; and
- the Employer has notified the Contractor that it *may* require payment of liquidated damages (or that it *may* withhold or deduct liquidated damages), before either:
  - the due date for the final payment under clause 4.24.5 (Final Statement and Final Payment); or
  - the due date for the Termination Payment under clause 8.7.5 (Termination by Employer: Consequences of termination under clauses 8.4 to 8.6), clause 8.8.2 (Termination by Employer: Employer's decision not to complete the Works), or clause 8.12.4 (Consequences of Termination under clauses 8.9 to 8.11, etc.), as applicable, where the Contractor's employment under the Contract has been terminated,

then the Employer may give a notice to the Employer in accordance with clause 2.29.2, not later than 5 days before the final date for payment of the amount payable under either:

- clause 4.24 (Final Statement and Final Payment); or
- clause 8.7.6 (Termination by Employer: Consequences of termination under clauses 8.4 to 8.6), clause 8.8.3 (Termination by Employer: Employer's decision not to complete the Works), or clause 8.12.5 (Consequences of Termination under clauses 8.9 to 8.11, etc.), as applicable, where the Contractor's employment under the Contract has been terminated.

Resource 89 page 464	Notice: Employer to Contractor	A notice from the Employer to the Contractor that the Employer may require payment of, or may withhold or deduct, liquidated damages.

2.29.2 *A notice from the Employer under clause 2.29.1 shall state that for the period between the Completion Date and the date of practical completion of the Works or that Section:*

2.29.2.1 *it requires the Contractor to pay liquidated damages at the rate stated in the Contract Particulars, or lesser rate stated in the notice, in which event the Employer may recover the same as a debt; and/or*

2.29.2.2 *that it will withhold or deduct liquidated damages at the rate stated in the Contract Particulars, or at such lesser stated rate, from sums due to the Contractor.*

Clause 2.29.2 includes the requirements for the Employer's notice to the Contractor referred to in the final paragraph of clause 2.29.1, and provides how the amount of liquidated damages will be calculated.

For the period between the Completion Date (when the Contractor *should* have completed the Works or Section) and the date of practical completion of the Works or Section, the notice must state that:

- the Employer requires the Contractor to pay liquidated damages to the Employer, either (i) at the rate stated against the reference to clause 2.29.2 in the Contract Particulars, or (ii) a lesser amount (and in both cases, clause 2.29.2.1 further provides that the Employer may recover those liquidated damages as a debt); and/or
- the Employer will withhold or deduct liquidated damages from payments due to the Contractor, again either (i) at the rate stated against the reference to clause 2.29.2 in the Contract Particulars, or (ii) a lesser amount.

If the Employer intends to withhold or deduct any of the liquidated damages payable, then, in addition to this notice, it must also give the appropriate Pay Less Notice under clause 4.9.5 (Interim and final payments – final date and amount), or (if applicable) clause 8.13.3 (Termination Payment – final date, notices and amount) or clause 8.13.5. (This is in order to comply with section 111(3) of the Housing Grants, Construction and Regeneration Act 1996.)

Resource 90 page 464	Notice: Employer to Contractor	A notice from the Employer to the Contractor that the Employer requires payment of, or will withhold or deduct, liquidated damages.

2.29.3 If the Employer fixes a later Completion Date for the Works or a Section, the Employer shall pay or repay to the Contractor any amounts recovered, allowed or paid under clause 2.29 for the period up to that later Completion Date.

Clause 2.29.3 applies in circumstances where the Employer had been paid liquidated damages, or had withheld or deducted liquidated damages from sums due to the Contractor, but subsequently gave an extension of time for the Works or relevant Section, with the effect that the Employer has now been paid, or has withheld or deducted, too much.

It therefore provides that the Employer must repay the relevant amount to the Contractor, for the period up to the revised Completion Date.

However, where applicable, the Employer may continue to withhold or deduct liquidated damages for the period between the revised Completion Date, and the date of practical completion of the Works or Section.

Resource 91 page 465	Reminder: Contractor to Employer	A reminder from the Contractor to the Employer to repay liquidated damages following an extension of time.

> 2.29.4 *If the Employer in relation to the Works or a Section has notified the Contractor in accordance with clause 2.29.1.2 that it may require payment of, or may withhold or deduct, liquidated damages, then, unless the Employer states otherwise in writing, clause 2.29.1.2 shall remain satisfied in relation to the Works or Section, notwithstanding the cancellation of the relevant Non-Completion Notice and issue of any further Non-Completion Notice.*

Clause 2.29.4 relates to the Employer's notice to the Contractor in accordance with clause 2.29.1.2, that it *may* require payment of (or *may* withhold or deduct) liquidated damages, in respect of the Works or relevant Section.

It provides that the Employer is not required to give that same notice twice, in respect of the Works or relevant Section, in circumstances where:

- the initial Non-Completion Notice is cancelled (by fixing a new Completion Date after it has been issued, in accordance with clause 2.28); and
- a subsequent Non-Completion Notice is issued.

This is provided that the Employer has not stated otherwise in writing to the Contractor.

Resource 92 page 466	Correspondence: Employer to Contractor	Correspondence from the Employer to the Contractor that the Employer no longer may require payment of, or may withhold or deduct, liquidated damages.

> 2.29.5 *If the Contractor's employment is terminated under this Contract:*
>
> 2.29.5.1 *where the date of termination occurs prior to the date of practical completion of the Works, the provisions of clauses 2.28 and 2.29 shall apply in respect of the period between the Completion Date and the date of termination, and the reference to practical completion of the Works or Section in clause 2.29.2 shall be deemed to be a reference to the date of termination;*
>
> 2.29.5.2 *in respect of the period after the date of termination, subject to clause 2.29.5.1, the Employer shall not be empowered to require the payment of or to withhold or deduct liquidated damages under clause 2.29 but the provisions of this clause 2.29.5.2 shall be without prejudice to and not in substitution of any other rights and remedies of the Employer.*

Clause 2.29.5 sets out the extent of the Employer's ability to withhold or deduct liquidated damages, in circumstances where the Contractor's employment under the Contract has been terminated.

(It has been introduced to take account of the Supreme Court's decision in *Triple Point Technology Inc v PTT Public Company Ltd*,[14] before which there was uncertainty in this regard.)

If the Contractor's employment under the Contract was terminated *before* the date of practical completion of the Works, clause 2.29.5.1 provides that the liquidated damages mechanism in clause 2.28 and clause 2.29 applies in respect of the period between (i) the Completion Date, and (ii) the date when the Contractor's employment under the Contract was terminated. (It further clarifies that the reference to '*practical completion*' in clause 2.29.2 will be deemed to be a reference to the date when the Contractor's employment under the Contract was terminated.)

In respect of the period *after* the date the Contractor's employment under the Contract was terminated, the Employer will *not* be entitled to be paid, or to withhold or deduct, liquidated damages under clause 2.29. However, this will not adversely affect any *other* rights and remedies of the Employer, including any right to bring a claim for general damages against the Contractor, arising as a result of the termination of the Contractor's employment under the Contract.

Resource 93 page 466	Checklist: Employer	A checklist for the deduction of liquidated damages by the Employer.

[14] *Triple Point Technology Inc v PTT Public Company Ltd* UKSC 2019/0074

Partial Possession by Employer

Overview

This sub-section relates to the potential for the Employer to take possession of part of the Works (or Section) *before* the Practical Completion Statement (or relevant Section Completion Statement) has been issued.

This is described as 'partial possession' by the Employer.

Key provisions

- clause 2.30 (Contractor's consent)
- clause 2.31 (Practical completion date)
- clause 2.32 (Defects etc. – Relevant Part)
- clause 2.33 (Insurance – Relevant Part)
- clause 2.34 (Liquidated damages – Relevant Part)

Clause 2.30: Contractor's consent

2.30 If at any time or times before the Practical Completion Statement or relevant Section Completion Statement the Employer wishes to take possession of any part or parts of the Works or a Section and the Contractor's consent has been obtained, then, notwithstanding anything expressed or implied elsewhere in this Contract, the Employer may take possession of such part or parts. The Contractor shall thereupon give the Employer notice identifying the part or parts taken into possession and giving the date when the Employer took possession ('the Relevant Part' and 'the Relevant Date' respectively).

Clause 2.30 entitles the Employer to take possession of part of the Works (or Section), *before* the Practical Completion Statement (or relevant Section Completion Statement) has been issued.

This is provided that the Employer has first obtained the Contractor's consent to such partial possession.

Once the Employer has taken partial possession, the Contractor must then give the Employer a notice both:

- identifying the part (or parts) taken into possession by the Employer, defined as the '*Relevant Part*'; and

DOI: 10.1201/9781003485919-25

- giving the date when the Employer took possession, defined as the '*Relevant Date*'.

Resource 94 page 467	Request: Employer to Contractor	A request from the Employer for the Contractor's consent for the Employer to take partial possession.
Resource 95 page 468	Consent given: Contractor to Employer	Consent given by the Contractor for the Employer to take partial possession.
Resource 96 page 468	Consent withheld: Contractor to Employer	Consent withheld by the Contractor for the Employer to take partial possession.
Resource 97 page 469	Notice: Contractor to Employer	A notice from the Contractor to the Employer of partial possession.

Clause 2.31: Practical completion date

> *2.31 For the purposes of clauses 2.35 and 4.18.2, practical completion of the Relevant Part shall be deemed to have occurred, and the Rectification Period in respect of the Relevant Part shall be deemed to have commenced, on the Relevant Date.*

Clause 2.31 provides that, on the Relevant Date, both:

- practical completion of the Relevant Part is deemed to have occurred; and consequently,
- the Rectification Period in respect of the Relevant Part is deemed to have commenced.

This is for the purposes of both:

- clause 2.35 (Schedules of defects and instructions), in respect of the making good of any defects, shrinkages, or other faults within the Relevant Part during the Rectification Period; and
- clause 4.18.2 (Retention – amounts and periods), relating to the deduction of half the Retention Percentage in respect of the Relevant Part.

Clause 2.32: Defects etc. – Relevant Part

> 2.32 *When any defects, shrinkages or other faults in the Relevant Part which the Employer has required to be made good under clause 2.35 have been made good, it shall issue a notice to that effect.*

Clause 2.32 should be read in conjunction with clause 2.35 (Schedules of defects and instructions) and clause 2.36 (Notice of Completion of Making Good), both considered below.

It relates to the issue of a '*Notice of Completion of Making Good*' (as defined in clause 2.36), in respect of the Relevant Part (rather than in respect of the Works or a Section).

If the Employer has required any defects, shrinkages, or other faults in the Relevant Part to be made good by the Contractor, in accordance with clause 2.35, then once they have been made good, the Employer is required to issue the Notice of Completion of Making Good in respect of the Relevant Part.

Resource 98 page 469	Notice: Employer to Contractor	A Notice of Completion of Making Good template for a Relevant Part, from the Employer to the Contractor.

Clause 2.33: Insurance – Relevant Part

> 2.33 *As from the Relevant Date the Works insurance obligation under Insurance Option A, B or C.2, whichever applies, shall cease in respect of the Relevant Part (but not otherwise) and the Existing Structures (if any) shall from that date be deemed to include the Relevant Part.*

Clause 2.33 relates to the Works insurance obligation in respect of the Relevant Part, whether under:

- Insurance Option A (New Buildings – All Risks Insurance of the Works by the Contractor);
- Insurance Option B (New Buildings – All Risks Insurance of the Works by the Employer); or
- paragraph C.2 (The Works – Joint Names Policy for All Risks) of Insurance Option C (Joint Names Insurance by the Employer of Existing Structures and Works in or Extensions to them).

From the Relevant Date (which is also the deemed date of practical completion of the Relevant Part, under clause 2.31):

- the applicable Works insurance obligation will cease, but only in respect of the Relevant Part; and
- the Existing Structures (if there are any) will be deemed to include the Relevant Part.

Clause 2.34: Liquidated damages – Relevant Part

> 2.34 *As from the Relevant Date, the rate of liquidated damages stated in the Contract Particulars in respect of the Works or Section containing the Relevant Part shall reduce by the same proportion as the value of the Relevant Part bears to the Contract Sum or to the relevant Section Sum, as shown in the Contract Particulars.*

Clause 2.34 relates to the proportionate reduction in the rate of liquidated damages for late completion (stated against the reference to clause 2.29 in the Contract Particulars), in the event that the Employer takes partial possession of the Relevant Part.

This is on the basis that the Relevant Date is the deemed date of practical completion of the Relevant Part (under clause 2.31), and so the Contractor should not be liable to pay the full rate of liquidated damages to the extent that it has already completed part of the Works or Section.

Clause 2.34 therefore provides that, from the Relevant Date, the rate of liquidated damages in respect of the Works or Section containing the Relevant Part will reduce proportionately, in line with the value of the Relevant Part to either the Contract Sum or the relevant Section Sum (stated against the reference to clause 2.34 in the Contract Particulars).

Recommendation: Agreeing the value of the Relevant Part

Clause 2.34 does not set out how the value of the Relevant Part is to be calculated.

Given the effect this value will have on the corresponding reduction in the rate of liquidated damages, the Parties are encouraged to agree the value of the Relevant Part before the Employer takes partial possession, such that the calculation under clause 2.34 can be readily performed.

Resource 99 page 470	Tracker: Contractor and Employer	A tracker template for recording partial possessions by the Employer.

Defects

Overview

This sub-section relates to defects and their making good, and includes:

- the Employer's options in respect of any defects which may appear in the Works (or Section), during the relevant Rectification Period; and
- the Employer's obligation to issue the Notice of Completion of Making Good, once all relevant defects have been made good.

Key provisions

- clause 2.35 (Defects)
- clause 2.36 (Notice of Completion of Making Good)

Clause 2.35: Schedules of defects and instructions

2.35 If any defects, shrinkages or other faults in the Works or a Section appear within the relevant Rectification Period due to any failure of the Contractor to comply with its obligations under this Contract:

2.35.1 such defects, shrinkages and other faults shall be specified by the Employer in a schedule of defects which it shall deliver to the Contractor as an instruction not later than 14 days after the expiry of that Rectification Period; and

2.35.2 prior to issue of that schedule, the Employer may whenever it considers it necessary issue instructions requiring any such defect, shrinkage or other fault to be made good, provided no instructions under this clause 2.35.2 shall be issued after delivery of that schedule or more than 14 days after the expiry of the relevant Rectification Period.

Within a reasonable time after receipt of such schedule or instructions, the defects, shrinkages and other faults shall at no cost to the Employer be made good by the Contractor unless the Employer shall otherwise instruct. If the Employer so instructs otherwise, an appropriate deduction may be made from the Contract Sum in respect of the defects, shrinkages or other faults not made good.

DOI: 10.1201/9781003485919-26

Clause 2.35 relates to any '*defects, shrinkages and other faults*' (typically shortened simply to 'defects') in the Works (or a Section).

It applies in respect of defects which are both caused by the Contractor's failure to comply with its obligations under the Contract, and which appear during the relevant Rectification Period (which begins upon practical completion of the Works or Section, and the duration for which is stated against the reference to clause 2.35 in the Contract Particulars).

Clause 2.35 provides the Employer with a number of options as to how such defects are to be '*made good*' (or remedied):

- Firstly, the Employer is required to prepare a schedule of all such defects, and issue it to the Contractor as an instruction, no later than 14 days after the expiry of the relevant Rectification Period.
- If it considers it necessary, the Employer may also issue instructions to the Contractor requiring individual defects to be made good. However, the Employer is not permitted to issue such instructions (i) after the issue of the schedule of defects under clause 2.35.1, or (ii) more than 14 days after the expiry of the relevant Rectification Period.

After receiving the schedule of defects, or any instructions relating to individual defects, the Contractor is then obliged to make good those defects, at no cost to the Employer, and within a '*reasonable time*'.

Alternatively, the Employer may instruct the Contractor *not* to make good a particular defect, and instead make an appropriate corresponding deduction from the Contract Sum (for example, in respect of costs incurred by the Employer in arranging for others to make good the defect).

Resource 100 page 470	Instruction: Employer to Contractor	An instruction by the Employer of a schedule of defects, shrinkages, and other faults in the Works, to be made good by the Contractor.
Resource 101 page 471	Instruction: Employer to Contractor	An instruction from the Employer requiring a particular defect, shrinkage, or other fault in the Works to be made good by the Contractor.
Resource 102 page 472	Instruction: Employer to Contractor	An instruction from the Employer requiring a particular defect, shrinkage, or other fault in the Works not to be made good by the Contractor.

Clause 2.36: Notice of Completion of Making Good

2.36 *When the defects, shrinkages or other faults in the Works or a Section which under clause 2.35 the Employer has required to be made good have been made good, it shall issue a notice to that effect (a 'Notice of Completion of Making Good'). That notice shall not be unreasonably delayed or withheld, and completion of that making good shall for the purposes of this Contract be deemed to have taken place on the date stated in that notice.*

Once all of the defects in the Works (or Section) which the Employer required to have been made good (in accordance with clause 2.35) have then been made good, clause 2.36 requires the Employer to issue a notice confirming this, defined as a '*Notice of Completion of Making Good*'.

It also requires that any such Notice of Completion of Making Good must not be unreasonably delayed or withheld by the Employer.

Clause 2.36 further provides that, for the purposes of the Contract, completion of such making good will be deemed to have taken place on the date stated in the Notice of Completion of Making Good.

This has a number of significant implications, relating to the following:

- the release of Retention to the Contractor, in accordance with clause 4.18 (Retention – amounts and periods); and
- the due date for the final payment, in accordance with clause 4.24.5 (Final Statement and Final Payment).

Recommendation: Record keeping for making good defects, shrinkages, or other faults

Given the implications relating to the release of the Retention and the due date for the final payment, the Contractor is encouraged (and incentivised) to make good all relevant defects in accordance with clause 2.35 as soon as possible, such that the Employer is then obliged to issue the Notice of Completion of Making Good.

The Contractor should therefore keep accurate, contemporaneous records of all such making good, and provide frequent updates to the Employer.

This is particularly necessary in circumstances where the Employer (or Employer's Agent) does not maintain as frequent or as active a presence on site during the Rectification Period, and where it may therefore take longer to coordinate any related site visits or inspections required to confirm that all such making good has taken place.

Resource 103 page 472	Reminder: Contractor to Employer	A reminder from the Contractor to the Employer of its obligation to issue a Notice of Completion of Making Good.
Resource 104 page 473	Notice: Employer to Contractor	A Notice of Completion of Making Good template, from the Employer to the Contractor.
Resource 105 page 473	Tracker: Contractor and Employer	A tracker template for the making good of defects, shrinkages, or other faults in the Works.

Contractor's Design Documents

Overview

This sub-section relates to the Contractor's Design Documents, and includes:

- the Contractor's obligation to provide the Employer with Contractor's Design Documents describing the Works as they have been built, and which relate to their maintenance and operation; and
- the Parties' rights and obligations in respect of the copyright and licences relating to the Contractor's Design Documents.

Key provisions

- clause 2.37 (As-built Drawings)
- clause 2.38 (Copyright and use)

Clause 2.37: As-built Drawings

2.37 The Contractor, in addition to its obligations under the CDM Regulations in relation to information for the health and safety file, shall, before practical completion of the Works or relevant Section and without further charge to the Employer, supply for retention and use by the Employer such Contractor's Design Documents and related information as is specified in the Contract Documents or as the Employer may reasonably require that show or describe the Works as built or relate to the maintenance and operation of them or their installations.

Clause 2.37 relates to documents which will assist the Employer in the maintenance and operation of the completed Works following practical completion, as they have been built.

This is in addition to the Contractor's obligations under the CDM Regulations relating to the provision of information for the health and safety file, considered under clause 3.16 (CDM Regulations and Part 2A of the Building Regulations) below.

Clause 2.37 requires the Contractor to supply to the Employer such Contractor's Design Documents and related information (i) as is stated in the Contract Documents, or (ii) as the Employer may reasonably require, that:

- show or describe the Works as built; or
- relate to the maintenance and operation of the Works, or their installations.

DOI: 10.1201/9781003485919-27

These must be provided before practical completion of the Works (or relevant Section), and will be at no further cost to the Employer.

Sufficient compliance with clause 2.37 is also effectively a precondition for the issue of the Practical Completion Statement (or Section Completion Statement) under clause 2.27 (Practical completion).

Recommendation: Collating as-built drawings and operation and maintenance manuals

Clause 2.37 potentially imposes a very significant administrative burden on the Contractor, particularly in the case of large or complex projects.

As far as possible, therefore, it is a task which is best performed consistently over the course of the project, as the relevant information and documents become available (for example, testing and commissioning certificates), rather than waiting until the point of practical completion (when the Contractor's focus is typically on the completion of the construction works).

This is particularly important, given that a failure to comply with clause 2.37 may delay the issue of the Practical Completion Statement (or Section Completion Statement) under clause 2.27.

The Contractor should therefore plan at a very early stage how it intends to identify, collate, and supply all of this information to the Employer, both by reviewing the relevant Contract Documents, and by establishing with the Employer what further information it may require.

Resource 106 page 474	Request for further documents: Employer to Contractor	A request from the Employer to the Contractor for further documents to show or describe the Works as built, or relate to the maintenance and operation of them, or their installations.
Resource 107 page 474	Tracker: Contractor and Employer	A tracker template for recording the collation of Contractor's Design Documents, as-built drawings, and documents for operation and maintenance.

Clause 2.38: Copyright and use

2.38.1 *Subject to any rights in any designs, drawings and other documents supplied to the Contractor for the purposes of this Contract by or on behalf of the Employer, all rights including (without limitation) copyright in all Contractor's Design Documents shall remain vested in the Contractor.*

Clause 2.38 relates to both (i) copyright in, and (ii) the use of, documents in relation to the Works.

Clause 2.38.1 provides that all rights (including copyright) in all Contractor's Design Documents will remain vested in the Contractor.

The only exception is in relation to any rights in any designs, drawings and other documents given to the Contractor by the Employer, for the purposes of the Contract.

2.38.2 *Subject to all sums due and payable under this Contract to the Contractor having been paid, the Employer shall have an irrevocable, royalty-free, non-exclusive licence to copy and use the Contractor's Design Documents and to reproduce the designs and content of them for any purpose relating to the Works including, without limitation, the construction, completion, maintenance, letting, sale, promotion, advertisement, reinstatement, refurbishment and repair of the Works. That licence shall enable the Employer to copy and use the Contractor's Design Documents for an extension of the Works but shall not include any right or licence to reproduce the designs contained in them for any such extension.*

Given that clause 2.38.1 provides that copyright in the Contractor's Design Documents will remain vested in the Contractor, the Employer will require a licence to copy and use them.

This licence is granted in clause 2.38.2, provided that the Employer has paid all sums due and payable to the Contractor.

The licence is irrevocable, royalty-free, and non-exclusive, and permits the Employer to:

- copy and use the Contractor's Design Documents; and
- to reproduce the designs and contents of Contractor's Design Documents, for any purpose (including maintenance, letting, sale, refurbishment, etc.).

The licence will also enable the Employer to copy and use Contractor's Design Documents for an extension of the Works, but this does not include any right or licence to *reproduce* the designs contained within those Contractor's Design Documents for this extension.

Market practice: Sums due and payable

Clause 2.38 is often amended to remove the condition that all sums due and payable must be paid to the Contractor, before the Employer will have the licence.

2.38.3 *The licence shall be assignable to any owner from time to time of the Works or any part of them and may be sub-licensed to any owner or tenant of the Works or part of them and to any person engaged for the purposes permitted by clause 2.38.2.*

Clause 2.38.3 entitles the Employer to assign the licence to any owner of the Works (or any part of the Works).

The licence may also be sub-licensed to:

- any owner or tenant of the Works (or part of the Works); and
- any person engaged to copy and use the Contractor's Design Documents, or to reproduce the designs and contents of the Contractor's Design Documents, for any purpose (including maintenance, letting, sale, refurbishment, etc.).

2.38.4 *The Contractor shall not be liable for any use by the Employer of any of the Contractor's Design Documents for any purpose other than that for which they were prepared.*

Clause 2.38.4 provides that the Contractor will only be liable to the Employer in respect of the Contractor's Design Documents to the extent that the Employer uses them for the purpose for which they were prepared (namely in relation to the design of the Works), but not otherwise.

Section 3: Control of the Works

Section overview

Section 3 covers the following four matters, all of which broadly relate to the control of the Works by the Parties:

- **Access and representatives:** the ability of the Employer (and the Employer's Agent) to monitor the progress of the Works, and the obligation on the Contractor (and its representative, the Site Manager), to control the carrying out and completion of the Works.
- **Sub-contracting:** the Contractor's obligation to obtain the Employer's consent before sub-contracting any of the design or construction of the Works, and, if consent is given by the Employer, what the Contractor must include in the corresponding sub-contracts.
- **Employer's instructions:** the Employer's authority to issue instructions in respect of certain matters, including (i) the postponement of work, (ii) Provisional Sums, (iii) materials, goods, work or workmanship not in accordance with the Contract, and (iv) the discovery on site of antiquities, asbestos, contaminated material, and unexploded ordnance, along with the Contractor's general obligation to comply with Employer's instructions (subject to a number of exceptions).
- **CDM Regulations and Part 2A of the Building Regulations:** the Parties' key obligations in respect of both the CDM Regulations and Part 2A of the Building Regulations, and in particular, the obligations relating to the roles of both the Principal Designer and the Principal Contractor (both for the purposes of the CDM Regulations and the Building Regulations).

Sub-sections

- Access and Representatives (clause 3.1 – clause 3.2)
- Sub-Contracting (clause 3.3 – clause 3.4)
- Employer's Instructions (clause 3.5 – clause 3.15)
- CDM Regulations and Part 2A of the Building Regulations (clause 3.16)

DOI: 10.1201/9781003485919-29

Access and Representatives

Overview

This sub-section includes a provision for each of the Parties' key representatives, namely:

- the Employer's Agent, on behalf of the Employer; and
- the Site Manager, on behalf of the Contractor.

These provisions relate to the monitoring of the Works, in the case of the Employer's Agent, and the control of the Works, in the case of the Site Manager.

Key provisions

- clause 3.1 (Access for Employer's Agent)
- clause 3.2 (Site Manager)

Clause 3.1: Access for Employer's Agent

3.1 The Employer's Agent and any person authorised by it or the Employer shall at all reasonable times have access to the Works and to the workshops or other premises of the Contractor where work is being prepared for this Contract. When work is to be prepared in workshops or other premises of a sub-contractor the Contractor shall by a term in the sub-contract secure so far as possible a similar right of access to those workshops or premises for the Employer and any person authorised by the Employer and take any steps reasonably necessary to make that right effective. Access under this clause 3.1 may be subject to such reasonable restrictions as are necessary to protect proprietary rights.

Clause 3.1 relates to the provision of access to the Works for the Employer's Agent, to help them monitor progress on the Employer's behalf.

It requires the Contractor to provide the Employer's Agent (along with any person authorised by the Employer's Agent or the Employer, for example, other Employer's Persons such as a clerk of works), access at all reasonable times, to both:

(i) the Works, and
(ii) the workshops or other premises where work is being prepared for the Contract.

DOI: 10.1201/9781003485919-30

In circumstances where the Contractor has appointed sub-contractors, which are preparing work in their workshops or other premises, the Contractor is further required to:

- obtain a similar right of access to those premises by the Employer (and anyone authorised by the Employer), so far as possible. This is to be achieved by including a term in the relevant sub-contracts to that effect. Such a provision is included in clause 3.9 (Access for Contractor and Employer) of the JCT Design and Build Sub-Contract Conditions (reference DBSub/C 2024); and
- to take any steps '*reasonably necessary*' to allow the Employer to access these premises (for example, by coordinating inspections by the Employer with the relevant sub-contractors).

The Employer's right of access under clause 3.1 is subject to the '*reasonable restrictions*' necessary to protect the proprietary rights of the Contractor, any relevant sub-contractors, and those who have an interest in the relevant premises.

Resource 108 page 475	Request: Employer to Contractor	A request from the Employer to the Contractor for access for the Employer's Agent and the Employer to the Contractor's workshops and other premises.

Clause 3.2: Site Manager

3.2 *The Contractor shall prior to the commencement of work on site appoint a full-time Site Manager, approved by the Employer, to act as the Contractor's representative there, in charge of the Works. The Contractor shall ensure that the appointee, or a competent deputy, is on site at all material times and, if the appointee ceases to hold the post, shall promptly appoint a replacement approved by the Employer. Any instructions issued to the Site Manager or its deputy shall be treated as issued to the Contractor.*

Clause 3.2 relates to the Contractor's management of the Works.

It requires the Contractor to appoint a full-time Site Manager, who will act as the Contractor's representative, in charge of the Works. For this reason, any instructions issued by the Employer or Employer's Agent to the Site Manager (or their '*competent*' deputy) will be treated as issued to the Contractor.

The Site Manager must be approved by the Employer, and must be appointed before the commencement of work on site.

The Site Manager (or their deputy) must also be present on site at all '*material*' times (in other words, during site operating hours).

If, for whatever reason, the original Site Manager leaves their position, the Contractor is then obliged to appoint a replacement Site Manager, who must also be approved by the Employer.

Recommendation: Distributing Employer's instructions issued to the Site Manager (or their deputy)

Given that any instructions issued to the Site Manager or their deputy (but not to any other staff) will be treated as issued to the Contractor, the Contractor should put measures in place to ensure that any such instructions issued to the Site Manager or their deputy are received and suitably distributed (for example, in case of any absence).

The contact details for the Site Manager and their deputy should be updated and shared with the Employer and the Employer's Agent (ideally within the communication protocol), along with any relevant Employer's Persons and Contractor's Persons.

The Employer and Employer's Agent should also ensure that they do not issue any instructions to anyone on site other than the Site Manager or their deputy, to avoid any uncertainty.

Resource 109 page 475	Request: Contractor to Employer	A request from the Contractor to the Employer for approval to appoint a Site Manager.
Resource 110 page 475	Request: Contractor to Employer	A request from the Contractor to the Employer for approval to replace the Site Manager.
Resource 111 page 476	Approval given: Employer to Contractor	Approval given by the Employer for the Contractor to appoint a Site Manager (or a replacement Site Manager).
Resource 112 page 476	Approval withheld: Employer to Contractor	Approval withheld by the Employer for the Contractor to appoint a Site Manager (or a replacement Site Manager).

Market practice: Key personnel

It is common for clause 3.2 to be expanded to require the Contractor to establish and maintain a team of 'key personnel' across key disciplines (for example, project management, commercial management, and design management) throughout the project, which may only be replaced with the Employer's consent. Such a team brings the benefits of consistency and experience to both Parties.

Sub-Contracting

Overview

This sub-section relates to any sub-contracting by the Contractor of the design or construction of the Works, and sets out both:

- the ability of the Contractor to sub-contract in principle; and
- to the extent that the Employer provides its consent to any such sub-contracting, what those corresponding sub-contracts must include.

Key provisions

- clause 3.3 (Consent to sub-contracting)
- clause 3.4 (Conditions of sub-contracting)

Clause 3.3: Consent to sub-contracting

3.3.1 Save for any sub-contract entered into in accordance with Supplemental Provision 1, where it applies:

3.3.1.1 the Contractor shall not without the Employer's consent sub-contract the whole or any part of the Works;

3.3.1.2 the Contractor shall not without the Employer's consent sub-contract the design for the Works or any part of them.

In no case shall any such consent or any sub-contracting in any way affect the Contractor's obligations under any other provision of this Contract.

Clause 3.3 relates to the Contractor's ability to sub-contract the Works.

The Contractor requires the Employer's consent to sub-contract:

- the whole, or any part, of the Works; and
- the whole, or any part, of the design for the Works (for example, by the appointment of design consultants).

The exception relates to Supplemental Provision 1 (Named Sub-Contractors), considered below, where it is stated in the Contract Particulars to apply (against the reference to the Seventh Recital and Part 1 of Schedule 2).

DOI: 10.1201/9781003485919-31

If the Employer provides its consent to the sub-contracting of any part of the design or construction of the Works, this consent will not affect the Contractor's obligations under the Contract. In other words, the Contractor will still remain responsible to the Employer for carrying out and completing the design and construction of the Works in accordance with the Contract Documents.

Resource 113 page 477	Request: Contractor to Employer	A request from the Contractor to the Employer for consent to sub-contract the whole or any part of the Works or the design for the Works.
Resource 114 page 477	Consent given: Employer to Contractor	Consent given by the Employer for the Contractor to sub-contract the whole or any part of the Works or the design for the Works.
Resource 115 page 478	Consent withheld: Employer to Contractor	Consent withheld by the Employer for the Contractor to sub-contract the whole or any part of the Works or the design for the Works.

3.3.2 *The provisions of this clause 3.3 and of clause 3.4 shall not apply to the execution of part of the Works by a Statutory Provider, which shall not in that capacity be a sub-contractor within the terms of this Contract.*

Clause 3.3.2 clarifies that both clause 3.3 and clause 3.4 do not apply to any part of the Works carried out by a Statutory Provider, which will not be classified as a sub-contractor.

'*Statutory Providers*' are defined in clause 1.1 (Definitions) as follows:

1.1 ***Statutory Providers:*** *any person executing work solely in pursuance of its statutory obligations, including any persons employed, engaged or authorised by such person upon or in connection with that work.*

Clause 3.4: Conditions of sub-contracting

3.4 *Where considered appropriate, the Contractor shall engage the sub-contractor using the JCT Design and Build Sub-Contract. It shall be a condition of any sub-contract that:*

Clause 3.4 is a detailed provision, setting out what must be included in any sub-contract (assuming the Employer has given its consent to sub-contracting, in accordance with clause 3.3 above).

3.4.1 *the sub-contractor's employment under the sub-contract shall terminate immediately upon the termination (for any reason) of the Contractor's employment under this Contract unless prior to the date of termination of the Contractor's employment the sub-contractor has conferred third party rights on a Funder and/or the Employer and/or executed and delivered any collateral warranty or warranties in favour of a Funder and/or the Employer which entitle the beneficiary, in the event of such termination, to issue a written notice to the sub-contractor to accept the instructions of the beneficiary or its appointee to the exclusion of the Contractor, in which case:*

3.4.1.1 *the sub-contractor may suspend the performance of its obligations under the sub-contract with effect from the date of termination of the Contractor's employment until receipt of any such written notice (if one is served), or if no such notice is served the sub-contractor's employment under the sub-contract shall immediately terminate on the day after the last date on which such written notice may be served;*

3.4.1.2 *any such suspension shall be deemed to be a suspension as a result of an instruction of the Employer under this Contract passed on to the sub-contractor as a direction by the Contractor under the sub-contract;*

Clause 3.4.1 requires a condition to be included in the sub-contract to the effect that if the Contractor's employment under the Contract is terminated, then the sub-contractor's employment under the sub-contract is also terminated immediately.

The exception is if, prior to the termination of the Contractor's employment under the Contract, the sub-contractor has (i) conferred third party rights on a Funder or the Employer, and/or (ii) delivered collateral warranties in favour of a Funder or the Employer, which entitle the beneficiary to notify the sub-contractor to accept the beneficiary's instructions, instead of the Contractor's instructions.

These are known as 'step-in' rights, where the beneficiary effectively 'steps in' to take the place of the Contractor under the sub-contract.

In such circumstances:

- the sub-contractor may suspend the performance of its obligations under the sub-contract, from the date of termination of the Contractor's employment, until they receive the beneficiary's notice. If the beneficiary does not serve a notice, then the sub-contractor's employment under the sub-contract will terminate immediately on the day after the last date when the beneficiary *could* have served that notice; and
- this suspension will be treated as though it was the result of an Employer's instruction to the Contractor under the Contract, which was then passed on to the sub-contractor by the Contractor under the sub-contract.

3.4.2 the sub-contract shall provide:

3.4.2.1 that, except for use on the Works, no Site Materials delivered to the Works by or for the sub-contractor shall be removed without the Contractor's written consent (such consent not to be unreasonably delayed or withheld) and (in addition to any provision for earlier vesting in the Contractor of title to any Listed Items for the purposes of clause 4.15.2.1 of these Conditions) that:

3.4.2.1.1 where, in accordance with clause 4.12 or 4.13 and clause 4.14 of these Conditions, the value of any Site Materials has been included in any Interim Payment and that Interim Payment has been paid by the Employer to the Contractor, they shall upon such payment become, and the sub-contractor shall not deny that they have become, the Employer's property;

3.4.2.1.2 if the Contractor pays the sub-contractor for any Site Materials before their value is included in an Interim Payment, they shall upon such payment become the Contractor's property;

3.4.2.2 that the sub-contractor shall give access to workshops or other premises in accordance with clause 3.1 of these Conditions;

3.4.2.3 that (without limiting either party's statutory and/or regulatory duties) each party to the sub-contract shall in relation to the Works and the site comply with applicable CDM Regulations and as applicable Part 2A of the Building Regulations;

3.4.2.4 in terms equivalent to those of clause 4.9.6 of these Conditions, that if by a final date for payment under the sub-contract the Contractor fails to pay the sub-contractor any amount that should properly have been paid, the Contractor shall, in addition to that amount, pay simple interest on it at the Interest Rate for the period from the final date for payment until such payment is made;

3.4.2.5 that where the Rights Particulars provide for the grant of third party rights from and/or for the execution and delivery of collateral warranties by the sub-contractor:

3.4.2.5.1 the sub-contract and, where applicable, those collateral warranties shall if those particulars require be executed as deeds;

3.4.2.5.2 any third party rights required shall vest on receipt of notice from the Contractor to that effect and any collateral warranty required shall be executed and delivered by the sub-contractor within 14 days of receipt of the Contractor's notice requiring execution;

3.4.2.5.3 the terms of and those governing such third party rights or collateral warranties shall in each case be consistent with those of clauses 2.26 and 2.27 and Schedule 5 of the JCT Design and Build Sub-Contract Conditions;

Clause 3.4.2 requires a number of further provisions to be included in the sub-contract, to the effect that:

- except for use on the Works, no Site Materials delivered to the Works by the sub-contractor may be removed without the Contractor's written consent (which should not be unreasonably delayed or withheld);
- in addition to any provision for the earlier vesting of title in the Contractor to any Listed Items (relating to materials and goods off site) for the purposes of clause 4.15.2.1 (Listed Items):
  - where the value of any Site Materials has been included in any Interim Payment, which has then been paid to the Contractor, those Site Materials will become the Employer's Property; and
  - if the Contractor pays the sub-contractor for Site Materials *before* their value is included in an Interim Payment, they will become the Contractor's property following such payment.
- the sub-contractor must give access to their workshops and other premises, in accordance with clause 3.1 (Access for Employer's Agent);
- the sub-contractor and the Contractor must comply with their respective obligations under the CDM Regulations and, as applicable, Part 2A of the Building Regulations, in relation to the Works and the site;
- the Contractor must pay simple interest on any amount that the Contractor has failed to pay to the sub-contractor by the relevant final date for payment under the sub-contract, from the period from the final date for payment (when payment *should* have been made) until when payment was actually made;
- where the Rights Particulars provide for (i) the grant of third party rights from, and/or (ii) the execution and delivery of collateral warranties by, the sub-contractor:
  - if the Rights Particulars require it, the sub-contract and any collateral warranties must be executed as deeds;
  - any third party rights will '*vest*' (or be granted to the relevant beneficiary) on receipt of the Contractor's notice;
  - any collateral warranty required must be executed and delivered by the sub-contractor within 14 days of receipt of the Contractor's notice; and
  - the form of such third party rights and collateral warranties, and the terms governing them, must be consistent with those of clause 2.26 (Third Party Rights and Collateral Warranties etc.: Particulars and notices), clause 2.27 (Third Party Rights and Collateral Warranties etc.: Rights of termination, variation etc.), and Schedule 5 (Third Party Rights) of the JCT Design and Build Sub-Contract Conditions (DBSub/C 2024).

3.4.3 where documents or information and/or the grant of licences are or may be required from the sub-contractor for the BIM Protocol, where applicable, and/or for the purposes of clauses 2.37 and 3.16 of these Conditions, the sub-contract shall also provide for the supply and grant to and by the sub-contractor of all information and licences reasonably necessary for that purpose.

Clause 3.4.3 applies where the sub-contractor is required to (i) provide documents or information, or (ii) grant licences, for the purposes of:

- the BIM Protocol (where applicable);
- clause 2.37 (As-built Drawings), relating to documents or information that show or describe the Works as built, or relate to the maintenance and operation of them, or their installations; or
- clause 3.16 (CDM Regulations and Part 2A of the Building Regulations), for example, relating to the health and safety file.

In such circumstances, the sub-contract must also provide for the supply and grant of all information and licences reasonably necessary to achieve this.

3.4 Continued *The Contractor shall not give such consent as is referred to in clause 3.4.2.1 without the Employer's prior consent under clause 2.21 of these Conditions.*

The final paragraph of clause 3.4 provides that the Contractor must not consent to the removal of Site Materials from the Works under clause 3.4.2.1, without having first obtained the Employer's corresponding consent under clause 2.21 (Materials and goods – on site).

Recommendation: JCT Design and Build Sub-Contract

The JCT Design and Build Sub-Contract Agreement (DBSub/A 2024) and Design and Build Sub-Contract Conditions (DBSub/C 2024) meet the requirements of this rather detailed clause 3.4, and so the Contractor is encouraged to adopt them.

Resource 116 page 478	Tracker: Contractor and Employer	A tracker template for the appointment of sub-contractors in accordance with clause 3.3 and clause 3.4.

Market practice: Further conditions of sub-contracting

Clause 3.4 is often amended to impose further conditions of sub-contracting, specific to the relevant project.

For example, sub-contractors with design responsibility may be required to take out and maintain a Professional Indemnity insurance policy, with limits of indemnity of certain types and amounts.

Employer's Instructions

Overview

This sub-section includes several provisions which relate to the Employer's instructions to the Contractor. It includes both:

- the rights of the Employer to issue instructions to the Contractor, across a wide range of matters; and
- the Contractor's rights and obligations in respect of those instructions.

Key provisions

- clause 3.5 (Compliance with instructions)
- clause 3.6 (Non-compliance with instructions)
- clause 3.7 (Instructions other than in writing)
- clause 3.8 (Provisions empowering instructions)
- clause 3.9 (Instructions requiring Changes)
- clause 3.10 (Postponement of work)
- clause 3.11 (Instructions on Provisional Sums)
- clause 3.12 (Inspection – tests)
- clause 3.13 (Work not in accordance with the Contract)
- clause 3.14 (Workmanship not in accordance with the Contract)
- clause 3.15 (Antiquities, asbestos, contaminated material and unexploded ordnance)

Clause 3.5: Compliance with instructions

3.5 The Contractor shall forthwith comply with all instructions issued to it by the Employer on any matter on which these Conditions expressly empower the Employer to issue instructions, save that:

3.5.1 where an instruction requires a Change of the type referred to in clause 5.1.2, the Contractor need not comply to the extent that it notifies a reasonable objection to the instruction to the Employer;

3.5.2 in the case of a notification by the Contractor under clause 2.20.2, the Contractor need not comply pending confirmation of the instruction.

DOI: 10.1201/9781003485919-32

Clause 3.5 requires the Contractor to immediately comply with all instructions which the Employer has the authority to issue under the Conditions.

There are two exceptions:

- **Changes under clause 5.1.2 (Definition of Changes)**: this relates to Changes defined in clause 5.1.2 (although not clause 5.1.1), which include the imposition, alteration, or omission by the Employer of any obligations or restrictions relating to (i) access to the site, (ii) limitations of working space, (iii) limitations of working hours, or (iv) the execution or completion of the work in any specific order. The Contractor is not obliged to comply with instructions requiring Changes of this nature, provided it notifies a reasonable objection to the Employer.
- **Patent Rights**: this relates to the potential infringement of patent rights, as set out in clause 2.20 (Patent rights – Instructions). Where, following receipt of an instruction, the Contractor notifies the Employer of a potential infringement of patent rights in accordance with clause 2.20.2, the Contractor is not obliged to comply with that instruction until the Employer confirms it under clause 2.20.2.

Resource 117 page 479	Notice: Contractor to Employer	A notice from the Contractor to the Employer of a reasonable objection to an Employer's instruction.

Market practice: Contractor's right to objection

Clause 3.5.1 is often amended to remove the Contractor's right to object to an instruction requiring a Change of the type referred to in clause 5.1.2.

Clause 3.6: Non-compliance with instructions

3.6 *Subject to clauses 3.5 and 3.9, if within 7 days after receipt of a notice from the Employer requiring compliance with an instruction the Contractor does not comply, the Employer may employ and pay other persons to execute work of any kind that may be necessary to give effect to that instruction. The Contractor shall be liable for all additional costs incurred by the Employer in connection with such employment and an appropriate deduction may be made from the Contract Sum.*

Clause 3.6 provides a remedy for the Employer, in the event that the Contractor does not comply with an instruction:

- If the Contractor fails to comply with an instruction, the Employer may give a notice to the Contractor, requiring it to comply with that instruction.

- If the Contractor still fails to comply with the instruction within 7 days after receipt of that notice, then the Employer may employ and pay others to execute work of any kind, in order to give effect to that instruction.
- In such circumstances, the Contractor will be liable to the Employer for all additional costs incurred in connection with the employment of those others, and the Employer may make an appropriate deduction from the Contract Sum.

The exceptions to this remedy are set out in the following clauses:

- clause 3.5 (Compliance with instructions), which sets out exceptions in relation to:
  - Changes referred to in clause 5.1.2 (Definition of Changes); and
  - the potential infringement of patent rights under clause 2.20 (Patent rights – Instructions); and
- clause 3.9 (Instructions requiring Changes), which sets out exceptions in relation to:
  - the alteration or modification of the design of the Works (clause 3.9.1);
  - again, Changes referred to in clause 5.1.2 (Definition of Changes) (clause 3.9.2), and
  - the Contractor's objections in its capacity as Principal Designer and/or Principal Contractor (whether for the purposes of either the CDM Regulations or the Building Regulations) (clause 3.9.4).

Resource 118 page 480	Notice: Employer to Contractor	A notice from the Employer to the Contractor requiring compliance with an instruction.
Resource 119 page 480	Reply: Contractor to Employer	A reply from the Contractor to the Employer following a notice from the Employer requiring compliance with an instruction.
Resource 120 page 481	Notice: Employer to Contractor	A further notice from the Employer to the Contractor, following the Employer's initial notice under clause 3.6, that the Employer will employ and pay other persons.
Resource 121 page 481	Notice: Employer to Contractor	A further notice from the Employer to the Contractor, following the Employer's initial notice under clause 3.6, that the Employer has employed and paid other persons.

Clause 3.7: Instructions other than in writing

3.7.1 *Where the Employer gives an instruction otherwise than in writing, it shall be of no immediate effect but the Contractor shall confirm the terms of the instruction in writing to the Employer within 7 days, and, if the Employer does not dissent by notice to the Contractor within 7 days from receipt of the Contractor's confirmation, the instruction shall take effect as from the expiry of the latter 7 day period.*

3.7.2 *If prior to or within 7 days from receipt of the Contractor's confirmation the Employer confirms the terms of the instruction in writing, it shall take effect from the date and in the terms of the Employer's confirmation.*

3.7.3 *If neither the Contractor nor the Employer confirms such an instruction in the manner and time stated but the Contractor nevertheless complies with it, the Employer may at any time prior to the due date for the final payment under clause 4.24.5 confirm it with retrospective effect.*

Clause 1.7.1 (Notices and other communications) requires that communications between the Parties, including instructions, are to be in writing.

However, clause 3.7 is relevant where the Employer gives an instruction which is *not* in writing (often referred to colloquially as a 'verbal' instruction).

In contrast to written instructions, these instructions will be of no immediate effect. In other words, the Contractor is not obliged to immediately comply with them.

Instead, the following process will apply:

- In accordance with clause 3.7.1, the Contractor must confirm the terms of the instruction to the Employer (this time in writing), within 7 days of receiving it. If the Employer does not notify the Contractor of its disagreement with the Contractor's confirmation within 7 days of receiving it, then the instruction will take effect from the expiry of the second 7-day period.
- Alternatively, under clause 3.7.2, the Employer may confirm the terms of the instruction in writing, either within 7 days from receipt of the Contractor's confirmation, or before even receiving the Contractor's confirmation. In such cases, the instruction will take effect from the date of the Employer's confirmation, and will be on the terms set out in that confirmation (as opposed to the Contractor's confirmation).
- Finally, under clause 3.7.3, if neither Party confirms the instruction, but the Contractor still complies with it, the Employer may retrospectively confirm it, at any time prior to the due date for the final payment under clause 4.24.5 (Final Statement and Final Payment).

Recommendation: Give instructions in writing

Whilst clause 3.7 may be useful in certain circumstances, it is still greatly preferable (for both Parties) for the Employer to give all instructions in writing, in accordance with clause 1.7.1 (Notices and other communications).

This reduces the potential for uncertainty over the terms of the instruction and when it takes effect, and it avoids unnecessary delay.

Resource 122 page 482	Confirmation: Contractor to Employer	Confirmation by the Contractor to the Employer of an instruction given otherwise than in writing.
Resource 123 page 482	Notice: Employer to Contractor	A notice from the Employer to the Contractor dissenting from the Contractor's confirmation of an instruction given otherwise than in writing.
Resource 124 page 483	Confirmation: Employer to Contractor	A notice from the Employer to the Contractor confirming the terms of an instruction given otherwise than in writing.
Resource 125 page 483	Confirmation: Employer to Contractor	A notice from the Employer to the Contractor retrospectively confirming the terms of an instruction given otherwise than in writing.

Clause 3.8: Provisions empowering instructions

3.8 On receipt of an instruction the Contractor may request the Employer to notify it which provision of these Conditions empowers issue of the instruction and the Employer shall forthwith comply with the request. If the Contractor thereafter complies with that instruction with neither Party then having invoked any dispute resolution procedure under this Contract to establish the Employer's powers in that regard, the instruction shall be deemed to have been duly given under the specified provision.

In most circumstances, it should be clear to the Contractor on what basis the Employer has issued a particular instruction. Indeed, the Employer will ideally have referenced the relevant provision within the instruction. For example, clause 3.9 (Instructions requiring Changes), considered below.

However, to the extent that it is *not* clear, clause 3.8 allows the Contractor to query the basis on which an instruction was issued, and then requires the Employer to immediately confirm which provision of the Conditions entitles the Employer to issue that instruction.

If the Employer answers the query, and the Contractor then complies with the instruction, it will be deemed to have been given in accordance with the provision identified in the Employer's answer. The only exception is if either Party (but in practice, most likely the Contractor) first commences any dispute resolution procedure (including negotiation, mediation, adjudication, arbitration, or litigation) to establish whether the Employer's answer is correct.

Resource 126 page 484	Request: Contractor to Employer	A request from the Contractor to the Employer to notify which of the Conditions empowers the Employer to issue a particular instruction.
Resource 127 page 484	Notice: Employer to Contractor	A notice from the Employer to the Contractor of the provision of the Conditions which empowers the Employer to issue a particular instruction.

Clause 3.9: Instructions requiring Changes

3.9.1 The Employer may issue instructions requiring a Change, subject to clause 3.9.4 and provided that the Employer may not effect a Change which is, or which makes necessary, an alteration or modification in the design of the Works without the Contractor's consent.

Clause 3.9.1 entitles the Employer to issue an instruction requiring a Change, which is defined in clause 5.1 (Definition of Changes).

This is subject to the following:

- the Contractor's right of objection set out in clause 3.9.4, in its capacity as Principal Designer and/or Principal Contractor (in both cases, for the purposes of both the CDM Regulations and the Building Regulations); and
- any Change which would alter or modify the *design* of the Works requires the Contractor's consent.

Resource 128 page 485	Request: Employer to Contractor	A request from the Employer for the Contractor's consent for the Employer to issue an instruction requiring a Change which is, or makes necessary, an alteration or modification in the design of the Works.
Resource 129 page 485	Consent given: Contractor to Employer	Consent given by the Contractor for the Employer to issue an instruction requiring a Change which is, or makes necessary, an alteration or modification in the design of the Works.
Resource 130 page 486	Consent withheld: Contractor to Employer	Consent withheld by the Contractor for the Employer to issue an instruction requiring a Change which is, or makes necessary, an alteration or modification in the design of the Works.

Market practice: Contractor's consent

Clause 3.9.1 is often amended to remove the requirement to obtain the Contractor's consent before instructing a Change which makes necessary an alternation or modification in the design of the Works.

3.9.2 *Any instruction of the type referred to in clause 5.1.2 shall be subject to the Contractor's right of reasonable objection set out in clause 3.5.1.*

Clause 3.9.2 reflects the position set out in clause 3.5.1 (Compliance with instructions), considered above.

Any instruction requiring a Change as defined in clause 5.1.2 (Definition of Changes), which includes the imposition, alteration, or omission by the Employer of any obligations or restrictions relating to (i) access to the site, (ii) limitations of working space, (iii) limitations of working hours, or (iv) the execution or completion of the work in any specific order, is subject to the Contractor's right to object to such instructions, included in clause 3.5.1.

3.9.3 *No Change required or sanctioned by the Employer shall vitiate this Contract.*

Clause 3.9.3 provides that the Contract cannot be vitiated (in other words, rendered void or unenforceable, or invalidated) by any Change which the Employer requires or sanctions.

3.9.4 The Contractor shall, within a reasonable time after receipt of an instruction effecting a Change or of an instruction in regard to the expenditure of a Provisional Sum included in the Employer's Requirements, notify the Employer whether in its capacity as Principal Designer and/or Principal Contractor it has any objection to such instruction. If the Contractor has any reasonable objection the Employer shall, to the reasonable satisfaction of the Contractor, vary the terms of the instruction so as to remove that objection; and, until the Employer has so varied the terms of the instruction, the Contractor shall not be required pursuant to clauses 2.1.1 to 2.1.4 to comply with such instruction.

Clause 3.9.4 relates to the Contractor in its capacity as Principal Designer and/or Principal Contractor (in both instances, both for the purposes of the CDM Regulations and the Building Regulations).

If the Contractor, in any of those four capacities, has an objection to either:

- an instruction effecting a Change; or
- an instruction in regard to the expenditure of a Provisional Sum included in the Employer's Requirements,

it must notify the Employer within a reasonable time after receiving the instruction.

The Employer is then required to vary the terms of the instruction to remove the Contractor's objection, to the Contractor's reasonable satisfaction.

Until this variation is made, the Contractor will not be required to comply with the instruction (as is otherwise required in accordance with clause 2.1.1 to clause 2.1.4 (Contractor's Obligations: General obligations)).

The Contractor should therefore consider every instruction of this type in each of its capacities, to determine whether it has any objections.

Resource 131 page 486	Notice: Contractor to Employer	A notice from the Contractor to the Employer of an objection to an instruction, in its capacity as Principal Designer or Principal Contractor.
Resource 132 page 487	Instruction: Employer to Contractor	A varied instruction from the Employer to the Contractor, following the Contractor's notice under clause 3.9.4.
Resource 133 page 487	Confirmation: Contractor to Employer	Confirmation by the Contractor to the Employer of the varied instruction, following the Contractor's notice under clause 3.9.4.

Clause 3.10: Postponement of work

> 3.10 *The Employer may issue instructions in regard to the postponement of any work to be executed under this Contract.*

Clause 3.10 entitles the Employer to instruct the Contractor to postpone carrying out any of the work to be executed under the Contract.

Whilst in most circumstances the Employer will want the Works to achieve practical completion as soon as possible, this provision can provide the Employer with more time, should it be needed (for example, to take a significant decision in respect of a proposed Change to the Works).

In other circumstances, it may not always be immediately obvious or clear that an Employer's instruction requires the Contractor to postpone a particular element of the works. (The verb 'postpone' may not be expressly stated.) For example, the Employer may instruct the Contractor not to proceed with a particular element of the Works, until certain conditions are met. In such instances, the Contractor should clarify whether the Employer wishes for it to postpone that element of the Works.

Instructions under clause 3.10 have further potential consequences. They constitute:

- a Relevant Event under clause 2.26.2.2 (Relevant Events), thereby potentially entitling the Contractor to an extension of time;
- a Relevant Matter under clause 4.21.2.1 (Relevant Matters), thereby potentially entitling the Contractor to reimbursement of loss and expense; and
- potential grounds for termination by either Party under clause 8.11.1.2 (Termination by either Party and regulation 73(1)(a) of the PC Regulations), for instructions issued as a result of the negligence or default of a Statutory Provider.

Resource 134 page 488	Instruction: Employer to Contractor	An instruction from the Employer to postpone any work to be executed under the Contract.
Resource 135 page 488	Confirmation: Contractor to Employer	Confirmation by the Contractor of an Employer's instruction regarding the postponement of any work to be executed under the Contract.

Clause 3.11: Instructions on Provisional Sums

> 3.11 *The Employer shall issue instructions in regard to the expenditure of Provisional Sums included in the Employer's Requirements.*

The term '*Provisional Sum*' is defined in clause 1.1 (Definitions) simply as '*a provisional sum for work included in the Employer's Requirements*'.

As the term suggests, it is a provisional (or perhaps 'interim' or 'temporary') allowance, for certain works which were not sufficiently defined at tender stage for the Parties to agree a fixed price.

Clause 3.11 requires the Employer to issue instructions regarding the expenditure of Provisional Sums.

Once the Employer does so, the scope of the relevant work will be established, and it follows that:

- the value of the Provisional Sum will be *deducted* from the Contract Sum, in accordance with clause 4.2.5 (Contract Sum and Adjustments: Items included in adjustments); and
- the work carried out by the Contractor in accordance with instructions under clause 3.11 will be valued in accordance with clause 5.2 (Valuation of Changes and provisional sum work), and then *added* to the Contract Sum, in accordance with clause 4.2.1.

Instructions under clause 3.11 have a number of further consequences. They constitute:

- a Relevant Event under clause 2.26.2.2 (Relevant Events), thereby potentially entitling the Contractor to an extension of time; and
- a Relevant Matter under clause 4.21.2.1 (Relevant Matters), thereby potentially entitling the Contractor to reimbursement of loss and expense.

Recommendation: Employer's Requirements

Understandably, Provisional Sums are often referenced in the Contract Sum Analysis.

However, in light of the express wording of clause 3.11 and the definition of Provisional Sums in clause 1.1, it is important to ensure that they have also been included in the Employer's Requirements, to avoid any uncertainty.

Resource 136 page 489	Instruction: Employer to Contractor	An instruction from the Employer regarding the expenditure of Provisional Sums.
Resource 137 page 489	Tracker: Contractor and Employer	A Provisional Sums tracker template.

Market practice: 'Defined' and 'undefined' provisional sums

The Contract does not distinguish between 'defined' provisional sums, which have been described in sufficient detail to enable the Contractor to allow for them in its preliminaries, and 'undefined' provisional sums, which have not.

Clause 3.11 is therefore sometimes amended to make reference to both 'defined' and 'undefined' Provisional Sums, determined by reference to the '*New rules of measurement*', published by the Royal Institution of Chartered Surveyors (RICS), and currently in its second edition (published in October 2021).

Parties often also agree deadlines for the instruction of the expenditure of each Provisional Sum, to help avoid delay to the progress of the Works. However, much will ultimately depend on the scope of the works instructed.

Clause 3.12: Inspection – tests

> 3.12 *The Employer may issue instructions requiring the Contractor to open up for inspection any work covered up or to arrange for or carry out any test of any materials or goods (whether or not already incorporated in the Works) or of any executed work. The cost of that opening up or testing (including the cost of making good) shall be added to the Contract Sum unless provided for in the Employer's Requirements or in the Contractor's Proposals or unless the inspection or test shows that the materials, goods or work are not in accordance with this Contract.*

Clause 3.12 provides the Employer with a way of determining whether materials, goods, or work are in accordance with the Contract.

It entitles the Employer to instruct the Contractor to both:

- open up any work previously covered up (for example, by floors, walls, or ceilings), to allow it to be inspected; and
- carry out any test (whether by the Contractor or a third party) of (i) any materials or goods, whether or not those materials or goods have been incorporated into the Works, or (ii) of any work carried out.

Given that these instructions will require the Contractor to carry out additional work, the cost of both (i) the opening up or testing, and (ii) making good following the inspection or testing, will be added to the Contract Sum, unless either:

- the Employer's Requirements or Contractor's Proposals already provide for that opening up or testing; or
- the inspection or test demonstrated that the materials, goods, or work were *not* in accordance with the Contract.

Recommendation: Covering up work

As far as possible, the Contractor is encouraged to give the Employer the opportunity to undertake any inspections of work before it is covered up, to reduce the need for instructions to be issued under clause 3.12.

Resource 138 page 490	Instruction: Employer to Contractor	An instruction from the Employer to open up work for inspection or testing.
Resource 139 page 490	Confirmation: Employer to Contractor	Confirmation from the Employer to the Contractor that the costs of opening up or inspection will be added to the Contract Sum, where the materials, goods or work were in accordance with the Contract.
Resource 140 page 491	Confirmation: Employer to Contractor	Confirmation from the Employer to the Contractor that the costs of opening up or inspection will not be added to the Contract Sum, where the materials, goods or work were not in accordance with the Contract.

Clause 3.13: Work not in accordance with the Contract

3.13 If any work, materials or goods are not in accordance with this Contract the Employer, in addition to its other powers, may:

3.13.1 issue instructions in regard to the removal from the site of all or any of such work, materials or goods;

Clause 3.13 provides the Employer with a number of options in the event that any work, materials, or goods are discovered not to be in accordance with the Contract.

Firstly, under clause 3.13.1, the Employer may instruct the Contractor to remove any such work, materials, or goods from the site.

Resource 141 page 491	Instruction: Employer to Contractor	An instruction from the Employer to remove from the site all or any work, materials or goods not in accordance with the Contract.
Resource 142 page 492	Confirmation: Contractor to Employer	Confirmation from the Contractor of compliance with the Employer's instruction to remove from the site all or any work, materials, or goods not in accordance with the Contract.

3.13.2 *after consultation with the Contractor, issue such Change instructions (to which the proviso in clause 3.9.1 applies) as are a reasonably necessary consequence of any instruction under clause 3.13.1 (but to the extent that such instructions are reasonably necessary, no addition shall be made to the Contract Sum and no extension of time shall be given); and/or*

Secondly, under clause 3.13.2, the Employer may issue Change instructions, which are a '*reasonably necessary*' consequence of any instruction issued by the Employer under clause 3.13.1, relating to the removal of any work, materials, or goods from the site.

These instructions are subject to the following:

- the Employer must first consult with the Contractor, before issuing such instructions;
- like other instructions requiring a Change, they are subject to the following conditions set out in clause 3.9.1 (Instructions requiring Changes):
  - the Contractor's right of objection set out in clause 3.9.4, in its capacity as Principal Designer and/or Principal Contractor (in both cases, for the purposes of both the CDM Regulations and the Building Regulations); and
  - any Change which would alter or modify the *design* of the Works requires the Contractor's consent; and
- to the extent that such instructions are '*reasonably necessary*', the Contractor will not be entitled to any addition to the Contract Sum, nor any extension of time.

Resource 143 page 492	Instruction: Employer to Contractor	An instruction from the Employer, as a reasonably necessary consequence of any instruction issued under clause 3.13.1.
Resource 144 page 493	Confirmation: Contractor to Employer	Confirmation from the Contractor of compliance with the Employer's instructions issued as a reasonably necessary consequence of any instruction under clause 3.13.1.

3.13.3 having due regard to the Code of Practice set out in Schedule 4, issue such instructions under clause 3.12 to open up for inspection or to test as are reasonable in all the circumstances to establish to its reasonable satisfaction the likelihood or extent, as appropriate to the circumstances, of any further similar non-compliance. To the extent that those instructions are reasonable, whatever the results of the opening up, no addition shall be made to the Contract Sum but clauses 2.25 and 2.26.2.3 shall apply unless the inspection or test shows that the work, materials or goods are not in accordance with this Contract.

Finally, under clause 3.13.3, the Employer may issue instructions to the Contractor in accordance with clause 3.12 (Inspection – tests), to open up works for inspection, or to test materials, goods, or work.

This is in order to help the Employer understand the probability or extent of any *further* similar instances of materials, goods, or work not in accordance with the Contract.

The Employer's rights under clause 3.13.3 are subject to the following:

- these instructions must be '*reasonable in all the circumstances*', to help the Employer establish the likelihood and extent of further non-compliance; and
- to assist with this, before issuing these instructions, the Employer must first have '*due regard*' to Schedule 4 (Code of Practice), which provides fifteen criteria to assist in the '*fair and reasonable operation of the requirements of clause 3.13.3*'.

To the extent that these instructions are reasonable in all the circumstances, and irrespective of the results of any further opening up, the Contractor will not be entitled to any addition to the Contract Sum. This is on the basis that these instructions stemmed from the original discovery of work, materials, or goods which were not in accordance with the Contract.

However, provided the inspections or tests demonstrated that the work, materials, or goods *were* in accordance with the Contract, the corresponding Relevant Event in clause 2.26.2.3 (Relevant Events) will apply, thereby potentially entitling the Contractor to an extension of time.

Resource 145 page 493	Instruction: Employer to Contractor	An instruction from the Employer to open up for inspection or to test the likelihood or extent of any further similar non-compliance.
Resource 146 page 494	Confirmation: Contractor to Employer	Confirmation by the Contractor of compliance with the Employer's instruction made under clause 3.13.3 and clause 3.12.

Clause 3.14: Workmanship not in accordance with the Contract

3.14 Where there is any failure to comply with clauses 2.1.1 to 2.1.4 in regard to the carrying out of work in a proper and workmanlike manner or in accordance with the Construction Phase Plan, the Employer, in addition to its other powers, may, after consultation with the Contractor, issue such instructions (whether requiring a Change or otherwise) as are in consequence reasonably necessary. To the extent that such instructions are reasonably necessary, no addition shall be made to the Contract Sum and no extension of time shall be given.

Similar to clause 3.13 above, clause 3.14 provides the Employer with a remedy in the event of the Contractor's failure to comply with clause 2.1.1 to clause 2.1.4 (Contractor's Obligations: General obligations), in regard to (i) the carrying out of work in a proper and workmanlike manner, or (ii) in accordance with the Construction Phase Plan.

In these circumstances, and provided it has first consulted with the Contractor, the Employer may issue such instructions which are consequently *'reasonably necessary'* to address the matter. These instructions may require a Change, in accordance with clause 3.9.1 (Instructions requiring Changes).

Similar to clause 3.13.2 above, to the extent that such instructions are *'reasonably necessary'*, the Contractor will not be entitled to any addition to the Contract Sum, nor any extension of time.

Recommendation: Record keeping in relation to clause 3.12 to clause 3.14

Given the potentially significant adverse consequences to the Contractor as a result of the Employer's instructions under clause 3.12 (Inspection – tests), clause 3.13 (Work not in accordance with the Contract), and clause 3.14 (Workmanship not in accordance with the Contract), the consultations required between the Parties under these provisions are important. The consultations should:

- give both Parties the opportunity to make representations, to better understand the matter;
- provide the Contractor with an opportunity to demonstrate that the relevant materials, goods, work, or workmanship may be in accordance with the Contract, such that no further instructions are necessary; and
- clarify the extent of any remedial work necessary.

Each Party should take an attendance note of each consultation, or better still, agree a joint attendance note.

Resource 147 page 494	Instruction: Employer to Contractor	An instruction from the Employer as a result of the Contractor's failure to comply with clause 2.1.1 to clause 2.1.4.
Resource 148 page 495	Confirmation: Contractor to Employer	Confirmation from the Contractor of compliance with an Employer's instruction issued as a result of the Contractor's failure to comply with clause 2.1.1 to clause 2.1.4.
Resource 149 page 496	Template: Attendance Note	An attendance note template for consultations between the Contractor and the Employer, relating to materials, goods, work, or workmanship not in accordance with the Contract.

Clause 3.15: Antiquities, asbestos, contaminated material and unexploded ordnance

3.15.1 All fossils, antiquities and other objects of interest or value which may be found on the site or in excavating it during the progress of the Works shall become the Employer's property. Upon discovery of any such object the Contractor shall forthwith:

3.15.1.1 use its best endeavours not to disturb the object and cease work if and insofar as continuing work would endanger the object or prevent or impede the object's excavation or removal;

3.15.1.2 take all steps necessary to preserve the object in the exact position and condition in which it was found; and

3.15.1.3 inform the Employer of the object's discovery and precise location.

Clause 3.15.1 relates to the discovery of fossils, antiquities (or artefacts), and other objects of interest or value, either on the site, or in excavating the site, during the course of carrying out the Works.

Any such object found will become the Employer's property, and the Contractor must immediately take the following precautions to protect the object:

- use its best endeavours not to disturb the object;
- stop any work which would otherwise (i) endanger the object, or (ii) prevent or impede the excavation or removal of the object;
- take all steps to preserve the object in precisely the same position and condition in which it was discovered; and
- inform the Employer of the object's discovery, and its location on site.

Resource 150 page 496	Notice: Contractor to Employer	A notice from the Contractor to the Employer of the discovery and precise location of any fossils, antiquities, or other objects of interest or value.

3.15.2 The Employer shall issue instructions as to action to be taken concerning any object reported under clause 3.15.1, which (without limiting its powers) may require the Contractor to permit the examination, excavation or removal of the object by a third party.

Once the Contractor has notified the Employer of the discovery of the object under clause 3.15.1, the Employer must then issue instructions as to the action to be taken.

This could include requiring the Contractor to permit a third party (for example, the Museum of London Archaeological Archive (MOLA)) to examine, excavate, or remove the object.

Any such third party will be considered an Employer's Person, in accordance with the definition of '*Employer's Persons*' in clause 1.1 (Definitions).

Compliance with these instructions constitutes both:

- a Relevant Event under clause 2.26.4 (Relevant Events), thereby potentially entitling the Contractor to an extension of time; and
- a Relevant Matter under clause 4.21.3 (Relevant Matters), thereby potentially entitling the Contractor to reimbursement of loss and expense.

Resource 151 page 497	Instruction: Employer to Contractor	An instruction from the Employer concerning the object reported by the Contractor under clause 3.15.1.

3.15.3 Upon discovery of any asbestos, contaminated material or unexploded ordnance on the site or in excavating it during the progress of the Works the Contractor shall forthwith:

3.15.3.1 use its best endeavours not to disturb the material or item and cease work if and insofar as continuing work would endanger health and safety and/or life and property or prevent or impede the disposal or removal of the material or item; and

3.15.3.2 report the discovery of such material or item and the steps it is taking under clause 3.15.3.1 to the Employer.

Clause 3.15.3 is similar to clause 3.15.1, but relates to the discovery of asbestos, contaminated material, and unexploded ordnance, either on the site, or in excavating the site, during the course of carrying out the Works.

The Contractor must immediately take the following precautions:

- use its best endeavours not to disturb the material or item;
- stop any work which would otherwise (i) endanger health and safety, or indeed, life (ii) endanger property, or (iii) prevent or impede the disposal or removal of the object; and
- inform the Employer of the discovery of the material or item, and the steps the Contractor has taken in respect of the points above.

Resource 152 page 497	Notice: Contractor to Employer	A notice from the Contractor to the Employer of the discovery of any asbestos, contaminated material, or unexploded ordnance.

3.15.4 The Employer shall issue instructions with respect to any material or item reported under clause 3.15.3, which (without limiting its powers) may require the Contractor to permit the investigation, disposal or removal of such material or item by a third party.

Similar to clause 3.15.2, once the Contractor has notified the Employer of the discovery of asbestos, contaminated material, or unexploded ordnance under clause 3.15.3, the Employer must then issue instructions as to the action to be taken.

This could include requiring the Contractor to permit a third party to investigate, dispose, or remove the material or item.

Again, any such third party will be considered an Employer's Person, in accordance with the definition of '*Employer's Persons*' in clause 1.1 (Definitions).

In general, compliance with these instructions constitutes both:

- a Relevant Event under clause 2.26.4 (Relevant Events), thereby potentially entitling the Contractor to an extension of time; and
- a Relevant Matter under clause 4.21.3 (Relevant Matters), thereby potentially entitling the Contractor to reimbursement of loss and expense.

However, these instructions will not constitute Relevant Events nor Relevant Matters to the extent that the presence of either asbestos or contaminated material (i) has been identified in the Contract Documents, or (ii) has been brought on to the site by the Contractor, or any Contractor's Persons.

Resource 153 page 498	Instruction: Employer to Contractor	An instruction from the Employer concerning the material or item reported by the Contractor under clause 3.15.3.
Resource 154 page 498	Tracker: Contractor and Employer	A tracker template for recording Employer's instructions.
Resource 155 page 499	Checklist: Contractor	A checklist for the Contractor upon receiving an Employer's instruction.

CDM Regulations and Part 2A of the Building Regulations

Overview

This sub-section relates to each of the Parties' obligations under both of the following:

- the Construction (Design and Management) Regulations 2015 (defined as the '*CDM Regulations*'); and
- Part 2A of the Building Regulations 2010 (defined as the '*Building Regulations*').

It includes several references to the roles of Principal Designer and Principal Contractor. This includes the roles of Principal Designer and Principal Contractor for the purposes of both:

- the CDM Regulations, in accordance with Article 6 (CDM Regulations – Principal Designer and Principal Contractor); *and*
- the Building Regulations, in accordance with Article 7 (Building Regulations – Principal Designer and Principal Contractor).

As stated above, Part 2A of the Building Regulations, entitled '*Dutyholders and Competence*', is known as the 'dutyholder regime'. The 'dutyholders' will have responsibility under the legislation for managing risks throughout the design, construction, and occupation phase of the project. During the construction phase, the dutyholders include the following:

- clients (which includes the Employer);
- the Principal Contractor (which potentially includes the Contractor);
- the Principal Designer (which potentially includes the Contractor);
- designers; and
- contractors (which includes the Contractor's sub-contractors).

These roles are similar to (albeit distinct from) those under the CDM Regulations.

Key provisions

- clause 3.16 (CDM Regulations and Part 2A of the Building Regulations)

DOI: 10.1201/9781003485919-33

Clause 3.16: CDM Regulations and Part 2A of the Building Regulations

> 3.16 *Without limiting either Party's statutory and regulatory duties and responsibilities, each Party undertakes to the other that in relation to the Works and site it will duly comply with applicable CDM Regulations and as applicable Part 2A of the Building Regulations, and in particular but without limitation:*

Clause 3.16 relates to each of the Parties' obligations under both (i) the CDM Regulations, and (ii) the Building Regulations.

Both Parties undertake to comply with both the applicable CDM Regulations and, as applicable, Part 2A of the Building Regulations, in relation to the Works and the site.

This general undertaking is stated not to limit (or 'cut across') either Party's statutory or regulatory duties and responsibilities.

The remainder of clause 3.16 sets out a number of specific obligations in that regard.

> 3.16.1 *where the Contractor is not the Principal Designer, the Employer shall ensure that the Principal Designer carries out its duties and, where the Contractor is not the Principal Contractor, shall ensure that the Principal Contractor carries out its duties under those regulations;*

The Employer is responsible for the appointment of the Principal Designer and the Principal Contractor, for the purposes of both the CDM Regulations and Part 2A of the Building Regulations.

In the event that the Contractor is not the Principal Designer (whether for the purposes of the CDM Regulations or the Building Regulations), clause 3.16.1 requires the Employer to ensure that the Principal Designer it has appointed carries out its duties under the CDM Regulations and Part 2A of the Building Regulations.

Similarly, in the event that the Contractor is not the Principal Contractor (whether for the purposes of the CDM Regulations or the Building Regulations), clause 3.16.1 requires the Employer to ensure that the Principal Contractor it has appointed carries out its duties under the CDM Regulations and Part 2A of the Building Regulations.

> 3.16.2 *where the Contractor is and while it remains the Principal Designer, it shall comply with the duties of a Principal Designer and (where it is the Principal Designer for the purposes of the CDM Regulations) it shall without charge prepare, and deliver to the Employer, the health and safety file;*

In the event that the Contractor is the Principal Designer (whether for the purposes of the CDM Regulations or the Building Regulations), it is required to comply with the corresponding duties of a Principal Designer, as set out in both the CDM Regulations and Part 2A of the Building Regulations.

In addition, in the event that the Contractor is the Principal Designer for the purposes of the CDM Regulations, it is responsible for preparing the health and safety file, and providing it to the Employer (at no cost).

Further reading: Health and Safety Executive Guidance on the health and safety file

The Health and Safety Executive's publication '*Managing health and safety in construction: Construction (Design and Management) Regulations 2015*'[15] includes the following sections relating to the health and safety file, which may be helpful for the Parties:

- Regulation 12: Construction phase plan and health and safety file;
- the guidance at paragraph 45, and paragraphs 110-115;
- Appendix 4 ('*The health and safety file*'); and
- Appendix 5 ('*How different types of information relate to and influence each other in a construction project involving more than one contractor: A summary*').

Resource 156 page 499	Correspondence: Contractor to Employer	The provision from the Contractor to the Employer of the health and safety file.

3.16.3 the Contractor shall comply with:

3.16.3.1 regulations 8 to 10 and 15 of the CDM Regulations and, where it is the Principal Contractor for the purposes of the CDM Regulations, with regulations 12 to 14 of those regulations; and

3.16.3.2 regulations 11F, 11J, 11K and 11L of the Building Regulations and, where it is the Principal Contractor for the purposes of the Building Regulations, with regulation 11N of those regulations;

[15] Available at: https://www.hse.gov.uk/pubns/books/l153.htm

Clause 3.16.3 requires the Contractor to comply with further specific provisions of both the CDM Regulations and the Building Regulations, namely the following:

In accordance with clause 3.16.3.1, relating to the CDM Regulations, the Contractor must comply with the following:

- regulation 8 (*General duties*);
- regulation 9 (*Duties of designers*);
- regulation 10 (*Designs prepared or modified outside Great Britain*); and
- regulation 15 (*Duties of contractors*).

Where it is the Principal Contractor for the purposes of the CDM Regulations, the Contractor must also comply with the following:

- regulation 12 (*Construction phase plan and health and safety file*);
- regulation 13 (*Duties of a principal contractor in relation to health and safety at the construction phase*); and
- regulation 14 (*Principal contractor's duties to consult and engage with workers*).

In accordance with clause 3.16.3.2, relating to the Building Regulations, the Contractor must comply with the following:

- regulation 11F (*Competence: general requirement*)
- regulation 11J (*Duties of dutyholders: General duty*)
- regulation 11K (*Duties of dutyholders: Additional duties of designers*)
- regulation 11L ((*Duties of dutyholders: Additional duties of contractors*)

Where it is the Principal Contractor for the purposes of the Building Regulations, the Contractor must also comply with regulation 11N (*Duties of dutyholders: Additional duties of a principal contractor*).

(Footnote [41] of the Contract further provides that, where the Employer is a '*domestic*' client, as defined in regulation 2(1) of both the CDM Regulations and the Building Regulations), the Principal Contractor may also be responsible for carrying out certain of the client's duties.)

3.16.4 if the Employer appoints a replacement for any Principal Designer or Principal Contractor, the Employer shall immediately upon that appointment notify the Contractor with details of the new appointee.

In the event that the Employer appoints a replacement Principal Designer or Principal Contractor (in both cases, whether for the purposes of the CDM Regulations or the Building Regulations), clause 3.16.4 requires the Employer to immediately notify the Contractor, with the details of the replacement Principal Designer or Principal Contractor.

Resource 157 page 500	Notice: Employer to Contractor	A notice from the Employer to the Contractor of the appointment of a replacement Principal Designer or Principal Contractor, for the purposes of either the CDM Regulations or the Building Regulations.

Section 4: Payment

Section overview

Section 4 sets out the key provisions in the Contract relating to payment, including:

- the circumstances in which the Contract Sum can be adjusted;
- VAT and the Construction Industry Scheme (CIS);
- the procedure for the calculation and making of Interim Payments, taking account of Listed Items (where applicable), and Retention;
- the mechanism for the reimbursement of direct loss and expense to the Contractor; and
- the procedure for the Final Payment.

Section 4 requires careful administration, as many of its provisions require several complicated calculations to be performed, and impose strict deadlines.

The Parties are particularly encouraged to ensure that the following documents are prepared accurately and issued on time:

- Interim Payment Applications, applications for reimbursement of loss and expense, and the Final Statement, in the case of the Contractor; and
- Payment Notices, Pay Less Notices, and the ascertainment of loss and expense, in the case of the Employer.

Given the complexities of tax law, the Parties should also consult with tax advisers when administering the provisions relating to VAT and the Construction Industry Scheme (CIS).)

Sub-sections

- Contract Sum and Adjustments (clause 4.1 – clause 4.3)
- Taxes (clause 4.4 – clause 4.5)
- Payments and Notices – general provisions (clause 4.6 – clause 4.11)
- Interim Payments – calculation of sums due (clause 4.12 – clause 4.14)
- Listed Items (clause 4.15)
- Retention (clause 4.16 – clause 4.18)
- Loss and Expense (clause 4.19 – clause 4.23)
- Final Statement and Final Payment (clause 4.24)

DOI: 10.1201/9781003485919-35

Contract Sum and Adjustments

Overview

This sub-section sets out the circumstances where the Contract Sum may be adjusted, and when those adjustments are to be taken into account.

Key provisions

- clause 4.1 (Adjustment only under the Conditions)
- clause 4.2 (Items included in adjustments)
- clause 4.3 (Taking adjustments into account)

Clause 4.1: Adjustment only under the Conditions

4.1 The Contract Sum shall not be adjusted or altered in any way other than in accordance with the express provisions of these Conditions.

Clause 4.1 provides that the Contract Sum may *only* be adjusted (whether increased or decreased) in circumstances where the Conditions expressly (in other words, specifically or explicitly) provide for it to be adjusted.

(For example, clause 4.2 below expressly provides for several circumstances where the Contract Sum may be adjusted, and begins with the phrase *'The Contract Sum shall be adjusted by…'*)

DOI: 10.1201/9781003485919-36

Clause 4.2: Items included in adjustments

4.2 *The Contract Sum shall be adjusted by:*

4.2.1 *any amount agreed by the Employer and the Contractor in respect of Changes and other work of the types referred to in clause 5.2 and the amount of each Valuation;*

4.2.2 *any amount agreed by Confirmed Acceptance of an Acceleration Quotation;*

4.2.3 *(where the Contract Particulars state that a Fluctuations Provision applies) any amounts payable or allowable under that provision;*

4.2.4 *any other amounts referred to in clause 4.12.2 or 4.13.2 (excluding any loss and/or expense to the extent included under clause 4.2.2) and any other deductions referred to in clause 4.12.3 or 4.13.3;*

4.2.5 *the deduction of all Provisional Sums included in the Employer's Requirements; and*

4.2.6 *any other amount which under this Contract is to be added to the Contract Sum or may be deducted from it.*

Following on from clause 4.1, clause 4.2 provides several matters by which the Contract Sum may be adjusted:

- amounts (i) agreed by the Employer and Contractor, or (ii) of Valuations, in respect of:
  - Changes required by Employer's instructions;
  - all work under the Conditions which is to be *treated* as a Change; and
  - work relating to Employer's instructions for the expenditure of Provisional Sums;
- amounts agreed by the Confirmed Acceptance of an Acceleration Quotation, in accordance with Supplemental Provision 4 (Acceleration Quotation), in circumstances where the Employer wishes to achieve practical completion before the Completion Date;
- amounts payable or allowable under any Fluctuations Provision, if they apply (as stated against the reference to clause 4.2, clause 4.12, and clause 4.13 in the Contract Particulars), including:
  - JCT Fluctuations Option A (Contribution, levy and tax fluctuations);
  - JCT Fluctuations Option B (Labour and materials cost and tax fluctuations); or
  - JCT Fluctuations Option C (Formula adjustment);
- other amounts referred to in either clause 4.12.2 (Gross Valuation – Alternative A) or clause 4.13.2 (Gross Valuation – Alternative B), but excluding any loss and/or expense which has already been included under clause 4.2.2;
- other deductions referred to in clause 4.12.3 or clause 4.13.3 (both considered below);

- the deduction of all Provisional Sums, included in the Employer's Requirements; and
- any other amount to be added or deducted from the Contract Sum.

Resource 158 page 500	Checklist: Contractor and Employer	A checklist for other amounts to be added to, or deducted from, the Contract Sum, as referred to in clause 4.2.6.

Market practice: Fluctuations Options

It is common for none of the Fluctuations Options to apply, on the basis that the Employer wishes to limit the circumstances where the Contract Sum may be adjusted, and to obtain as 'fixed' a price as possible.

Clause 4.3: Taking adjustments into account

4.3 Where these Conditions provide that an amount is to be added to, deducted from or dealt with by adjustment of the Contract Sum, then, as soon as the amount is ascertained in whole or in part, the ascertained amount shall be taken into account in the next Interim Payment.

Clause 4.3 sets out when any adjustments to the Contract Sum are to be taken into account.

As soon as the amount of any such adjustment has been ascertained (or established), whether in whole or in part, that ascertained amount will be taken into account in the next Interim Payment.

Each Interim Payment may therefore need to take account of several separate adjustments of the Contract Sum.

Resource 159 page 501	Reminder: Contractor to Employer	A reminder from the Contractor to the Employer to take adjustments to the Contract Sum into account in the next Interim Payment.

Taxes

Overview

This sub-section contains two provisions, both of which relate to tax matters. The Parties are therefore encouraged to consult a tax adviser to ensure compliance with their obligations in this regard.

Key provisions

- clause 4.4 (VAT)
- clause 4.5 (Construction Industry Scheme (CIS))

Clause 4.4: VAT

4.4.1 *The Contract Sum is exclusive of VAT and in relation to each payment to the Contractor under this Contract, the Employer shall in addition pay the amount of any VAT properly chargeable in respect of it.*

Clause 4.4 clarifies that the Contract Sum is exclusive of VAT. (This is reflected in Article 2 (Contract Sum), which refers to the '*VAT-exclusive sum*'.)

The Employer must therefore also pay the amount of any VAT chargeable in respect of each payment to the Contractor under the Contract.

4.4.2 *If after the Base Date the supply of any goods or services to the Employer becomes exempt from VAT there shall be paid to the Contractor an amount equal to the input tax on the supply to the Contractor of goods and services that contribute to the Works which as a consequence of that exemption the Contractor cannot recover.*

Clause 4.4.2 protects the Contractor in the event that a VAT exemption for the supply of any goods or services to the Employer is introduced after the Base Date.

In such circumstances, the Employer must also pay to the Contractor the amount equal to the '*input*' tax on the supply of goods and services to the Contractor (in other words, the VAT paid by the Contractor in respect of those goods and

DOI: 10.1201/9781003485919-37

services) that contribute to the Works, which, following the VAT exemption, the Contractor can no longer recover.

In other words, this payment will ensure that the Contractor is no worse off than it would have been, had the VAT exemption not been introduced.

Resource 160 page 502	Reminder: Contractor to Employer	A reminder from the Contractor to the Employer in respect of a VAT exemption.

Market practice: VAT domestic reverse charge

Clause 4.4 is often amended to provide for the VAT domestic reverse charge, which relates to the supply of certain building and construction services, and which came into force on 1 March 2021. In summary, it requires the recipient, rather than the supplier, to account for VAT due on the relevant services, to help ensure that VAT is being correctly reported.

Whilst HMRC has published guidance,[16] the Parties are encouraged to consult tax advisors in respect of the VAT domestic reverse charge.

Clause 4.5: Construction Industry Scheme (CIS)

4.5 If the Employer is or at any time up to the final payment under clause 4.9 becomes a 'contractor' for the purposes of the CIS, its obligation to make any payment under this Contract is subject to the provisions of the CIS.

Clause 4.5 relates to the Construction Industry Scheme (or 'CIS'), which is a tax deduction scheme designed to prevent tax avoidance.[17]

The CIS requires '*contractors*' (as defined under the CIS, and to be distinguished from the 'Contractor' under the Contract) to deduct money from payments to be made to '*subcontractors*' (again, as defined under the CIS) and pass it to HMRC, which then count as advance payments towards the subcontractors' tax and National Insurance.

'*Contractors*' are required to register for the CIS.

Clause 4.5 therefore provides that, if the Employer either (i) is a '*contractor*' under the CIS, or (ii) becomes a '*contractor*' under the CIS, its obligations to make payments to the Contractor under the Contract is subject to the CIS.

[16] Available at: https://www.gov.uk/guidance/vat-domestic-reverse-charge-for-building-and-construction-services

[17] Available at: https://www.gov.uk/what-is-the-construction-industry-scheme

This relates to the Fourth Recital and the corresponding entry in the Contract Particulars, which requires the Employer to state whether or not it is a '*contractor*' for the purposes of the CIS, at the Base Date.

Resource 161 page 502	Confirmation: Employer to Contractor	Confirmation from the Employer to the Contractor of the Employer becoming a 'contractor' for the purposes of the Construction Industry Scheme.

Payments and Notices – general provisions

Overview

This sub-section sets out the key provisions in respect of notices relating to payment, including:

- the Contractor's obligation to issue Interim Payment Applications;
- the Employer's obligation to issue corresponding Payment Notices; and
- the Employer's right to issue Pay Less Notices.

These provisions have been drafted to comply with the Housing Grants, Construction and Regeneration Act 1996.

This sub-section also provides for advance payments to be made to the Contractor, along with the Contractor's right to suspend for non-payment.

Key provisions

- clause 4.6 (Advance payment)
- clause 4.7 (Interim Payments – Contractor's Interim Payment Applications, due dates and Payment Notices)
- clause 4.8 (Relevant statement and Final Payment Notice)
- clause 4.9 (Interim and final payments – final date and amount)
- clause 4.10 (Pay Less Notices and other general provisions)
- clause 4.11 (Contractor's right of suspension)

Clause 4.6: Advance payment

4.6 *Where the Contract Particulars state that clause 4.6 applies, and an advance payment is to be made, it shall be paid to the Contractor on the date and reimbursed to the Employer on the terms stated in the Contract Particulars, save that, if the Contract Particulars state that an advance payment bond is required, payment shall only be made if the Contractor has provided to the Employer a bond in the terms set out in Part 1 of Schedule 6 from a surety approved by the Employer.*

DOI: 10.1201/9781003485919-38

Resource 162 page 503	Tracker: Contractor and Employer	A tracker template to record the reimbursement to the Employer of an advance payment made to the Contractor.

Clause 4.6 provides for an advance payment to be made by the Employer to the Contractor, to help the Contractor's cashflow. (For example, in circumstances where the Contractor needed to place a high-value order for goods or materials at an early stage in the project, in order to ensure practical completion was achieved by the Completion Date.)

Clause 4.6 is optional, and whether it applies will be stated against the reference to clause 4.6 in the Contract Particulars. (However, it is not applicable where the Employer is a Local or Public Authority.)

If it applies, then the Employer must pay the advance payment to the Contractor on the terms stated in the Contract Particulars, which include the following:

- the amount of the advance payment (whether as a monetary amount, or as a percentage of the Contract Sum);
- the date on which it will be paid to the Contractor;
- the manner in which it will be reimbursed to the Employer; and
- whether or not an advance payment bond is required, as security for the Employer.

If an advance payment bond is required, then the Employer will only be obliged to make the advance payment to the Contractor if the Contractor has provided that advance payment bond, in the form included in Part 1 of Schedule 6 (Forms of Bonds), and from a surety approved by the Employer.

Recommendation: Surety under the advance payment bond

If the Parties have agreed for an advance payment to be made to the Contractor, and for an advance payment bond to be provided to the Employer, then the Contractor should obtain the Employer's approval of the surety for the bond prior to the execution of the Contract, to avoid any delay in receiving the advance payment.

The advance payment bond should also be executed simultaneously with the Contract.

Clause 4.7: Interim Payments – Contractor's Interim Payment Applications, due dates and Payment Notices

4.7.1 *Interim Payments shall be made by the Employer to the Contractor in accordance with section 4 and whichever of Alternative A (Stage Payments) or Alternative B (Periodic Payments) is stated in the Contract Particulars to apply.*

Clause 4.7 requires the Employer to make Interim Payments to the Contractor in accordance with the provisions of section 4 and either:

- clause 4.12 (Alternative A – Stage Payments), upon completion of the relevant stages; or
- clause 4.13 (Alternative B (Periodic Payments), upon the expiry of every period.

The applicable Alternative will be stated against the reference to clause 4.7.1 in the Contract Particulars.

4.7.2 *During the period up to the due date for the final payment fixed under clause 4.24.5 and subject to clause 4.7.3, the monthly due dates for Interim Payments by the Employer shall in each case be the date 7 days after the relevant Interim Valuation Date.*

Clause 4.7.2 provides the monthly due dates for Interim Payments, namely the date 7 days after the relevant Interim Valuation Date.

The Interim Valuation Dates are set out against the reference to clause 4.7.2 in the Contract Particulars.

This applies to the period up to the due date for the final payment, under clause 4.24.5 (Final Statement and Final Payment), considered below.

It is subject to the exception in clause 4.7.3.

Recommendation: Interim Valuation Dates

The first Interim Valuation Date will be stated against the reference to clause 4.7.2 in the Contract Particulars.

All subsequent Interim Valuation Dates will be either on (i) the same date in each month, or (ii) the *nearest* (rather than the *next*) Business Day in that month.

This will affect the corresponding due date, as well as the 5-day period for the Employer to issue the corresponding Payment Notice, in accordance with clause 4.7.5 (considered below).

The Parties are therefore encouraged to make note on their project calendar of the relevant months where the Interim Valuation Date, and corresponding due date, will be adjusted.

4.7.3 *In relation to each Interim Payment, the Contractor shall make an application to the Employer (an 'Interim Payment Application'), stating the sum that the Contractor considers to be due to it at the due date and the basis on which that sum has been calculated. Where the Interim Payment Application is received no later than the relevant Interim Valuation Date, the due date shall be the date that would apply under clause 4.7.2; if the Interim Payment Application is received later, the due date shall be 7 days after the date of receipt by the Employer.*

Clause 4.7.3 defines the term '*Interim Payment Application*'.

For each Interim Payment, clause 4.7.3 requires the Contractor to make an Interim Payment Application to the Employer, stating both:

- the sum the Contractor considers to be due, at the due date; and
- the basis on which that sum was calculated.

The due date will depend on whether the Interim Payment Application is received on or before the Interim Valuation Date:

- if it is, the due date under clause 4.7.2 will apply, namely the date 7 days after the relevant Interim Valuation Date; or
- if it is not, the due date will be 7 days after receipt by the Employer of the Interim Payment Application.

For cashflow purposes, the Contractor will want the (earlier) due date in clause 4.7.2 to apply, and should therefore make its Interim Payment Application by the Interim Valuation Date.

Resource 163 page 503	Template: Interim Payment Application	An Interim Payment Application template, from the Contractor to the Employer.

4.7.4 Each Interim Payment Application shall be accompanied by such further information as may be specified in the Employer's Requirements.

Clause 4.7.4 requires the Contractor to support each Interim Payment Application with any further information specified in the Employer's Requirements. For example, evidence showing the completion of design work or construction work.

4.7.5 Not later than 5 days after each due date the Employer shall give a notice (a 'Payment Notice') to the Contractor, stating the sum that it considers to be or have been due to the Contractor at the due date, calculated in accordance with clause 4.12 or 4.13 and clause 4.14, and the basis on which that sum has been calculated.

Clause 4.7.5 defines the term '*Payment Notice*', and requires the Employer to give a Payment Notice to the Contractor in respect of each Interim Payment Application.

Each Payment Notice must:

- be given not later than 5 days after each due date;
- state the sum that the Employer considers to be (or have been) due to the Contractor, at the due date. This is achieved by performing the calculations in

either clause 4.12 (Alternative A – Stage Payments) or clause 4.13 (Alternative B – Periodic Payments), and clause 4.14 (Sums due as Interim Payments); and
- state the basis on which that sum was calculated.

Resource 164 page 504	Template: Payment Notice	A Payment Notice template, from the Employer to the Contractor.

Clause 4.8: Relevant statement and Final Payment Notice

> 4.8 *Not later than 5 days after the due date for the final payment fixed under clause 4.24.5 and notwithstanding any dispute regarding the relevant statement (as defined in clause 4.24.5.3), the Party by which the statement shows the final payment as payable ('the paying Party') shall give a notice (a 'Final Payment Notice') to the other Party, stating the sum that the paying Party considers to be or have been due under clause 4.24.2 at the due date and the basis on which that sum has been calculated.*

Clause 4.8 defines the term '*Final Payment Notice*', and should be read in conjunction with clause 4.24 (Final Statement and Final Payment).

The Final Payment Notice must be given by the Party for whom the final payment is stated as payable, in either (i) the Final Statement, under clause 4.24.1, or (ii) the Employer's Final Statement, under clause 4.24.4.

It must be issued not later than 5 days after the due date for the final payment, which is determined in accordance with clause 4.24.5.

This is irrespective of whether the Parties are in dispute regarding the Final Statement or the Employer's Final Statement.

The Final Payment Notice must state:

- the sum the paying Party considers to be due under clause 4.24.2, at the due date; and
- the basis on which that sum was calculated.

Resource 165 page 505	Final Payment Notice: Contractor or Employer to the other Party	A Final Payment Notice template.

Clause 4.9: Interim and final payments – final date and amount

> 4.9.1 *The final date for payment of each Interim Payment and the final payment shall be 14 days from its due date.*

Clause 4.9.1 provides the final date for payment (in other words, the date by when payment must be made), in respect of both (i) each Interim Payment, and (ii) the final payment under clause 4.24 (Final Statement and Final Payment).

In each case, the final date for payment will be 14 days from its corresponding due date.

Market practice: Extended final date for payment and implications for cashflow

Clause 4.9.1 is often amended to increase the period between the due date and the final date for payment, to give the Employer longer to pay the Contractor. However, the Contractor should have regard to any corresponding negative effects on its cashflow, before agreeing to any such increase.

The Contractor is also encouraged to prepare and regularly update a cashflow forecast in respect of the Works, to ensure that it has sufficient resources and working capital to meet its obligations under the Contract.

> 4.9.2 *Subject to any Pay Less Notice given by the Employer under clause 4.9.5, the Employer shall pay the sum stated as due in the Payment Notice on or before the final date for payment.*

Clause 4.9.2 requires the Employer to pay the Contractor the sum stated as due in the Payment Notice, by no later than the final date for payment (included in clause 4.9.1).

The exception is where the Employer gives a Pay Less Notice, in accordance with clause 4.9.5 below.

> 4.9.3 *If a Payment Notice is not given in accordance with clause 4.7.5, the Employer shall, subject to any Pay Less Notice under clause 4.9.5, pay the Contractor the sum stated as due in the Interim Payment Application.*

Clause 4.9.3 provides a default position, if the Employer fails to give a Payment Notice in accordance with clause 4.7.5 above.

In these circumstances, the Employer will be obliged to pay the Contractor the sum stated as due by the Contractor in its Interim Payment Application.

Again, the exception is where the Employer gives a Pay Less Notice, in accordance with clause 4.9.5 below.

> 4.9.4 *In the case of the final payment, subject to any Pay Less Notice under clause 4.9.5, the paying Party shall pay the sum stated as due in the Final Payment Notice or, if such notice is not given, the balance stated in the relevant statement on or before the final date for payment.*

Clause 4.9.4 provides what must be paid in respect of the final payment, as set out in clause 4.24 (Final Statement and Final Payment).

The paying Party (whether the Employer or the Contractor) must pay either:

- the sum stated as due in the Final Payment Notice; or
- if no Final Payment Notice has been given, the balance stated as due in either the Final Statement or the Employer's Final Statement (as applicable).

In either case, this payment must be made by no later than the final date for payment, provided in clause 4.9.1 above.

Again, the exception is where the paying Party gives a Pay Less Notice, in accordance with clause 4.9.5 below.

> 4.9.5 *Where:*
>
> 4.9.5.1 *the Employer intends to pay less than the sum stated as due from it in a Payment Notice or Interim Payment Application; or*
>
> 4.9.5.2 *the paying Party intends to pay less than the sum so stated as due in the Final Payment Notice or, in default of such notice, less than the amount stated as due in the relevant statement,*
>
> *the Party by which the payment is stated to be payable shall not later than 5 days before the final date for payment give the other Party notice of that intention in accordance with clause 4.10.1 (a 'Pay Less Notice'). Where a Pay Less Notice is given, the payment to be made on or before the final date for payment shall not be less than the amount stated in it as due.*

Clause 4.9.5 defines the term '*Pay Less Notice*', where the paying Party notifies the other Party that it intends to pay less than the sum currently stated as due.

Clause 4.9.5 provides that Pay Less Notices may be given in the following circumstances:

- **For Interim Payments:** by the Employer, where it intends to pay less than the sum stated as due from it in either:

  - a Payment Notice from the Employer; or
  - an Interim Payment Application from the Contractor; and
- **For the Final Payment:** by either the Employer or the Contractor, where the paying Party intends to pay less than the sum stated as due in either:
  - the Final Payment Notice; or, if this has not been given,
  - the Final Statement or the Employer's Final Statement (as applicable).

In each case, the Pay Less Notice must:

- be given not later than 5 days before the final date for payment; and
- in accordance with clause 4.10 (Pay Less Notices and other general provisions), specify:
  - the sum considered to be due at the time the Pay Less Notice is given, and
  - the basis on which that sum was calculated.

In circumstances where a Pay Less Notice is given, the paying Party must pay *at least* the sum stated in it as due (it may pay more, if it wishes), by the final date for payment.

Resource 166 page 506	Pay Less Notice: Employer and Contractor	A Pay Less Notice template.

Market practice: Extended period for giving a Pay Less Notice

Clause 4.9.5 is often amended to extend the period for giving a Pay Less Notice, by reducing the 5-day period. Whilst this assists the paying Party (usually the Employer), it creates greater uncertainty for the recipient's cashflow, and its financial planning.

4.9.6 If either Party fails to pay a sum, or any part of it, due to the other Party under these Conditions by the final date for payment, it shall, in addition to any unpaid amount that should properly have been paid, pay the other Party simple interest on that amount at the Interest Rate for the period from the final date for payment until payment is made.

Clause 4.9.6 imposes an obligation on both Parties to pay interest on any principal amounts which either of them should previously have paid, but have failed to do so.

This applies where the Party has failed to pay the principal sum in its entirety, or only in part.

In such circumstances, the paying Party must, in addition to the payment of the principal sum, pay simple interest (rather than compound interest) on that principal sum from:

- when payment should have been made, namely the final date for payment; to
- when payment was actually made.

The '*Interest Rate*' is defined in clause 1.1 (Definitions) as 5% per annum above the official bank rate of the Bank of England, current at the date when the relevant payment becomes overdue. This information is available from the Bank of England website.[18]

The following information is therefore required to perform the calculation of the interest due:

- the principal sum;
- the Interest Rate;
- the final date for payment; and
- the date when payment was made.

> 4.9.7 *Any such unpaid amount and any interest under clause 4.9.6 shall be recoverable as a debt. Acceptance of a payment of interest shall not in any circumstances be construed as a waiver either of the recipient's right to proper payment of the principal amount due or of the Contractor's rights to suspend performance under clause 4.11 or terminate its employment under section 8.*

Clause 4.9.7 provides that both (i) any unpaid principal amount, and (ii) any interest due under clause 4.9.6, may be recovered as a debt. In other words, the Party to whom payment is owed may bring a debt claim against the Party which should have made payment.

Clause 4.9.7 further provides that, just because one Party accepts payment of interest from the other Party, it does *not* mean that the Party accepting payment waives:

- its right to proper payment of the principal amount due; or
- in the case of the Contractor, its rights to either:
  - suspend the performance of its obligations under clause 4.11 (Contractor's right of suspension); or
  - terminate its employment under the Contract under section 8 (Termination), and clause 8.9.1.1 in particular.

Clause 4.10: Pay Less Notices and other general provisions

> 4.10.1 *A Pay Less Notice given by either Party shall specify the sum it considers to be due to the other Party at the date the notice is given and the basis on which that sum has been calculated.*

[18] Available at: https://www.bankofengland.co.uk/boeapps/database/Bank-Rate.asp

Clause 4.10.1 sets out what must be included in a Pay Less Notice (whether given by the Employer or the Contractor).

The Party giving a Pay Less Notice must specify both:

- the sum it considers to be due to the other Party, at the date the Pay Less Notice is given; and
- the basis on which that sum was calculated.

> 4.10.2 *A Payment Notice, a Final Payment Notice or a Pay Less Notice to be given by the Employer may be given on its behalf by the Employer's Agent or by any other person which the Employer notifies the Contractor as being authorised to do so.*

Clause 4.10.2 relates to Payment Notices, Final Payment Notices, and Pay Less Notices given by or on behalf of the Employer.

It provides that such notices may be given on the Employer's behalf by either:

- the Employer's Agent; or
- any other person which the Employer notifies the Contractor as being authorised by the Employer to do so.

(There is no corresponding provision in respect of the Contractor.)

> 4.10.3 *In relation to the requirements for the giving of Payment Notices, the Final Payment Notice, Pay Less Notices and the submission of a Final Statement, it is immaterial that the amount then considered to be due may be zero.*

Clause 4.10.3 relates to the giving of:

- Payment Notices;
- the Final Payment Notice;
- Pay Less Notices; and
- the Final Statement.

In each case, it is immaterial if the amount then considered to be due is zero. (In other words, this is not prohibited.)

> 4.10.4 *The Employer's fiduciary interest in the Retention referred to in clause 4.16 shall not prevent the Employer from exercising any right under this Contract to withhold or deduct from a sum due to the Contractor, subject to clause 4.9.5, even if that sum includes any Retention due for release under clause 4.18.*

As considered further in clause 4.16 (Rules on treatment of Retention), the Employer's interest in the Retention is fiduciary, as trustee for the Contractor.

However, clause 4.10.4 provides that this fiduciary interest will not prevent the Employer from exercising any right it has under the Contract to withhold or deduct from a sum due to the Contractor, even if that sum includes any Retention which is due to be paid to the Contractor under clause 4.18 (Retention – amounts and periods).

(For example, any right the Employer may have to withhold or deduct liquidated damages, in accordance with clause 2.29 (Payment or allowance of liquidated damages).)

This is subject to the Employer giving a Pay Less Notice, in accordance with clause 4.9.5.

Market practice: Employer's fiduciary interest

Clause 4.10.4 is often amended (in conjunction with clause 4.16 (Rules on treatment of Retention)) to remove any references to the Employer's fiduciary interest in the Retention.

In other words, to remove the Employer's obligation to hold the Retention on trust for the Contractor.

This may have significant adverse implications in the event of the Employer's insolvency, and may impact the Contractor's ability to recover any Retention.

Clause 4.11: Contractor's right of suspension

4.11.1 *If the Employer fails to pay a sum payable to the Contractor in accordance with clause 4.9 (together with any VAT properly chargeable in respect of that payment) by the final date for payment and the failure continues for 7 days after the Contractor has given notice to the Employer of its intention to suspend the performance of its obligations under this Contract and the grounds for such suspension, the Contractor, without affecting its other rights and remedies, may suspend performance of any or all of its obligations until payment is made in full.*

Clause 4.11 entitles the Contractor to suspend the performance of its obligations under the Contract, in the event of non-payment by the Employer.

(This is also a statutory right, under section 112 (*Right to suspend performance for non-payment*) of the Housing Grants, Construction and Regeneration Act 1996.)

Paying the Contractor is one of the Employer's most important obligations under the Contract, and clause 4.11 protects the Contractor from having to continue carrying out the Works without payment.

If the Employer is obliged in accordance with clause 4.9 to pay a sum (along with any applicable VAT) to the Contractor, and fails to do so by the final date for payment, the Contractor may give a notice to the Employer stating:

- that the Contractor intends to suspend the performance of its obligations under the Contract; and
- the grounds for such suspension. (In other words, the particulars of the failure to pay by the Employer.)

If the Employer still fails to make payment for 7 days after the Contractor has given that notice, then the Contractor may suspend the performance of any or all of its obligations (for example, the design or construction of the Works), until the Employer makes payment in full.

Exercising this right of suspension will not affect any of the Contractor's other rights and remedies, whether under the Contract, or generally at law.

Resource 167 page 507	Notice: Contractor to Employer	A notice from the Contractor to the Employer of the Contractor's intention to suspend the performance of its obligations under the Contract.
Resource 168 page 507	Reply: Employer to Contractor	A reply from the Employer to the Contractor's notice of intention to suspend the performance of its obligations under the Contract.
Resource 169 page 508	Confirmation: Contractor to Employer	Confirmation from the Contractor to the Employer of the Contractor's intention to suspend the performance of its obligations under the Contract.

4.11.2 *Where the Contractor exercises its right of suspension under clause 4.11.1, it shall be entitled to a reasonable amount in respect of costs and expenses reasonably incurred by it as a result of exercising the right.*

If the Contractor does suspend the performance of any of its obligations under clause 4.11.1, then clause 4.11.2 provides that the Contractor will be entitled to a '*reasonable amount*' in respect of the costs and expenses the Contractor reasonably incurs as a result.

4.11.3 *Applications in respect of any such costs and expenses shall be made to the Employer and the Contractor shall with its application or on request submit such details of them as are reasonably necessary for ascertaining the amount in question.*

Clause 4.11.3 requires the Contractor to make applications to the Employer for reimbursement of the costs and expenses referred to in clause 4.11.2, supported (either with the application, or at the Employer's request) by the details reasonably necessary to allow the amount to be ascertained.

Recommendation: Record keeping of costs and expenses associated with suspension

In accordance with clause 4.11.2, the Contractor is entitled to a '*reasonable*' amount in respect of the costs and expenses '*reasonably incurred*' by it.

The Contractor is therefore encouraged to keep records of the costs and expenses incurred by it in connection with any such suspension, separate from any other costs and expenses it may have incurred. For example, 'standing time' costs, cancellation costs, extended hire charges, costs associated with missed delivery slots, other preliminary costs, etc.

Resource 170 page 509	Application: Contractor to Employer	An application from the Contractor to the Employer for payment of costs and expenses incurred as a result of the Contractor's suspension of the performance of its obligations under the Contract.
Resource 171 page 509	Request: Employer to Contractor	A request from the Employer to the Contractor for details of costs and expenses incurred as a result of the Contractor's suspension of the performance of its obligations under the Contract.

Interim Payments – calculation of sums due

Overview

This sub-section governs the calculation of both the Gross Valuation, and thereafter sums due as Interim Payments to the Contractor, whether payment is to be made either:

- by stages, in accordance with clause 4.12 (Gross Valuation – Alternative A); or
- periodically, in accordance with clause 4.13 (Gross Valuation – Alternative B).

The applicable Alternative will be stated against the reference to clause 4.7.1 in the Contract Particulars.

Whilst a summary is provided below, Part 2 contains more detailed checklists for the calculation of the Gross Valuation under both Alternative A (Stage Payments) and Alternative B (Periodic Payments).

Key provisions

- clause 4.12 (Gross Valuation – Alternative A)
- clause 4.13 (Gross Valuation – Alternative B)
- clause 4.14 (Sums due as Interim Payments)

Clause 4.12: Gross Valuation – Alternative A

4.12 The Gross Valuation for each Interim Payment shall be the total of the amounts referred to in clauses 4.12.1 and 4.12.2, less the deductions referred to in clause 4.12.3, each calculated as at the Interim Valuation Date:

Clause 4.12 applies where Interim Payments are to be made in stages (rather than periodically, in accordance with clause 4.13 below), and sets out how the Gross Valuation is to be calculated.

DOI: 10.1201/9781003485919-39

In summary, the calculation is as follows:

(the total of the amounts in clause 4.12.1) + (the total of the amounts in clause 4.12.2) – (the total of the amounts referred to in clause 4.12.3)

In each case, these amounts are calculated as at the relevant Interim Valuation Date (which is set out against the reference to clause 4.7.2 in the Contract Particulars).

4.12.1 *the total of the following, which are subject to Retention:*

4.12.1.1 *the cumulative value of stages completed;*

4.12.1.2 *the value of any Changes or other work referred to in clause 5.2 that are relevant to the Interim Payment (whether agreed pursuant to clause 5.2 or valued under the Valuation Rules) but excluding any amounts referred to in clause 4.12.2.4;*

4.12.1.3 *the value of any Listed Items for which the conditions set out in clause 4.15 are satisfied; and*

4.12.1.4 *where JCT Fluctuations Option C is applicable and where in accordance with the JCT Formula Rules amounts in the Value of Work are to be allocated to lift installations, structural steelwork installations or catering equipment installations, the total value of Site Materials of those descriptions, provided that their value shall only be included if they are adequately protected against weather and other casualties and they are not on the Works prematurely;*

those values shall be adjusted, where appropriate, in accordance with any applicable Fluctuations Provision or any Acceleration Quotation for which there has been Confirmed Acceptance;

Clause 4.12.1 lists values under four categories, all of which are subject to Retention (considered further in clause 4.18 (Retention – amounts and periods)).

Where appropriate, these values will be adjusted in accordance with:

- any applicable Fluctuations Provision (considered below); or
- any Acceleration Quotation for which there has been Confirmed Acceptance, in accordance with Supplemental Provision 4 (Acceleration Quotation).

4.12.2 *the total of the following, which are not subject to Retention:*

4.12.2.1 *any amounts to be included in accordance with clause 4.3 by the Employer as a result of payments made or costs incurred by the Contractor under clause 2.5.2, 2.20, 3.12, 6.10.2 or 6.10.3, 6.11.3, 6.12.2 or 6.20;*

4.12.2.2 *any amounts payable under clause 4.11.2;*

4.12.2.3 *the amount of any loss and/or expense to which the Contractor is entitled under clause 4.19.1 or by a Confirmed Acceptance;*

4.12.2.4 *where Insurance Option B or C applies or to the extent that the work is under clause 6.13.5.3 to be treated as a Change, any amounts in respect of reinstatement work under clause 6.13.4; and*

4.12.2.5 *any amount payable to the Contractor under any applicable Fluctuations Provision, other than by means of an adjustment made under clause 4.12.1;*

Clause 4.12.2 lists values under a further five categories, which (unlike clause 4.12.1) are *not* subject to Retention.

4.12.3 *the following deductions:*

4.12.3.1 *any amounts deductible under clause 2.35, 3.6, 6.12.2 or 6.19.2; and*

4.12.3.2 *any amount allowable by the Contractor under clause 6.10.2 or under any applicable Fluctuations Provision, other than by means of an adjustment made under clause 4.12.1.*

Clause 4.12.3 sets out a number of deductions to be made, across several provisions.

Resource 172 page 510	Checklist: Contractor and Employer	A checklist for calculating the Gross Valuation in accordance with Alternative A (Stage Payments).

Clause 4.13: Gross Valuation – Alternative B

4.13 The Gross Valuation for each Interim Payment shall be the total of the amounts referred to in clauses 4.13.1 and 4.13.2, less the deductions referred to in clause 4.13.3, each calculated as at the Interim Valuation Date:

Clause 4.13 applies where Interim Payments are to be made periodically (rather than in stages, in accordance with clause 4.12 above), and sets out how the Gross Valuation is to be calculated.

In summary, the calculation is as follows:

(the total of the amounts in clause 4.13.1) + (the total of the amounts in clause 4.13.2) – (the total of the amounts referred to in clause 4.13.3)

In each case, these amounts are calculated as at the relevant Interim Valuation Date (which is set out against the reference to clause 4.7.2 in the Contract Particulars).

4.13.1 the total values of the following, which are subject to Retention:

4.13.1.1 work properly executed including any design work carried out by the Contractor and work so executed for which a value has been agreed pursuant to clause 5.2 or which has been valued under the Valuation Rules, but excluding any amounts referred to in clause 4.13.2.4;

4.13.1.2 Site Materials provided they are adequately protected against weather and other casualties and they are not on the Works prematurely; and

4.13.1.3 Listed Items (if any) for which the conditions set out in clause 4.15 are satisfied;

those values shall be adjusted, where appropriate, in accordance with any applicable Fluctuations Provision or any Acceleration Quotation for which there has been Confirmed Acceptance;

Clause 4.13.1 lists values under three categories, all of which are subject to Retention (considered further in clause 4.18 (Retention – amounts and periods)).

Where appropriate, these values will be adjusted in accordance with:

- any applicable Fluctuations Provision (considered below); or
- any Acceleration Quotation for which there has been Confirmed Acceptance, in accordance with Supplemental Provision 4 (Acceleration Quotation).

4.13.2 the total of the following, which are not subject to Retention:

4.13.2.1 any amounts to be included in accordance with clause 4.3 by the Employer as a result of payments made or costs incurred by the Contractor under clause 2.5.2, 2.20, 3.12, 6.10.2 or 6.10.3, 6.11.3, 6.12.2 or 6.20;

4.13.2.2 any amounts payable under clause 4.11.2;

4.13.2.3 the amount of any loss and/or expense to which the Contractor is entitled under clause 4.19.1 or by a Confirmed Acceptance;

4.13.2.4 where Insurance Option B or C applies or to the extent that the work is under clause 6.13.5.3 to be treated as a Change, any amounts in respect of reinstatement work under clause 6.13.4; and

4.13.2.5 any amount payable to the Contractor under any applicable Fluctuations Provision, other than by means of an adjustment made under clause 4.13.1;

Clause 4.13.2 lists values under a further five categories, which (unlike clause 4.13.1) are *not* subject to Retention.

4.13.3 the following deductions:

4.13.3.1 any amounts deductible under clause 2.35, 3.6, 6.12.2 or 6.19.2; and

4.13.3.2 any amount allowable by the Contractor under clause 6.10.2 or under any applicable Fluctuations Provision, other than by means of an adjustment made under clause 4.13.1.

Clause 4.13.3 sets out a number of deductions to be made, across several provisions.

Resource 173 page 512	Checklist: Contractor and Employer	A checklist for calculating the Gross Valuation in accordance with Alternative B (Periodic Payments).

Clause 4.14: Sums due as Interim Payments

4.14 *The sum due as an Interim Payment shall in each case be an amount equal to the Gross Valuation under clause 4.12 where the Contract Particulars state that Alternative A applies, or clause 4.13 where Alternative B applies, in either case less the following deductions:*

4.14.1 *any amount which may be deducted and retained by the Employer under clauses 4.16 to 4.18 ('the Retention');*

4.14.2 *the cumulative total of the amounts of any advance payment that have then become due for reimbursement to the Employer in accordance with the terms stated in the Contract Particulars for clause 4.6; and*

4.14.3 *the amounts paid in previous Interim Payments.*

Clause 4.14 sets out the calculation of sums due as Interim Payments to the Contractor, as follows:

Firstly, the Gross Valuation must be calculated, whether under clause 4.12 (if Alternative A: Stage Payments applies), or under clause 4.13 (if Alternative B: Periodic Payments applies). The applicable Alternative will be stated against the reference to clause 4.7.1 in the Contract Particulars.

The amount equal to the Gross Valuation is then subject to the following further three deductions:

- the amount which the Employer is entitled to deduct and retain under clause 4.16 to clause 4.18 below, defined as the '*Retention*';
- if clause 4.6 (Advance Payment) applies, the cumulative total of the amounts of any advance payment that has become due for reimbursement to the Employer (as set out against the reference to clause 4.6 in the Contract Particulars); and
- the total amount paid to the Contractor in all previous Interim Payments.

The balance will be the sum due as an Interim Payment to the Contractor.

Resource 174 page 513	Tracker: Contractor and Employer	A tracker template for recording Interim Payments.

Listed Items

Overview

This sub-section contains only one provision, relating to '*Listed Items*', which are defined in clause 1.1 (Definitions) as *'materials, goods and/or items pre-fabricated for inclusion in the Works which are listed as such items by the Employer in a list supplied to the Contractor and annexed to the Employer's Requirements'*.

Clause 4.15 provides for the possibility that the Contractor may be paid for such Listed Items, *before* they have been delivered to the Works.

Whilst this helps protect the Contractor's cashflow, it potentially creates a greater commercial risk for the Employer, so such Listed Items must be agreed between the Parties and included within the Employer's Requirements.

Furthermore, clause 4.15 sets out a number of conditions which must be met before the value of any Listed Item may be included in the Gross Valuation, before it has been delivered to the site.

However, in many circumstances, it makes good commercial sense for Listed Items to be included. For example, where the relevant materials or goods are subject to volatile price changes, or are on long lead times.

Key provisions

- clause 4.15 (Listed Items)

Clause 4.15: Listed Items

4.15 The conditions for inclusion of the value of a Listed Item in the Gross Valuation before its delivery to or adjacent to the Works as referred to in clause 4.12.1.3 or 4.13.1.3 are:

Clause 4.15 provides for the circumstances where the value of a particular Listed Item may be included in the Gross Valuation, whether calculated in accordance with clause 4.12 (if Alternative A: Stage Payments applies), or clause 4.13 (if Alternative B: Periodic Payments applies).

Provided the relevant conditions set out in clause 4.15.1 to clause 4.15.5 have been satisfied, this would allow the value of that Listed Item to be paid to the

DOI: 10.1201/9781003485919-40

Contractor, before it has been delivered to the Works (in other words, when it is not yet on site).

> 4.15.1 *the Listed Item is in accordance with this Contract;*

Clause 4.15.1 simply provides that the Listed Item must be in accordance with the Contract.

> 4.15.2 *the Contractor has provided the Employer with reasonable proof that:*
>
> 4.15.2.1 *property in the Listed Item is vested in the Contractor; and*
>
> 4.15.2.2 *the Listed Item is and will remain insured against loss or damage for its full value under a policy of insurance protecting the interests of the Employer and the Contractor in respect of the Specified Perils until delivered to, or adjacent to, the Works;*

Clause 4.15.2 requires the Contractor to provide reasonable proof to the Contractor that both:

- the property in the Listed Item has '*vested*' in the Contractor (rather than, for example, the supplier from which the Contractor purchased the Listed Item); and
- the Listed Item is (and will remain) insured until it is delivered to the Works. This insurance must be against loss or damage in respect of the '*Specified Perils*' (which are defined in clause 6.8 (Insurance of the Works and Existing Structures: Related definitions)), for the full value of the Listed Item, under a policy protecting the interests of both the Employer and the Contractor.

> 4.15.3 *at the premises where the Listed Item is situated pending delivery, there is in relation to that item clear identification of:*
>
> 4.15.3.1 *the Employer as the person to whose order it is held; and*
>
> 4.15.3.2 *the Works as its destination,*
>
> *each item being either set apart or clearly and visibly marked, individually or as a set, by letters, figures or a pre-determined code; and*

Wherever the Listed Item is currently being stored prior to its delivery to the Works, clause 4.15.3 requires that the following must be clearly identified:

- that the Listed Item is being held to the order of the Employer; and
- the Works, as the delivery address.

Each Listed Item must also be clearly identified and distinguished from other goods or materials, either by setting it apart, or clearly marking it, with the use of an appropriate referencing system.

> 4.15.4 *for uniquely identified Listed Items, the Contractor, if it is stated in the Contract Particulars as required, has provided a bond in favour of the Employer from a surety approved by the Employer in the terms set out in Part 2 of Schedule 6 ('the required bond') in the amount specified in the Contract Particulars for this clause 4.15.4; or*

Clause 4.15.4 relates to '*uniquely identified*' Listed Items. In other words, Listed Items that are readily identifiable or discernible from the list included in the Employer's Requirements.

In respect of such Listed Items, the Contractor must provide a bond in favour of the Employer (as a form of security), if it is stated as required against the reference to clause 4.15.4 in the Contract Particulars.

This bond must be:

- from a surety approved by the Employer;
- in the terms included in Part 2 of Schedule 6 (Forms of Bonds); and
- in the amount specified in the Contract Particulars (against the reference to clause 4.15.4).

Resource 175 page 514	Request: Contractor to Employer	A request from the Contractor for the Employer's approval to provide a bond from a particular surety.
Resource 176 page 514	Approval given: Employer to Contractor	Approval given from the Employer for the Contractor to provide a bond from a particular surety.
Resource 177 page 515	Approval withheld: Employer to Contractor	Approval withheld from the Employer for the Contractor to provide a bond from a particular surety.

> 4.15.5 *for Listed Items that are not uniquely identified, the Contractor has provided the required bond in the amount specified in the Contract Particulars for this clause 4.15.5.*

Unlike clause 4.15.4, clause 4.15.5 relates to Listed Items which are not '*uniquely identified*'. In other words, Listed Items that are not readily identifiable or discernible from the list included in the Employer's Requirements, and are more generic in nature.

In respect of such Listed Items, and if clause 4.15.5 applies, the Contractor must also provide a bond in favour of the Employer.

Again, this bond must be:

- from a surety approved by the Employer;
- in the terms included in Part 2 of Schedule 6 (Forms of Bonds); and
- in the amount specified in the Contract Particulars (against the reference to clause 4.15.5).

Resource 178 page 515	Checklist: Employer	A checklist for the Employer to determine whether the conditions for inclusion of the value of a Listed Item in the Gross Valuation before its delivery to or adjacent to the Works have been met.

Retention

Overview

This sub-section relates to the Retention, which may be deducted and retained by the Employer from Interim Payments made to the Contractor. It provides for:

- how that Retention is to be treated;
- the circumstances where a Retention Bond may apply, as an alternative to the Retention being deducted and retained by the Employer;
- the amounts of Retention which may be deducted and retained; and
- for how long the Retention may be deducted and retained.

The deduction of Retention is effectively another form of performance security, for the Employer's benefit (similar to the provision of a performance bond or a parent company guarantee), as the Contractor will need to comply with its key obligations under the Contract before it will be released.

Key provisions

- clause 4.16 (Rules on treatment of Retention)
- clause 4.17 (Retention Bond)
- clause 4.18 (Retention – amounts and periods)

Clause 4.16: Rules on treatment of Retention

4.16 *The Retention which the Employer may deduct and retain as referred to in clause 4.14 shall be subject to the following rules:*

4.16.1 *the Employer's interest in the Retention is fiduciary as trustee for the Contractor (but without obligation to invest);*

4.16.2 *except where the Employer is a Local or Public Authority, the Employer, to the extent that it exercises its right under clause 4.18 and if the Contractor so requests, shall at the time of each Interim Payment place the Retention in a separate bank account (so designated as to identify the amount as the Retention held by the Employer on trust as provided in clause 4.16.1) and notify the Contractor that the amount has been so placed. The Employer shall be entitled to the full beneficial interest in any interest accruing on the separate bank account and under no duty to account for any such interest to the Contractor.*

DOI: 10.1201/9781003485919-41

Clause 4.14.1 (Sums due as Interim Payments) provides for the deduction of Retention by the Employer from Interim Payments.

However, this is subject to a number of rules in clause 4.16, relating to the treatment of that Retention:

- **Fiduciary duty:** clause 4.16.1 provides that the Employer does not have the legal or beneficial interest in the Retention; it is not the Employer's money to do with as it wishes. Instead, the Employer's interest is '*fiduciary*', as trustee for the Contractor. In other words, the Employer owes a duty to the Contractor to hold the Retention on the Contractor's behalf, and in the Contractor's best interests. (However, this obligation does not extend to requiring the Employer to invest the Retention on the Contractor's behalf.)
- **Separate bank account:** if the Employer exercises its right under clause 4.18 to deduct Retention, and if the Contractor's requests it, clause 4.16.2 provides that the Employer must place the Retention in a *separate* bank account, at the same time of each Interim Payment, so as to clearly identify the amount of the Retention held on trust by the Employer under clause 4.16.1 above. The Employer must also notify the Contractor of any such deposits. However, the Employer is entitled to the full beneficial interest in any interest accruing on that separate bank account, and is not required to pay it to the Contractor. Clause 4.16.2 therefore serves to protect both the Employer and the Contractor, in the event of the other's insolvency. (However, it does not apply where the Employer is a Local or Public Authority.)

Resource 179 page 517	Request: Contractor to Employer	A request from the Contractor to the Employer to place the Retention in a separate bank account.
Resource 180 page 517	Notice: Employer to Contractor	A notice from the Employer to the Contractor that the Retention has been placed in a separate bank account.

Market practice: Treatment of Retention

Clause 4.16 is often amended to remove the Employer's fiduciary duties, given the additional administrative burden it can require.

Clause 4.17: Retention Bond

4.17 Where the Contract Particulars state that clause 4.17 applies, then:

If it applies (as stated against the reference to clause 4.17 in the Contract Particulars), clause 4.17 provides an alternative to the deduction of Retention by the Employer, in the form of a Retention Bond.

This aims to provide the same performance security to the Employer, but it better protects the Contractor's cashflow.

4.17.1 *subject to clauses 4.17.3 and 4.17.4, the provisions of clauses 4.14 and 4.18 permitting the deduction of the Retention shall not apply;*

As stated above, where clause 4.17 applies, clause 4.17.1 provides that clause 4.14 and clause 4.18 (both of which permit the deduction of Retention) will *not* apply. This is subject to clause 4.17.3 and clause 4.17.4 below.

Recommendation: Record keeping of Retention amounts

Even in circumstances where a Retention Bond is provided, footnote [44] of the Contract encourages the Employer to keep updated records of the amount of the Retention which *would* have been retained by the Employer, had clause 4.14 and clause 4.18 applied.

This is because any demand by the Employer under the Retention Bond needs to state the amount of Retention which would have been held by the Employer at the date of the demand, had Retention been deductible.

4.17.2 *on or before the Date of Possession the Contractor shall provide to the Employer and thereafter maintain a bond ('the Retention Bond') in favour of the Employer from a surety approved by the Employer ('the Surety') in the terms set out in Part 3 of Schedule 6, incorporating in clauses 2 (maximum aggregate sum) and 6.3 (expiry date) of the bond the sum and date stated in the Contract Particulars;*

Clause 4.17.2 requires the Contractor to provide and maintain the Retention Bond, in favour of the Employer.

The Retention Bond must be provided:

- on or before the Date of Possession (as stated against the reference to clause 2.3 (Date of Possession – progress) in the Contract Particulars);
- from a surety approved by the Employer (defined as '*the Surety*');
- in the terms set out in Part 3 of Schedule 6 (Forms of Bonds); and
- incorporating both the sum and date stated against the reference to clause 4.17 in the Contract Particulars in the following provisions of the Retention Bond:
  - clause 2, relating to the '*maximum aggregate sum*'; and
  - clause 6.3, relating to the expiry date of the bond.

Resource 181 page 517	Request: Contractor to Employer	A request from the Contractor for the Employer's approval to provide a Retention Bond from a particular surety.
Resource 182 page 518	Approval given: Employer to Contractor	Approval given from the Employer for the Contractor to provide a Retention Bond from a particular surety.
Resource 183 page 518	Approval withheld: Employer to Contractor	Approval withheld from the Employer for the Contractor to provide a Retention Bond from a particular surety.

4.17.3 if the Contractor fails to provide or maintain the Retention Bond in accordance with clause 4.17.2, the provisions of clauses 4.14 and 4.18 permitting the deduction of the Retention shall apply in respect of Interim Payments issued after the date of the failure, save that if the Contractor subsequently provides the required bond, any Retention deducted during the period of failure shall become due for release to the Contractor on the next due date thereafter;

Clause 4.17.3 relates to where the Contractor either fails to provide or maintain the Retention Bond, in accordance with clause 4.17.2.

In these circumstances, the provisions both clause 4.14 and clause 4.18 (both of which permit the deduction of Retention) will apply again, in respect of Interim Payments made after the date of the Contractor's failure to provide or maintain the Retention Bond.

However, if the Contractor subsequently *does* provide the Retention Bond, then any Retention previously deducted will become due to be released to the Contractor on the next due date.

Resource 184 page 519	Confirmation: Employer to Contractor	Confirmation from the Employer of the deduction of Retention by the Employer, if the Contractor fails to provide or maintain the Retention Bond.

4.17.4 *if at any time the amount of the Retention that would have been deducted had the provisions of clauses 4.14 and 4.18 applied exceeds the aggregate sum stated in the Retention Bond, then either the Contractor shall arrange with the Surety for the aggregate sum to equate to such amount or the amount not covered by the bond may be deducted as Retention;*

Clause 4.17.4 provides for where the Contractor has provided a Retention Bond, but the aggregate sum of that Retention Bond is *less* than the amount of the Retention which would have been deducted, had the provisions of clause 4.14 and clause 4.18 applied. In other words, there is a shortfall, thereby lessening the Employer's protection.

In these circumstances, there are two options:

- the Contractor may arrange with the Surety to increase the aggregate sum under the Retention Bond, to remove the shortfall; or
- the shortfall may be deducted as Retention by the Employer.

Resource 185 page 519	Request: Employer to Contractor	A request from the Employer for the Contractor to confirm how the additional Retention will be accounted for.

4.17.5 *where the Contractor has provided a performance bond or guarantee of the type referred to in clause 7.3.1, then, in respect of any default for which the Employer is entitled to make a demand under both that performance bond or guarantee and the Retention Bond, the Employer shall first have recourse to the Retention Bond.*

Clause 4.17.5 provides for where:

- the Contractor has provided both (i) a Retention Bond, and (ii) a performance bond under clause 7.3.1 (Performance Bonds and Guarantees); and
- the Employer is entitled to make a demand under both the Retention Bond and the performance bond.

In these circumstances, clause 4.17.5 requires the Employer to make the demand under the Retention Bond in the first instance.

Clause 4.18: Retention – amounts and periods

4.18 During the period prior to issue of the Notice of Completion of Making Good (or last such notice) the Retention which may be deducted and retained by the Employer shall be the following percentages of the total amount (or proportion of that amount) included in the Gross Valuation for any Interim Payment under clause 4.12.1 or 4.13.1, whichever is applicable, for work and (where applicable) Site Materials and Listed Items:

Clause 4.18 sets out both:

- the amount of the Retention which may be deducted and retained by the Employer; and
- for how long it may be deducted and retained.

The Retention may be deducted during the period before the issue of the Notice of Completion of Making Good (or, where the Works are divided into Sections, the issue of the last such Notice of Completion of Making Good), but not after. In other words, once the final Notice of Completion of Making Good has been issued, the Employer will not be entitled under clause 4.18 to deduct any Retention.

The Retention may be deducted as a percentage of the total amount included in the Gross Valuation for any Interim Payment under either (i) clause 4.12.1 (Gross Valuation – Alternative A), in the case of Stage Payments, or (ii) clause 4.13.1 (Gross Valuation – Alternative B), in the case of Periodic Payments, for:

- work; and, where applicable,
- Site Materials; and
- Listed Items.

The relevant percentages are set out in clause 4.18.1 and clause 4.18.2 below, and relate to the Retention Percentage, which is stated against the reference to clause 4.18.1 in the Contract Particulars (and the default percentage for which is 3%).

4.18.1 the Retention Percentage may be deducted from the total amount where the Works have not reached practical completion or (where there are Sections) from that proportion of the total amount that relates to uncompleted Sections (in either case excluding from the total amount any proportion of it attributable to a Relevant Part);

Clause 4.18.1 applies during the period *before* the Works have reached practical completion.

During this period, the *full* Retention Percentage may be deducted.

If the Works are divided into Sections, then the full Retention Percentage may be deducted from that proportion of the total amount that relates to those Sections which have not yet reached practical completion.

In both cases, this excludes any proportion of the total amount which relates to a Relevant Part, for which the Employer has taken partial possession. This is because, in accordance with clause 2.31 (Partial Possession by Employer: Practical completion date), practical completion of the Relevant Part is deemed to have occurred on the Relevant Date, when the Employer took partial possession.

4.18.2 *half the Retention Percentage may be deducted:*

4.18.2.1 *from the total amount where the Works as a whole have reached practical completion but the Notice of Completion of Making Good has not been issued; or*

4.18.2.2 *from the proportion of the total amount that relates to each Section that has reached practical completion but for which such a notice has not been issued or relates to a Relevant Part for which a notice has not been issued under clause 2.32.*

Clause 4.18.2 applies during the period where (i) the Works *have* reached practical completion, but (ii) the Notice of Completion of Making Good has not been issued.

During this period, *half* the Retention Percentage may be deducted.

If the Works are divided into Sections, then half the Retention Percentage may be deducted from the proportion of the total amount that relates to each Section that has reached practical completion, but for which the corresponding Notices of Completion of Making Good have not yet been issued.

Similarly, half the Retention Percentage may be deducted where it relates to a Relevant Part (for which the Employer has taken partial possession), but for which a Notice of Completion of Making Good has not yet been issued under clause 2.32 (Partial Possession by Employer: Defects etc. – Relevant Part).

In summary, the first half of the Retention should therefore be included within the Interim Payment made on or immediately after practical completion, with the second half included in the payment following the issue of the relevant Notice of Completion of Making Good.

Resource 186 page 520	Checklist: Contractor and Employer	A checklist for the calculation of the Retention which may be deducted.

Market practice: Additional requirements for the release of the Retention

Clause 4.18 is often amended to include further requirements for the release of the Retention, in addition to the issue of the relevant Notice of Completion of Making Good.

These include compliance by the Contractor with its obligations relating to the provision of documents to the Employer, such as collateral warranties required under clause 7.11 (Third Party Rights and Collateral Warranties from Sub-Contractors).

Although these documents should likely have been provided prior to practical completion, these additional requirements give the Employer the option of issuing the Practical Completion Statement (or Section Completion Statement), whilst retaining an incentive for the Contractor to provide those documents afterwards.

Loss and Expense

Overview

This sub-section relates to loss and expense incurred by the Contractor.

In certain instances, the Contractor will be entitled to reimbursement of that loss and expense from the Employer. In others, it will not.

Similar to the extension of time mechanism, considered in clause 2.23 to clause 2.26 (Adjustment of Completion Date) above, it is for the Employer to ascertain the Contractor's entitlement to reimbursement of loss and expense.

Again, in that commercial context, it is unsurprising that disputes over the Contractor's entitlement to reimbursement of loss and expense are very common.

This sub-section sets out the following:

- the circumstances where the Contractor is entitled to reimbursement of loss and expense;
- the steps the Contractor must take to realise that entitlement; and
- the Employer's corresponding obligations in assessing the Contractor's entitlement.

Key provisions

- clause 4.19 (Matters materially affecting regular progress)
- clause 4.20 (Notification and ascertainment)
- clause 4.21 (Relevant Matters)
- clause 4.22 (Amounts ascertained – addition to Contract Sum)
- clause 4.23 (Reservation of Contractor's rights and remedies)

Clause 4.19: Matters materially affecting regular progress

4.19.1 If in the execution of this Contract the Contractor incurs or is likely to incur any direct loss and/or expense as a result of any deferment of giving possession of the site or part of it under clause 2.4 or because regular progress of the Works or any part of them has been or is likely to be materially affected by any Relevant Matter, it shall, subject to clause 4.19.2 and compliance with the provisions of clause 4.20 be entitled to reimbursement of that loss and/or expense.

DOI: 10.1201/9781003485919-42

Clause 4.19 sets out the extent of the Contractor's entitlement to be reimbursed for direct loss and/or expense it incurs.

The Contractor will be entitled to the reimbursement of that direct loss and expense, if it is incurred in the course of the performance of its obligations under the Contract, and as a result of either:

- any deferment by the Employer of giving possession of the site (or part of it) under clause 2.4 (Deferment of possession); or
- because regular progress of the Works (or any part of the Works) either has been, or is likely to be, materially affected by a Relevant Matter (listed in clause 4.21 below).

The Contractor's entitlement to reimbursement is subject to both:

- the exceptions in clause 4.19.2; and
- compliance with the provisions of clause 4.20.

Furthermore, the Contractor is only entitled to recover '*direct*' loss and expense, rather than 'indirect' or 'consequential' loss and expense. Briefly stated, indirect or consequential loss is more 'remote' in nature, and relates to those losses which could not reasonably be considered to arise naturally, or could not reasonably be supposed to have been foreseen or contemplated by the Parties.

Whilst the types of loss and expense are not expressly set out in clause 4.19, they may potentially include loss and expense relating to prolongation, disruption, loss of profit, and finance charges.

> *4.19.2 No such entitlement arises where these Conditions provide that there shall be no addition to the Contract Sum or otherwise exclude the operation of this clause 4.19 or to the extent that the Contractor is reimbursed for such loss and/or expense under another provision of these Conditions.*

Clause 4.19.2 provides a number of exceptions to the Contractor's entitlement to reimbursement of loss and expense under clause 4.19.1.

The Contractor will have no entitlement in circumstances where:

- the Conditions provide that there will be *no* addition to the Contract Sum (for example, in clause 3.13 (Work not in accordance with the Contract), or clause 3.14 (Workmanship not in accordance with the Contract)); or
- the Conditions otherwise exclude the operation of clause 4.19 (for example, in clause 6.13.5.4 (Loss or damage – insurance claims and reinstatement)); or
- to the extent that the Contractor is reimbursed for the same loss and expense under *another* provision of the Conditions, to prevent the Contractor from being reimbursed twice (for example, in clause 4.11.2 (Contractor's right of suspension)).

Clause 4.20: Notification and ascertainment

> 4.20.1 *The Contractor shall notify the Employer as soon as the likely effect of a Relevant Matter on regular progress or the likely nature and extent of any loss and/or expense arising from a deferment of possession becomes (or should have become) reasonably apparent to it.*

Clause 4.20 sets out the procedure for both:

- the Contractor's application for reimbursement of loss and expense, and
- the Employer's ascertainment of the amount of that loss and expense.

Clause 4.20.1 requires the Contractor to notify the Employer, as soon as either:

- the likely effect of a Relevant Matter on regular progress of the Works; or
- the likely nature and extent of any loss and expense arising from a deferment of giving possession, becomes (or should have become) reasonably apparent to the Contractor.

> 4.20.2 *That notification shall be accompanied or, as soon as reasonably practicable, followed by the Contractor's initial assessment of the loss and/or expense incurred and any further amounts likely to be incurred, together with such information as is reasonably necessary to enable the Employer to ascertain the loss and/or expense incurred.*

Clause 4.20.2 requires the Contractor to also provide the Employer with:

- the Contractor's *initial* assessment of the loss and expense incurred;
- any further amounts of loss and expense likely to be incurred; and
- the information reasonably necessary to enable the Employer to ascertain the loss and expense incurred.

This must be provided either (i) with the Contractor's notice under clause 4.20.1, or (ii) as soon as reasonably practicable after that notice.

Resource 187 page 521	Notice: Contractor to Employer	A notice from the Contractor to the Employer of loss and expense incurred by the Contractor, arising from a deferment of giving possession of the site.
Resource 188 page 522	Notice: Contractor to Employer	A notice from the Contractor to the Employer of loss and expense incurred by the Contractor, arising from a Relevant Matter.

Market practice: 'Time bars' to reimbursement of loss and expense

Clause 4.20 is often amended to include a specific 'time bar' to the Contractor's entitlement to reimbursement of loss and expense.

In other words, if the Contractor fails to comply with clause 4.20.1 (or clause 4.20.2) within a certain prescribed time period, then the Contractor may have *no* entitlement to reimbursement of the associated loss and expense under clause 4.19.

Whilst this is potentially very onerous on the Contractor, it promotes such applications being dealt with at an earlier stage.

Furthermore, such provisions may actually help alleviate disagreements between the Parties relating to such claims, as they create certainty, and the Contractor will have little choice but to make its applications within the relevant time period.

4.20.3 The Contractor shall thereafter, in such form and manner as the Employer may reasonably require, update that assessment and information at monthly intervals until all information reasonably necessary to allow ascertainment of the total amount of such loss and expense has been supplied.

Clause 4.20.3 recognises that applications for reimbursement of loss and expense may relate to an extended period.

It therefore requires the Contractor to update both (i) its initial assessment of the loss and expense incurred, and (ii) supporting information, every month, until it has provided all the information reasonably necessary to allow the *total* amount of loss and expense to be ascertained.

These updates must be provided in both the form and manner the Employer reasonably requires.

It is therefore important that the Contractor maintains contemporaneous, accurate records of all such loss and expense, to assist with the Employer's ascertainment.

Resource 189 page 523	Requirement: Employer to Contractor	A requirement by the Employer of further information from the Contractor to ascertain loss and expense incurred.

4.20.4 *Within 28 days of receipt of the initial assessment and information and 14 days of each subsequent update of them the Employer shall notify the Contractor of the ascertained amount of the loss and/or expense incurred, each ascertainment being made by reference to the information supplied by the Contractor and in sufficient detail to enable the Contractor to identify differences between it and the Contractor's assessment.*

Once the Contractor has made the relevant notification and supplied all of the information reasonably necessary under clause 4.20.1 to clause 4.20.3, clause 4.20.4 then requires the Employer to notify the Contractor of the ascertained (or 'determined' or 'established') amounts of the loss and expense incurred.

These notices must be given:

- within 28 days of receipt of the *initial* assessment and information from the Contractor, under clause 4.20.1 and clause 4.20.2; *and*
- within 14 days of each subsequent update, under clause 4.20.3.

Clause 4.20.4 also requires the Employer to help the Contractor understand how the Employer has ascertained the amount of loss and expense. Each ascertainment must be made:

- by reference to the information provided by the Contractor; and
- in sufficient detail to enable the Contractor to identify any differences between the Contractor's assessment and the Employer's ascertainment.

Resource 190 page 523	Notice: Employer to Contractor	A notice from the Employer to the Contractor of the ascertained amount of loss and expense incurred.

Clause 4.21: Relevant Matters

> 4.21 *The following are the Relevant Matters:*

Clause 4.21 sets out the Relevant Matters which will, in principle, entitle the Contractor to reimbursement of loss and expense, in accordance with clause 4.19.

The Relevant Matters partly overlap with the Relevant Events considered in clause 2.26 (Relevant Events) above, which relate to the Contractor's entitlement to extensions of time.

> 4.21.1 *Changes and any other matters or instructions which under these Conditions are to be treated as a Change;*

Changes are one of the most common Relevant Matters in practice, and are defined in clause 5.1 (Definition of Changes).

This Relevant Matter applies:

- where the Employer instructs a Change which amends the Employer's Requirements, in accordance with clause 3.9 (Instructions requiring Changes); and
- to matters or Employer's instructions which are to be *treated* as a Change.

However, this is provided that such loss and expense is not already accounted for in the valuation of such Changes, in accordance with clause 5.2 (Valuation of Changes and provisional sum work). For example, where the value of the Change is agreed in accordance with Supplemental Provision 2 (Valuation of Changes – Contractor's estimates), which includes loss and expense within that value.

> 4.21.2 *Employer's instructions:*
>
> 4.21.2.1 *under clause 3.10 or 3.11; or*
>
> 4.21.2.2 *for the opening up for inspection or testing of any work, materials or goods under clause 3.12 (including making good), unless the inspection or test shows that the work, materials or goods are not in accordance with this Contract;*

The Relevant Matter in clause 4.21.2.1 relates to Employer's instructions under both:

- clause 3.10 (Postponement of work), where the Employer instructs the Contractor to postpone any work to be carried out under the Contract; and

- clause 3.11 (Instructions on Provisional Sums), where the Employer issues instructions regarding the expenditure of any Provisional Sums included in the Employer's Requirements.

The Relevant Matter in clause 4.21.2.2 relates to the Employer's instructions for the Contractor to either (i) open up for inspection, or (ii) test, any work, materials, or goods (along with any associated making good), in accordance with clause 3.12 (Inspection – tests). However, it will not apply in circumstances where the relevant work, materials, or goods are found not to be in accordance with the Contract.

4.21.3 compliance with clause 3.15.1 or with Employer's instructions under clause 3.15.2 or (except to the extent that the presence of asbestos or contaminated material has been identified in the Contract Documents and/or any such material has been brought on to the site by the Contractor or any Contractor's Person) compliance with clause 3.15.3 or with Employer's instructions under clause 3.15.4;

The Relevant Matter in clause 4.21.3 relates to compliance with:

- clause 3.15.1 (Antiquities, asbestos, contaminated material and unexploded ordnance), and the Employer's instructions under clause 3.15.2, both in relation to fossils, antiquities, and other objects of interest or value found on the site; or
- clause 3.15.3, and the Employer's instructions under clause 3.15.4, both in relation to asbestos, contaminated material, and unexploded ordnance. However, this does not apply to the extent that (i) the presence of asbestos or contaminated material was identified in the Contract Documents, or (ii) any asbestos or contaminated material was brought on to the site by either the Contractor or Contractor's Persons.

4.21.4 delay in receipt of any permission or approval for the purposes of Development Control Requirements necessary for the Works to be carried out or proceed, which delay the Contractor has taken all practicable steps to avoid or reduce;

The Relevant Matter in clause 4.21.4 relates to delay in receiving any permission or approval, for the purposes of Development Control Requirements (thought to relate primarily to planning), which is necessary for the Works to be carried out or proceed.

This is provided that the Contractor has taken '*all practicable steps*' to avoid or reduce this delay. For example, by ensuring that any associated applications for which the Contractor is responsible are made accurately and in good time, and supported by all relevant information.

The Contractor is therefore encouraged to copy the Employer into any correspondence in respect of Development Control Requirements, to demonstrate compliance with this obligation.

4.21.5 *any impediment, prevention or default, whether by act or omission, by the Employer or any Employer's Person, except to the extent caused or contributed to by any default, whether by act or omission, of the Contractor or any Contractor's Person.*

Along with Changes (in clause 4.21.1 above), acts of prevention by the Employer (or Employer's Persons) are a common Relevant Matter in practice.

Clause 4.21.5 applies to any '*impediment, prevention or default*' by either the Employer or any Employer's Persons, and whether this was caused by an act or an omission (in other words, a failure to do something).

This is broadly drafted, to reflect the wide range of potentially applicable acts or omissions by or on behalf of the Employer.

However, it does not apply to the extent that such impediment, prevention, or default was caused, or contributed to, by any default of the Contractor (or any Contractor's Person), again, whether by act or omission.

In other words, the Contractor will not be entitled to reimbursement of loss and expense for this Relevant Matter to the extent that it has caused or contributed to the underlying problem.

4.21.6 *(where the Contract Particulars state that clause 4.21.6 applies) an epidemic:*

4.21.6.1 *first occurring after the Base Date which affects the execution of the Works; or*

4.21.6.2 *first occurring before the Base Date whose effects change after the Base Date and any such change affects the execution of the Works*

by limiting the availability or use of labour or the availability of persons engaged in providing services for the Works where such labour and/or persons are necessary for the proper carrying out of the Works, or preventing the Contractor from, or delaying the Contractor in, securing such goods or materials or such services as are necessary for the proper carrying out of the Works;

Clause 4.21.6 includes an optional Relevant Matter, and whether it applies will be stated against the reference to clause 4.21.6 and clause 4.21.7 in the Contract Particulars.

It relates to epidemics either (i) first occurring *after* the Base Date, which affects the execution of the Works, or (ii) first occurring *before* the Base Date, but whose effects change *after* the Base Date, and any such change affects the execution of the Works, by:

- limiting (i) the availability or use of labour, or (ii) the availability of persons engaged in providing services for the Works, where such labour or persons are necessary to properly carry out the Works; or
- either (i) preventing the Contractor from, or (ii) or delaying the Contractor in, securing goods, materials, or services as are necessary to properly carry out the Works.

4.21.7 *(where the Contract Particulars state that clause 4.21.7 applies) the occurrence after the Base Date of any of the following which affects the execution of the Works:*

4.21.7.1 *the passing into law of any statute, statutory instrument or other subordinate legislation, regulation or bye-law, whether to make a new law or change or repeal an existing law;*

4.21.7.2 *the exercise of any statutory power, except to the extent caused or contributed to by any default, whether by act or omission, of the Contractor or any Contractor's Person; or*

4.21.7.3 *the publication of any guidance,*

by the United Kingdom Government or any of the devolved administrations, or any Local or Public Authority or any equivalent authority governed by public law in any of the devolved administrations, or in the case of guidance by the Construction Leadership Council or its successor.

Clause 4.21.7 includes another optional Relevant Matter, and whether it applies will also be stated against the reference to clause 4.21.6 and clause 4.21.7 in the Contract Particulars.

It relates to the occurrence after the Base Date of any of the following, which affects the execution of the Works:

- the passing into law of legislation, regulations, or bye-laws (whether to make a new law, or to change or repeal an existing law);
- the exercise of any statutory power, except to the extent that the exercise of the power has been caused or contributed to by any default (whether by act or omission) of the Contractor, or any Contractor's Person; or
- the publication of guidance,

by (i) the United Kingdom Government (or any devolved administration), or (ii) any Local or Public Authority (or any equivalent authority governed by public law in any devolved administration), or (iii) in the case of guidance, by the Construction Leadership Council, or its successor.

Clause 4.22: Amounts ascertained – addition to Contract Sum

> 4.22 *Amounts ascertained under clause 4.20 shall be added to the Contract Sum.*

Clause 4.22 simply provides that the ascertained amount of loss and expense incurred under clause 4.20 will be added to the Contract Sum.

Furthermore, clause 4.3 (Taking adjustments into account) provides that, once ascertained, that amount will be taken into account in the next Interim Payment.

Resource 191 page 524	Tracker: Contractor and Employer	A tracker template for recording applications for reimbursement of loss and expense.

Clause 4.23: Reservation of Contractor's rights and remedies

> 4.23 *The provisions of clauses 4.19 to 4.22 shall not limit or affect any other rights and remedies of the Contractor.*

Clause 4.23 provides that the Contractor's entitlement to reimbursement of loss and expense under clause 4.19 to clause 4.22 will not adversely affect any other rights and remedies available to the Contractor.

This includes any right of the Contractor to bring a claim for damages, as a result of a breach of contract by the Employer.

Market practice: Exclusive remedy

Clause 4.23 is often amended to the effect that the loss and expense provisions contained in clause 4.19 to clause 4.22 constitute the Contractor's sole and exclusive remedy for loss and expense it may incur.

In other words, it is amended in a way which *will* limit and adversely affect any other rights and remedies of the Contractor.

Final Statement and Final Payment

Overview

This sub-section contains only one provision, and sets out the process relating to the final payment to be made under the Contract, whether to the Contractor or to the Employer.

Key provisions

- clause 4.24 (Final Statement and Final Payment)

Clause 4.24: Final Statement and Final Payment

4.24.1 Following practical completion of the Works the Contractor shall submit the Final Statement to the Employer and supply such supporting documents as the Employer may reasonably require.

Following practical completion, clause 4.24.1 requires the Contractor to submit the Final Statement to the Employer, along with any supporting documents the Employer may reasonably require.

Resource 192 page 524	Correspondence: Contractor to Employer	The submission by the Contractor to the Employer of the Final Statement.
Resource 193 page 525	Request: Employer to Contractor	A request by the Employer for further supporting documents required from the Contractor, in respect of the Final Statement.

DOI: 10.1201/9781003485919-43

4.24.2 *The Final Statement shall set out the adjustments to the Contract Sum to be made in accordance with clause 4.2 and shall state:*

4.24.2.1 *the Contract Sum, as so adjusted; and*

4.24.2.2 *the sum of amounts already paid by the Employer to the Contractor,*

and the final payment shall be the difference (if any) between the two sums, which shall be shown as a balance due to the Contractor from the Employer or to the Employer from the Contractor, as the case may be. The Final Statement shall state the basis on which that amount has been calculated, including details of all such adjustments.

Clause 4.24.2 provides what the Final Statement must contain.

It must set out the adjustments required to be made to the Contract Sum in accordance with clause 4.2 (Contract Sum and Adjustments: Items included in adjustments), considered above, and then state the following two amounts:

(i) the Contract Sum, as adjusted; and
(ii) the total amount already paid by the Employer to the Contractor.

The final payment will be the difference (if there is one) between these two amounts, shown as a balance due either to the Contractor from the Employer, or to the Employer from the Contractor.

The Final Statement must also state the basis on which the amount of the final payment has been calculated, including details of all adjustments made to the Contract Sum.

Resource 194 page 525	Template: Final Statement	A Final Statement template, from the Contractor to the Employer.

4.24.3 *If the Contractor does not submit the Final Statement within 3 months of practical completion of the Works, the Employer may on or after the expiry of that period give the Contractor notice that unless that statement is submitted within 2 months from the date of the notice the Employer may itself issue a final statement in lieu of that from the Contractor.*

Clause 4.24.3 and clause 4.24.4 provide for a fallback position if the Contractor fails to submit the Final Statement to the Employer within 3 months of practical completion of the Works.

In such circumstances, clause 4.24.3 allows the Employer to give the Contractor a notice to the effect that, unless the Contractor submits the Final Statement within 2 months from the date of that notice, the Employer may issue a final statement (defined in clause 4.24.4 as the '*Employer's Final Statement*'), in place of the Final Statement from the Contractor.

This effectively gives the Contractor a further opportunity to submit the Final Statement.

If the Employer wishes to give this notice, it must do so on or after the expiry of the 3-month period after practical completion of the Works.

Resource 195 page 526	Notice: Employer to Contractor	A notice from the Employer to the Contractor of the Employer's intention to issue the Employer's Final Statement.

4.24.4 *Unless the Contractor submits the Final Statement prior to the Employer exercising that right, the Employer may at any time after the 2 month notice period issue a final statement to the Contractor ('the Employer's Final Statement') in the form and with the details required by clause 4.24.2, so far as the Employer, on the information in its possession, is reasonably able to do so.*

Clause 4.24.4 provides for where the Contractor still fails to submit the Final Statement to the Employer, despite having received the Employer's notice under clause 4.24.3.

In these circumstances, the Employer may issue the Employer's Final Statement, in place of the Final Statement from the Contractor.

If the Employer wishes to issue the Employer's Final Statement, it may do so at any time after the expiry of the 2-month notice period referred to in clause 4.24.3, provided the Contractor has not submitted the Final Statement beforehand.

The Employer's Final Statement must take the same form and include the same details as are required in respect of the Final Statement under clause 4.24.2, to the extent that the Employer is reasonably able to do so, on the basis of the information it possesses.

Resource 196 page 526	Correspondence: Employer to Contractor	The submission by the Employer to the Contractor of the Employer's Final Statement.
Resource 197 page 527	Template: Employer's Final Statement	An Employer's Final Statement template, from the Employer to the Contractor.

4.24.5 *The due date for the final payment shall be the date one month after whichever of the following occurs last:*

4.24.5.1 *the end of the Rectification Period in respect of the Works or (where there are Sections) the last such period to expire;*

4.24.5.2 *the date stated in the Notice of Completion of Making Good under clause 2.36 or (where there are Sections) in the last such notice to be issued; or*

4.24.5.3 *the date of submission to the other Party of the Final Statement or, if issued first, the Employer's Final Statement ('the relevant statement').*

Clause 4.24.5 provides the due date for the final payment.
The due date will be the date one month after the latest of the following:

- the end of the Rectification Period in respect of the Works, or (if the Works are divided into Sections) the last Rectification Period to expire;
- the date stated in the Notice of Completion of Making Good under clause 2.36 (Notice of Completion of Making Good), or (if the Works are divided into Sections) in the last Notice of Completion of Making Good to be issued; or
- the date of submission of the Final Statement to the Employer (under clause 4.24.1), or, if issued first, the date of submission of the Employer's Final Statement to the Contractor (under clause 4.24.4).

In practice, it is likely that these will occur in the following order:

1. clause 4.24.5.3, in the expectation that the Contractor will issue the Final Statement within 3 months of practical completion of the Works, in accordance with clause 4.24.1;
2. clause 4.24.5.1, on the assumption that the Rectification Period will be longer than 3 months; and
3. clause 4.24.5.2, on the assumption that the Employer will not issue the Notice of Completion of Making Good until after the expiry of the associated Rectification Period.

4.24.6 *Except to the extent that prior to the due date for the final payment the Employer gives notice to the Contractor disputing anything in the Final Statement or the Contractor gives notice to the Employer disputing anything in the Employer's Final Statement, and subject to clause 1.8.2, the relevant statement shall upon the due date become conclusive as to the sum due under clause 4.24.2 and have the further effects stated in clause 1.8.*

Clause 4.24.6 provides that, upon the due date determined under clause 4.24.5, either the Final Statement or the Employer's Final Statement (as applicable) will:

- become conclusive as to the sum due under clause 4.24.2; and
- have the further effects set out in clause 1.8 (Effect of Final Statement), considered above.

This is subject to the following:

- if, prior to the due date for the final payment, either (i) the Employer gives a notice to the Contractor disputing something in the Final Statement, or (ii) the Contractor gives a notice to the Employer disputing something in the Employer's Final Statement; and
- the conclusion of any adjudication, arbitration, or litigation, under clause 1.8.2.

Resource 198 page 527	Notice: Contractor or Employer to the other Party	A notice disputing the Final Statement / Employer's Final Statement.

Section 5: Changes

Section overview

Section 5 of the Contract governs what constitutes a Change, and provides how Changes are to be valued.

Whether or not something constitutes a Change has a number of important implications. As well as potential adjustments to the Contract Sum, a Change (and any other matters or instructions which under the Conditions are to be treated as, or as requiring, a Change) constitutes both:

- a Relevant Event under clause 2.26.1 (Relevant Events), which may entitle the Contractor to an extension of time; and
- a Relevant Matter under clause 4.21.1 (Relevant Matters), which may entitle the Contractor to reimbursement of loss and expense.

Section 5 also contains the Valuation Rules, which provide general principles to allow the Valuations of Changes to be undertaken.

The starting point for undertaking Valuations is clause 5.4 (Measurable Work), and thereafter, to the extent that they are relevant:

- clause 5.5 (Daywork), in relation to additional or substituted work;
- clause 5.6 (Change of conditions for other work); and, as a last resort,
- clause 5.7 (Additional provisions).

Sub-sections

- General (clause 5.1 – clause 5.3)
- The Valuation Rules (clause 5.4 – clause 5.7)

DOI: 10.1201/9781003485919-45

General

Overview

This sub-section includes a detailed definition of the term '*Change*', which is used extensively throughout the Conditions. This is helpful, given the potentially significant commercial implications if something constitutes a Change.

It also provides how those Changes (as well as work in relation to Provisional Sums), are to be valued, and for the consequent adjustment of the Contract Sum.

Key provisions

- clause 5.1 (Definition of Changes);
- clause 5.2 (Valuation of Changes and provisional sum work); and
- clause 5.3 (Giving effect to Valuations, agreements etc.)

Clause 5.1: Definition of Changes

5.1 The term 'Change' means:

5.1.1 a change in the Employer's Requirements which makes necessary the alteration or modification of the design, quality or quantity of the Works, otherwise than such as may be reasonably necessary for the purposes of rectification pursuant to clause 3.13, including:

5.1.1.1 the addition, omission or substitution of any work;

5.1.1.2 the alteration of the kind or standard of any of the materials or goods to be used in the Works;

5.1.1.3 the removal from the site of any work executed or Site Materials other than work, materials or goods which are not in accordance with this Contract;

5.1.2 the imposition by the Employer of any obligations or restrictions in regard to the following matters or any addition to or alteration or omission of any such obligations or restrictions that are so imposed or are imposed in the Employer's Requirements in regard to:

5.1.2.1 access to the site or use of any specific parts of the site;

5.1.2.2 limitations of working space;

5.1.2.3 limitations of working hours; or

5.1.2.4 the execution or completion of the work in any specific order.

DOI: 10.1201/9781003485919-46

Clause 5.1 sets out the definition of a Change, which is divided into two broad categories.

Firstly, under clause 5.1.1, it applies to any change in the Employer's Requirements which results in the alteration or modification of either (i) the design, (ii) the quality, or (iii) the quantity of the Works. Three examples are provided:

- the addition, omission or substitution of work for the Contractor to carry out;
- the alteration of the kind or standard of any materials or goods to be used in the Works; and
- the removal from the site of (i) any work executed or (ii) Site Materials, other than work, materials or goods which are not in accordance with the Contract.

The exception is in respect of any alteration or modification, which may be reasonably necessary for the purposes of rectification in accordance with clause 3.13 (Work not in accordance with the Contract), in circumstances where work, materials, or goods are not in accordance with the Contract.

Secondly, under clause 5.1.2, it applies to the imposition by the Employer of any obligations or restrictions (or the alteration or omission of any such obligations or restrictions), relating to the following four matters:

- the Contractor's access to the site, or use of any specific parts of the site;
- limitations of the Contractor's working space;
- limitations of the Contractor's working hours; or
- the execution or completion of the work in any specific order.

Broadly speaking, the first category relates to *what* is being designed and built, the second category relates to *how* the works are designed and built.

The Contractor also has a right of reasonable objection to Changes under the second category, in accordance with clause 3.5.1 (Compliance with instructions), considered above.

Whether or not something constitutes a Change has a number of other potentially significant consequences, including the fact that a Change constitutes both:

- a Relevant Event, in accordance with clause 2.26.1 (Relevant Events), which may entitle the Contractor to an extension of time (considered further in clause 2.23 to clause 2.26 (Adjustment of Completion Date)); and
- a Relevant Matter, in accordance with clause 4.21.1 (Relevant Matters), which may entitle the Contractor to reimbursement of direct loss and expense (considered further in clause 4.19 to clause 4.23 (Loss and Expense)).

Resource 199 page 528	Checklist: Contractor and Employer	A checklist for establishing whether an instruction is a Change.

Clause 5.2: Valuation of Changes and provisional sum work

5.2 *The value of:*

5.2.1 *all Changes required by Employer's instructions;*

5.2.2 *all work which under these Conditions is to be treated as a Change; and*

5.2.2 *all work executed by the Contractor in accordance with Employer's instructions as to the expenditure of Provisional Sums included in the Employer's Requirements*

shall be such amount as is agreed by the Employer and the Contractor or, where not agreed, the amount valued (a 'Valuation') in accordance with clauses 5.4 to 5.7 ('the Valuation Rules').

Clause 5.2 provides that the values of each of the following three matters are either (i) the amount agreed by the Parties, or, where the Parties do not reach agreement, (ii) the amount valued by or on behalf of the Employer (described as a 'Valuation'), in accordance with the Valuation Rules (considered below):

- all Changes required by Employer's instructions (for example, under clause 3.9 (Instructions requiring Changes));
- all work which is to be *treated* as a Change, under the Conditions (for example, under clause 5.6 (Change of conditions for other work)); and
- all work in accordance with Employer's instructions as to the expenditure of Provisional Sums (under clause 3.11 (Instructions on Provisional Sums)).

Resource 200 page 528	Tracker: Contractor and Employer	A tracker template for recording Changes, work to be treated as a Change, and the expenditure of Provisional Sums.

Clause 5.3: Giving effect to Valuations, agreements etc.

5.3 *The Contract Sum shall be adjusted for each agreement by the Employer and the Contractor under clause 5.2 and for each Valuation.*

Clause 5.3 provides that the Contract Sum will be adjusted to reflect the values of the matters set out in clause 5.2 above, namely:

- any amounts agreed by the Parties, or
- where not agreed, any Valuations.

Recommendation: Cashflow implications relating to the value of Changes and provisional sum work

Clause 4.3 (Contract Sum and Adjustments: Taking adjustments into account) provides that, where the Conditions (in this case, clause 5.3) provide that an amount is to be dealt with by adjustment of the Contract Sum, then as soon as the amount is ascertained in whole or in part, the ascertained amount will be taken into account in the next Interim Payment.

Some projects involve a great many Changes, and the expenditure of many Provisional Sums.

It is important for the Contractor's cashflow to ensure that, as soon as possible, it has provided the Employer with all the information the Employer reasonably requires either (i) in order to reach agreement with the Contractor under clause 5.2, or failing that, (ii) to undertake a Valuation.

This can often impose a very significant administrative burden on both Parties, and so it is therefore all the more important that they work with each other in a collaborative manner, as required by Article 3 (Collaborative working). This may also require support from the relevant Contractor's Persons (for example, design consultants, sub-contractors, and suppliers).

Once the values of the matters listed in clause 5.2 have been ascertained, the Contractor is reminded to take account of this in preparing its Interim Payment Applications, considered in clause 4.7.3 (Interim Payments – Contractor's Interim Payment Applications, due dates and Payment Notices) below.

The Valuation Rules

Overview

This sub-section sets out the Valuation Rules.

As considered above in relation to clause 5.2 (Valuation of Changes and provisional sum work), if the value of:

- any Change required by an Employer's instruction,
- any work which under the Conditions is to be treated as a Change, or
- any work executed by the Contractor in accordance with Employer's instructions as to the expenditure of Provisional Sums included in the Employer's Requirements,

is *not* agreed between the Employer and the Contractor, then the value will be ascertained by way of a Valuation, undertaken by or on behalf of the Employer, in accordance with the Valuation Rules.

The Valuation Rules are contained in clause 5.4 to clause 5.7. Given their potentially very wide application, they are not overly prescriptive, but are instead broadly drafted to provide general principles.

Key provisions

- clause 5.4 (Measurable Work)
- clause 5.5 (Daywork)
- clause 5.6 (Change of conditions for other work)
- clause 5.7 (Additional provisions)

DOI: 10.1201/9781003485919-47

Clause 5.4: Measurable Work

5.4 *Valuations shall be made in accordance with this clause 5.4 and, so far as is relevant, clauses 5.5 to 5.7.*

5.4.1 *Allowance shall be made in such Valuations for the addition or omission of the relevant design work.*

5.4.2 *The valuation of additional or substituted work shall be consistent with the values of work of a similar character set out in the Contract Sum Analysis, making due allowance for any change in the conditions under which work is carried out and/or any significant change in the quantity of the work so set out. Where there is no work of a similar character set out in the Contract Sum Analysis a fair valuation shall be made.*

5.4.3 *The valuation of the omission of work set out in the Contract Sum Analysis shall be in accordance with the values therein for such work.*

5.4.4 *Any valuation of work under clauses 5.4.2 and 5.4.3 shall include allowance for any necessary addition to or reduction of the provision of site administration, site facilities and temporary works.*

5.4.5 *Allowance shall be made for any percentage or lump sum adjustments in the Contract Sum Analysis.*

Clause 5.4 provides that it serves as the starting point for undertaking a Valuation, and only then should clause 5.5, clause 5.6, and clause 5.7 be considered.

It relates to measurable work in the following categories:

- **Design work:** allowance must be made for both the addition and omission of design work.
- **Additional or substituted work:** the valuation of additional or substituted work must be '*consistent*' with the values of work of a similar character set out in the Contract Sum Analysis, making appropriate allowance for (i) any change in the *conditions* under which work is carried out, and (ii) any significant change in the *quantity* of the work so set out. In the event there is no work of a similar character set out in the Contract Sum Analysis, a '*fair valuation*' must be made.
- **Omission of work:** the valuation of omissions is more straightforward than additional or substituted work, as it must be in accordance with the values for such work contained in the Contract Sum Analysis.
- **Site administration, site facilities, and temporary works:** any valuation in relation to either (i) additional or substituted work, or (ii) omitted work, must also include allowance for any required addition to, or reduction of (more likely in the case of an omission), the provision of the following (sometimes referred to as 'preliminaries'):
  - site administration;
  - site facilities; and
  - temporary works.

- **Percentage or lump sum adjustments:** allowance must also be made for any percentage adjustments or '*lump sum*' adjustments included in the Contract Sum Analysis.

Clause 5.5: Daywork

5.5 Where the execution of additional or substituted work cannot be valued in accordance with clause 5.4, the Valuation shall comprise:

5.5.1 the prime cost of such work (calculated in accordance with the 'Definition of Prime Cost of Daywork carried out under a Building Contract' issued by The Royal Institution of Chartered Surveyors (RICS) and Construction Industry Publications Ltd as current at the Base Date) together with Percentage Additions to each section of the prime cost at the rates stated in the document identified in the Contract Particulars or, if they apply in respect of labour, at the All- Inclusive Rates stated in such document; or

5.5.2 where the work is within the province of any specialist trade and the RICS and the appropriate body representing the employers in that trade have agreed and issued a definition of prime cost of daywork, the prime cost of such work calculated in accordance with that definition current at the Base Date, together with Percentage Additions on the prime cost at the rates stated in the document identified in the Contract Particulars or, if they apply in respect of labour, at the All-Inclusive Rates stated in such document.

Provided that in any case vouchers specifying the time daily spent upon the work, the operatives' names, the plant and the materials employed shall be delivered for verification to the Employer not later than 7 Business Days after the work has been executed.

Clause 5.5 relates to the execution of additional or substituted work (rather than omitted work), and applies in circumstances where such work cannot be valued in accordance with clause 5.4 above.

Under both clause 5.5.1 and clause 5.5.2, the Valuation is composed of two separate elements:

1. the '*prime cost*' of such work, which, generally speaking, relates to the direct cost of materials, labour and plant; and
2. an additional amount, referred to as the '*Percentage Additions*', to cover overheads and profit.

Whether clause 5.5.1 or clause 5.5.2 applies depends on whether the work relates to any '*specialist trade*', and the Royal Institution of Chartered Surveyors (RICS) and the appropriate body representing the employers in that trade have jointly issued a definition of prime cost of daywork. If both these conditions are met, then clause 5.5.2 applies.

Footnote [48] of the Contract states that there are currently three definitions to which clause 5.5.2 refers, namely those agreed between:

- the RICS and The Electrical Contractors' Association and SELECT;
- the RICS and The Building and Engineering Services Association; and
- the RICS and The Association of Plumbing and Heating Contractors and The Scottish and Northern Ireland Plumbing Employers' Federation.

The Valuation will be either:

Clause 5.5.1

- the prime cost of such work (in other words, the direct cost of materials, labour and plant, to be calculated in accordance with the '*Definition of Prime Cost of Daywork carried out under a Building Contract*'[19] issued by the RICS and Construction Industry Publications Ltd as current at the Base Date), plus
- the Percentage Additions to each section of the prime cost, at the rates stated in the document identified in the Contract Particulars (against the reference to clause 5.5) or, if they apply in respect of labour, at the All-Inclusive Rates stated in that document,

or

Clause 5.5.2

- where the work relates to any '*specialist trade*' and the RICS and the appropriate body representing the employers in that trade have agreed and issued a definition of prime cost of daywork, the prime cost of such daywork calculated in accordance with that definition, current at the Base Date, plus
- the Percentage Additions on the prime cost at the rates stated in the document identified in the Contract Particulars (against the reference to clause 5.5) or, if they apply in respect of labour, at the All-Inclusive Rates stated in that document.

In either case, this is provided that '*vouchers*' (or other documentary evidence) have been provided to the Employer for it to verify, not later than 7 Business Days after the work has been executed, specifying the following:

- the time daily spent upon the work;
- the operatives' names;
- the plant employed; and
- the materials employed.

[19] Available (with a subscription) at: https://www.isurv.com/downloads/download/1377/definition_of_prime_cost_of_daywork

Clause 5.6: Change of conditions for other work

5.6 *If as a result of:*

5.6.1 *compliance with any instruction requiring a Change; or*

5.6.2 *compliance with any instruction as to the expenditure of a Provisional Sum,*

there is a substantial change in the conditions under which any other work is executed, that other work shall be treated as a Change and shall be valued in accordance with the provisions of this section 5.

Clause 5.6 provides for the potential wider consequences of the Contractor complying with either (i) an instruction requiring a Change, or (ii) an instruction for the expenditure of a Provisional Sum.

If compliance with those instructions results in a '*substantial change*' in the conditions under which any *other* work is executed, that other work will *also* be treated as a Change, and will therefore be valued in accordance with section 5, again using clause 5.2 (Valuation of Changes and provisional sum work) as a starting point.

Clause 5.7: Additional provisions

5.7.1 *To the extent that a Valuation does not relate to the execution of additional or substituted work or the omission of work or to the extent that the valuation of any work or liabilities directly associated with a Change cannot reasonably be effected in the Valuation by the application of clauses 5.4 to 5.6, a fair valuation shall be made.*

Clause 5.7.1 effectively operates as a last resort, in circumstances where:

- a Valuation does not relate to either (i) the execution of additional or substituted work, or (ii) the omission of work; or
- to the extent that the valuation of any work or liabilities directly associated with a Change '*cannot reasonably be effected*' (or achieved) in the Valuation by applying:
  - clause 5.4 (Measurable Work);
  - clause 5.5 (Daywork); or
  - clause 5.6 (Change of conditions for other work).

In such circumstances, it provides only that a '*fair valuation*' must be made, which is much more subjective than the rules set out in clause 5.4, clause 5.5, and clause 5.6.

> 5.7.2 *No allowance shall be made under the Valuation Rules for any effect upon the regular progress of the Works or of any part of them or for any other direct loss and/or expense for which the Contractor would be reimbursed by payment under any other provision in these Conditions.*

Clause 5.7.2 provides that a Valuation must *not* include either:

- any effect upon the regular progress of the Works, or any part of them; or
- any other direct loss and/or expense,

for which the Contractor would be reimbursed by payment under any other provision in the Conditions.

This is most likely to apply in relation to clause 4.19 to clause 4.23 (Loss and Expense), considered above, in circumstances where the Contractor incurs direct loss and/or expense as a result of any deferment of giving possession of the site, or because regular progress of the Works has been materially affected by a Relevant Matter.

Resource 201 page 529	Checklist: Contractor and Employer	A checklist for carrying out a Valuation in accordance with the Valuation Rules.

Section 6: Injury, Damage and Insurance

Section overview

Section 6 sets out the key provisions relating to injury to persons, damage to property, the various insurances required under the Contract, and the Joint Fire Code.

The insurances include the following:

- public liability insurance (and employers' liability insurance);
- 'non-negligent loss' insurance, under clause 6.5.1;
- the insurance of the Works, and Existing Structures;
- Terrorism Cover; and
- Professional Indemnity insurance.

Given the commercial risks involved, and the fact that policies issued by insurers are not standardised (and may contain various restrictions, provisos, sub-limits, and exclusions), the Parties are encouraged to consult with insurance advisers when negotiating and administering these provisions, and when placing, maintaining, and administering the relevant insurances required.

Sub-sections

- Personal Injury and Property Damage (clause 6.1 – clause 6.3)
- Insurance against Personal Injury and Property Damage (clause 6.4 – clause 6.6)
- Insurance of the Works and Existing Structures (clause 6.7 – clause 6.14)
- Professional Indemnity Insurance (clause 6.15 – clause 6.16)
- Joint Fire Code – compliance (clause 6.17 – clause 6.20)

DOI: 10.1201/9781003485919-49

Personal Injury and Property Damage

Overview

This sub-section relates to the Contractor's liability to the Employer for loss or damage relating to personal injury, property, and (where applicable) Existing Structures or their contents.

Key provisions

- clause 6.1 (Contractor's liability – personal injury or death)
- clause 6.2 (Contractor's liability – loss, injury or damage to property)
- clause 6.3 (Loss or damage to Existing Structures or their contents)

Clause 6.1: Contractor's liability – personal injury or death

6.1 *The Contractor shall be liable for, and shall indemnify the Employer against, any expense, liability, loss, claim or proceedings whatsoever in respect of personal injury to or the death of any person arising out of or in the course of or caused by the carrying out of the Works, except to the extent that the same is due to any act or neglect of the Employer, any Employer's Person or any Statutory Provider.*

Clause 6.1 makes the Contractor liable to the Employer for any loss in respect of personal injury (or death) to any person, arising out of the carrying out of the Works.

It also includes an indemnity from the Contractor to the Employer in that regard, which is effectively a separate promise by the Contractor to be responsible for the Employer's loss under clause 6.1.

The exception is if that loss was due to any '*act or neglect*' of either (i) the Employer, (ii) any Employer's Persons, or (iii) any Statutory Provider.

DOI: 10.1201/9781003485919-50

Market practice: Obligations under clause 2.35 (Schedules of defects and instructions)

Clause 6.1 is often expanded to apply not only to the carrying out of the Works, but also to the Contractor's obligations under clause 2.35 (Schedules of defects and instructions), in relation to the making good of any defects during the Rectification Period.

Clause 6.2: Contractor's liability – loss, injury or damage to property

6.2 *Subject to clause 6.3, the Contractor shall be liable for, and shall indemnify the Employer against, any expense, liability, loss, claim or proceedings in respect of any loss, injury or damage whatsoever to any property real or personal in so far as such loss, injury or damage arises out of or in the course of or by reason of the carrying out of the Works and to the extent that the same is due to any negligence, breach of statutory duty, omission or default of the Contractor or any Contractor's Person.*

Clause 6.2 makes the Contractor liable to the Employer for any loss in respect of damage to any property, both:

- arising out of the carrying out of the Works, and
- to the extent that such loss was due to any negligence, breach of statutory duty, omission (in other words, a failure to do something), or default by either (i) the Contractor, or (ii) any Contractor's Persons.

Similar to clause 6.1, clause 6.2 also includes an indemnity from the Contractor to the Employer in that regard, which again is effectively a separate promise by the Contractor to be responsible for the Employer's loss under clause 6.2.

This is subject to the exclusions identified in clause 6.3 below.

Market practice: Obligations under clause 2.35 (Schedules of defects and instructions)

Similar to clause 6.1 above, clause 6.2 is often also expanded to apply not only to the carrying out of the Works, but also to the Contractor's obligations under clause 2.35 (Schedules of defects and instructions), in relation to the making good of any defects during the Rectification Period.

Clause 6.3: Loss or damage to Existing Structures or their contents

> 6.3.1 *Where paragraph C.1 of Insurance Option C applies, the Contractor's liability and indemnity under clause 6.2 excludes any loss or damage to Existing Structures or to any of their contents required to be insured under that option that is caused by any of the risks or perils required or agreed to be insured against under that option.*

Clause 6.3.1 relates to the Contractor's liability for loss or damage to '*Existing Structures*', or to their contents.

The definition of '*Existing Structures*' in clause 1.1 (Definitions) includes the following:

- any and all existing structures within which the Works, or part of them, are to be executed, or to which they are to form an extension, together with:
- any Section for which a Section Completion Statement has been issued (in other words, has achieved practical completion), and
- relating to partial possession by the Employer, from the Relevant Date, any Relevant Part taken into the Employer's possession under clause 2.30 (Partial Possession by Employer: Contractor's consent)

Clause 6.3.1 is relevant where paragraph C.1 of Insurance Option C (Joint Names Insurance by the Employer of Existing Structures and Works in or Extensions to them) applies.

Paragraph C.1 requires the Employer to place and maintain a Joint Names Policy in respect of the Existing Structures, along with the contents of the Existing Structures which are either (i) owned by the Employer, or (ii) for which the Employer is responsible. This Joint Names Policy is for the full cost of reinstatement, repair, or replacement of loss or damage, due to any of the '*Specified Perils*' (which are defined in clause 6.8 (Insurance of the Works and Existing Structures: Related definitions)).

Therefore, where paragraph C.1 applies, clause 6.3.1 provides that the Contractor's liability and indemnity to the Employer under clause 6.2 *excludes* the loss or damage to Existing Structures (or their contents) required to be insured under Insurance Option C, and which is caused by any of the '*risks or perils*' required or agreed to be insured against.

In other words, the Contractor is not liable to the Employer for these matters, as they are required to be covered by the Joint Names Policy under paragraph C.1.

> 6.3.2 *The exclusion in clause 6.3.1 shall apply notwithstanding that the loss or damage is or may be due in whole or in part to the negligence, breach of statutory duty, omission or default of the Contractor or any Contractor's Person.*

Clause 6.3.2 provides that the exclusion in clause 6.3.1 applies even in circumstances where the loss or damage is due (whether in whole or in part) to the negligence, breach of statutory duty, omission, or default, of either (i) the Contractor, or (ii) any Contractor's Persons.

> 6.3.3 *Where Insurance Option C applies but a C.1 Replacement Schedule applies in lieu of paragraph C.1, the Contractor's liability and indemnity under clause 6.2 shall, in respect of loss, injury or damage to the Existing Structures and their contents due to the causes specified in that clause, be subject only to such limitations or exclusions as are specified in that schedule.*

Clause 6.3.3 again is relevant where Insurance Option C applies, but in this instance, where a '*C.1 Replacement Schedule*' applies instead of paragraph C.1.

Footnote [63] of the Contract explains why a C.1 Replacement Schedule may be required, in circumstances where the Joint Names Policy required by paragraph C.1 is either not available or unsuitable:

'However, the Joint Names Policy required by paragraph C.1 or the extension of a subsisting structure and contents policy to being a Joint Names Policy may not be readily available – and that provision is often not now appropriate for – refurbishment projects or alterations by tenant Employers where Existing Structures insurance is the landlord's responsibility.

Joint Names cover may also not be readily available to some domestic owner occupiers looking to undertake refurbishments or extensions to their property.

The Contract Particulars for clause 6.7 and Schedule 3 therefore expressly allow the Parties in those circumstances to disapply paragraph C.1 and, by means of a C.1 Replacement Schedule, to include in place of that paragraph provisions that are tailored to their particular requirements.'

In other words (and as the name suggests), the C.1 Replacement Schedule replaces paragraph C.1.

Therefore, similar to clause 6.3.1 above, clause 6.3.3 provides that the Contractor's liability and indemnity to the Employer under clause 6.2 for loss, injury or damage to the Existing Structures (or their contents), due to the causes identified in clause 6.2, is subject to any limitations or exclusions stated in the C.1 Replacement Schedule.

> 6.3.4 *The reference in clause 6.2 to 'property real or personal' does not include the Works, work executed or Site Materials up to and including the date of issue of the Practical Completion Statement or, if earlier, the date of termination of the Contractor's employment, except that:*
>
> 6.3.4.1 *after the date of issue of a Section Completion Statement, the Section to which it relates shall no longer be regarded as 'the Works' or 'work executed' for these purposes; and*
>
> 6.3.4.2 *if partial possession is taken under clause 2.30, the Relevant Part shall no longer be so regarded after the Relevant Date.*

Clause 6.3.4 clarifies the meaning of the term '*property real or personal*' in clause 6.2. In particular, it excludes:

- the Works,
- work executed, or
- Site Materials,

up to and including either (i) the date of issue of the Practical Completion Statement or, if earlier, (ii) the date of termination of the Contractor's employment under the Contract.

This is because these matters are required to be covered by the Joint Names Policy for All Risks Insurance (considered below under clause 6.7 (Insurance of the Works and Existing Structures: Insurance Options and period)).

The exceptions are in relation to any areas which have achieved practical completion:

- If the Works are divided into Sections, a Section will no longer be regarded as '*the Works*' or '*work executed*' under clause 6.3.4, after the date of the corresponding Section Completion Statement.
- Similarly, if partial possession is taken by the Employer under clause 2.30 (Partial Possession by Employer: Contractor's consent), the corresponding Relevant Part will no longer be regarded as '*the Works*' or '*work executed*' under clause 6.3.4, after the Relevant Date.

Insurance against Personal Injury and Property Damage

Overview

Following on from the Contractor's liability to the Employer in respect of personal injury and property damage considered in the previous sub-section, this sub-section relates to the Contractor's obligations to take out and maintain insurance relating to those matters, including:

- personal injury (and death);
- property damage; and
- the liability of the Employer for injury or damage to property.

Key provisions

- clause 6.4 (Contractor's insurance of its liability)
- clause 6.5 (Contractor's insurance of liability of Employer)
- clause 6.6 (Excepted Risks)

Clause 6.4: Contractor's insurance of its liability

6.4.1 *Without limiting or affecting its indemnities to the Employer under clauses 6.1 and 6.2, the Contractor shall effect and maintain insurance in respect of claims arising out of the liabilities referred to in those clauses which:*

6.4.1.1 *in respect of claims for personal injury to or the death of any employee of the Contractor arising out of and in the course of such person's employment, shall comply with all relevant legislation; and*

6.4.1.2 *for all other claims to which clause 6.4.1 applies, shall indemnify the Employer in like manner to the Contractor (but only to the extent that the Contractor may be liable to indemnify the Employer under the terms of this Contract) and shall for any one occurrence or series of occurrences arising out of one event be in a sum not less than that stated in the Contract Particulars for clause 6.4.1.*

DOI: 10.1201/9781003485919-51

Clause 6.4.1 requires the Contractor to take out and maintain insurance (known as 'public liability insurance') in respect of any claims arising out of the liabilities referred to in both:

- clause 6.1 (Contractor's liability – personal injury or death); and
- clause 6.2 (Contractor's liability – loss, injury or damage to property).

This insurance is subject to the following further requirements:

- In respect of claims for personal injury (or death) of any of the Contractor's employees, arising in the course of their employment, the insurance (known as 'employers' liability insurance') must comply with all relevant legislation.
- In respect of all other claims under clause 6.4.1, the insurance must indemnify the Employer in the same way as the Contractor (to the extent that the Contractor is liable to indemnify the Employer under the Contract). Furthermore, that insurance must be for '*any one occurrence or series of occurrences arising out of one event*', and must have a limit of indemnity of at least the amount stated against the reference to clause 6.4.1 in the Contract Particulars. (The Contractor may insure for a greater amount, if it wishes.)

The Contractor's obligations under clause 6.4.1 will not limit or adversely affect the Contractor's indemnities to the Employer under either of clause 6.1 or clause 6.2.

Furthermore, footnote [49] of the Contract states: '*It should be noted that the cover granted under Public Liability policies taken out pursuant to clause 6.4.1 may not be co-extensive with the indemnity given to the Employer in clauses 6.1 and 6.2: for example, each claim may be subject to an excess and cover may not be available in respect of loss or damage due to gradual pollution.*'

In other words, there may be a gap or shortfall between the Contractor's indemnities to the Employer under clause 6.1 and clause 6.2, and the Contractor's ability to obtain insurance for those liabilities. This will be at the Contractor's risk.

6.4.2 *As to evidence that such insurances have been effected and are being maintained and the consequences of failure to comply, clause 6.12 shall apply.*

Clause 6.4.2 relates to the provision by the Contractor of evidence that the insurances required under clause 6.4.1 have been placed, and are being maintained (along with the consequences if the Contractor fails to do so). It simply directs the Parties to clause 6.12 (Evidence of insurance), which governs these matters in respect of a number of separate insurances.

Clause 6.5: Contractor's insurance of liability of Employer

6.5.1 *If the Employer's Requirements state that insurance under clause 6.5.1 is required, the Contractor shall effect and maintain a policy of insurance in the names of the Employer and the Contractor for the amount of indemnity stated in the Contract Particulars in respect of any expense, liability, loss, claim or proceedings which the Employer may incur or sustain by reason of injury or damage to any property caused by collapse, subsidence, heave, vibration, weakening or removal of support or lowering of ground water arising out of or in the course of or by reason of the carrying out of the Works, excluding injury or damage:*

6.5.1.1 *for which the Contractor is liable under clause 6.2;*

6.5.1.2 *which is attributable to errors or omissions in the designing of the Works;*

6.5.1.3 *which can reasonably be foreseen to be inevitable having regard to the nature of the work to be executed and the manner of its execution;*

6.5.1.4 *(if Insurance Option C applies) which it is the responsibility of the Employer to insure under paragraph C.1 of Schedule 3;*

6.5.1.5 *to the Works and Site Materials except where the Practical Completion Statement has been issued or in so far as any Section is the subject of a Section Completion Statement;*

6.5.1.6 *which arises from any consequence of war, invasion, act of foreign enemy, hostilities (whether war is declared or not), civil war, rebellion or revolution, insurrection or military or usurped power;*

6.5.1.7 *which is directly or indirectly caused by or contributed to by or arises from the Excepted Risks;*

6.5.1.8 *which is directly or indirectly caused by or arises out of pollution or contamination of buildings or other structures or of water or land or the atmosphere happening during the period of insurance, save that this exception shall not apply in respect of pollution or contamination caused by a sudden identifiable, unintended and unexpected incident which takes place in its entirety at a specific moment in time and place during the period of insurance (all pollution or contamination which arises out of one incident being considered for the purpose of this insurance to have occurred at the time such incident takes place); or*

6.5.1.9 *which results in any costs or expenses being incurred by the Employer or in any other sums being payable by the Employer in respect of damages for breach of contract, except to the extent that such costs or expenses or damages would have attached in the absence of any contract.*

Clause 6.5.1 relates to what is informally referred to as 'non-negligent loss' insurance, or simply '6.5.1' insurance.

It is intended to apply in respect of certain incidents or matters occurring during the course of carrying out the Works which cause injury or damage to property, and which cannot be attributed to either Party (so they are therefore not negligent), but for which the Employer may still have strict liability.

These incidents or matters include collapse, subsidence, heave, vibration, weakening or removal of support, or lowering of ground water.

If it is stated as required under the Employer's Requirements, then the Contractor must place and maintain an insurance policy, in the names of both the Employer and the Contractor, for at least the amount of indemnity stated against the reference to clause 6.5.1 in the Contract Particulars.

Clause 6.5.1.1 to clause 6.5.1.9 set out several other matters which may also cause injury or damage to property, but which are expressly excluded from the obligation to insure under clause 6.5.1.

In many instances, these matters are covered under other provisions within section 6. For example, clause 6.5.1.1 relates to injury or damage for which the Contractor is liable under clause 6.2 (Contractor's liability – loss, injury or damage to property), for which the Contractor is obliged to take out and maintain insurance under clause 6.4 (Contractor's insurance of its liability).

Recommendation: Employer's Requirements

Clause 6.5.1 refers to this insurance being required within the Employer's Requirements, rather than in any other Contract Document. The Parties should therefore ensure that, if this insurance is required, it is expressly provided for within the Employer's Requirements.

6.5.2 *Any insurance under clause 6.5.1 shall be placed with insurers approved by the Employer, and the Contractor shall upon its issue deposit the policy with the Employer.*

Clause 6.5.2 requires the Contractor to obtain the Employer's approval to place the insurance required under clause 6.5.1 with a particular insurer.

Once this insurance has been placed, the Contractor must then provide the Employer with the policy.

Resource 202 page 531	Request: Contractor to Employer	A request from the Contractor for the Employer's approval to place the insurance required under clause 6.5.1 with a particular insurer.
Resource 203 page 532	Approval given: Employer to Contractor	Approval given by the Employer for the Contractor to place the insurance required under clause 6.5.1 with a particular insurer.
Resource 204 page 532	Approval withheld: Employer to Contractor	Approval withheld by the Employer for the Contractor to place the insurance required under clause 6.5.1 with a particular insurer.
Resource 205 page 532	Correspondence: Contractor to Employer	The provision by the Contractor to the Employer of the policy for the insurance required under clause 6.5.1.

6.5.3 *As to evidence that such insurance has been effected and is being maintained and the consequences of failure to comply, clause 6.12 shall apply.*

Similar to clause 6.4.2 above, clause 6.5.3 relates to the provision by the Contractor of evidence that the insurance required under clause 6.5.1 has been placed, and is being maintained (along with the consequences if the Contractor fails to do so). It simply directs the Parties to clause 6.12 (Evidence of insurance), which governs these matters in respect of a number of separate insurances.

Clause 6.6: Excepted Risks

6.6 *Notwithstanding clauses 6.1, 6.2 and 6.4.1, the Contractor shall neither be liable to indemnify the Employer nor obliged to insure against any personal injury to or the death of any person or any damage, loss or injury to the Works, Site Materials, work executed, the site or any other property caused by the effect of an Excepted Risk.*

Clause 6.6 provides that the Contractor will not be required to indemnify the Employer, nor take out insurance against:

- any personal injury to (or death of) any person; or
- any damage, loss, or injury to:
  - the Works,
  - Site Materials,
  - work executed,
  - the site, or
  - any other property,

which has been caused by the effect of an '*Excepted Risk*'.

This is irrespective of the Contractor's obligations under each of:

- clause 6.1 (Contractor's liability – personal injury or death);
- clause 6.2 (Contractor's liability – loss, injury or damage to property); and
- clause 6.4.1 (Contractor's insurance of its liability).

'*Excepted Risks*' are defined in clause 6.8 (Insurance of the Works and Existing Structures: Related definitions), and include:

- contamination relating to nuclear radioactivity;
- pressure waves caused by aircraft travelling up to supersonic speeds; and
- acts of terrorism not included within the Terrorism Cover (considered below).

Insurance of the Works and Existing Structures

Overview

This sub-section primarily relates to the insurance of the Works and (if applicable) Existing Structures.

It also includes provisions governing:

- Terrorism Cover (namely insurance against loss or damage caused by terrorism); and
- the consequences in the event of loss or damage affecting executed work, Site Materials, and Existing Structures.

Key provisions

- clause 6.7 (Insurance Options and period)
- clause 6.8 (Related definitions)
- clause 6.9 (Sub-contractors – Specified Perils cover under Works Insurance Policies)
- clause 6.10 (Terrorism Cover – policy extensions and premiums)
- clause 6.11 (Terrorism Cover – non-availability – Employer's options)
- clause 6.12 (Evidence of insurance)
- clause 6.13 (Loss or damage – insurance claims and reinstatement)
- clause 6.14 (Loss or damage to Existing Structures – right of termination)

Clause 6.7: Insurance Options and period

6.7.1 *Insurance Options A, B and C are set out in Schedule 3. The Insurance Option that applies to this Contract is that stated in the Contract Particulars.*

Clause 6.7.1 introduces the three Insurance Options, which relate to the insurance of the Works, and which are set out in Schedule 3 (Insurance Options):

- **Insurance Option A** (New Buildings – All Risks Insurance of the Works by the Contractor): this is for use in the case of new buildings, where the Contractor is

DOI: 10.1201/9781003485919-52

required to take out a Joint Names Policy for All Risks Insurance of the Works (or alternatively, to include them within its 'Annual Construction' policy).

- **Insurance Option B** (New Buildings – All Risks Insurance of the Works by the Employer): again, this is for use in the case of new buildings, but where the Employer is required to take out the Joint Names Policy for All Risks Insurance of the Works.
- **Insurance Option C** (Joint Names Insurance by the Employer of Existing Structures and Works in or Extensions to them): this is for use in the case of either alterations of, or extensions to, Existing Structures. In this case, the Employer is required to take out:
  - a Joint Names Policy for All Risks Insurance for the Works, and
  - if paragraph C.1 applies, a Joint Names Policy to insure the Existing Structures and their contents owned by the Employer, or for which the Employer is responsible, against loss or damage by the Specified Perils.

The applicable Insurance Option will be stated against the reference to clause 6.7 and Schedule 3 in the Contract Particulars.

'*All Risks Insurance*' is defined in clause 6.8 (Insurance of the Works and Existing Structures: Related definitions) as follows:

> 6.8 *All Risks Insurance: insurance which provides cover against any physical loss or damage to work executed and Site Materials and against the reasonable cost of the removal and disposal of debris and of any shoring and propping of the Works which results from such physical loss or damage but excluding the cost necessary to repair, replace or rectify:*
>
> *a property which is defective due to:*
> *i wear and tear,*
> *ii obsolescence, or*
> *iii deterioration, rust or mildew;*
>
> *b any work executed or any Site Materials lost or damaged as a result of its own defect in design, plan, specification, material or workmanship or any other work executed which is lost or damaged in consequence thereof where such work relied for its support or stability on such work which was defective;*
>
> *c loss or damage caused by or arising from:*
> *i any consequence of war, invasion, act of foreign enemy, hostilities (whether war be declared or not), civil war, rebellion, revolution, insurrection, military or usurped power, confiscation, commandeering, nationalisation or requisition or loss or destruction of or damage to any property by or under the order of any government de jure or de facto or public, municipal or local authority,*
> *ii disappearance or shortage if such disappearance or shortage is only revealed when an inventory is made or is not traceable to an identifiable event, or*
> *iii an Excepted Risk.*

As previously stated, the Parties are encouraged to consult with insurance advisers when placing and maintaining this insurance.

Footnote [52] of the Contract warns that *'The risks and costs that All Risks Insurance is required to cover are defined by exclusions. Policies issued by insurers are not standardised; the way in which insurance for these risks is expressed varies and in some cases it may not be possible for insurance to be taken out against certain of the risks required to be covered...'*

> 6.7.2 *In each case the Party responsible for effecting a Joint Names Policy under the Insurance Option that applies (the 'Works Insurance Policy') shall maintain that policy up to and including the date of issue of the Practical Completion Statement, or last Section Completion Statement, or (if earlier) the date of termination of the Contractor's employment, except that the obligation to maintain a Works Insurance Policy:*
>
> 6.7.2.1 *shall not apply in relation to a Section after the date of issue of its Section Completion Statement; and*
>
> 6.7.2.2 *if partial possession is taken under clause 2.30, shall not as from the Relevant Date apply in relation to the Relevant Part.*

Clause 6.7.2 requires the Party responsible for placing the Joint Names Policy under either Insurance Option A, Insurance Option B, or Insurance Option C (defined as the '*Works Insurance Policy*') to maintain it up to:

- the date of issue of the Practical Completion Statement; or
- if the Works are divided into Sections, the date of issue of the final Section Completion Statement; or,
- if earlier, the date of termination of the Contractor's employment under the Contract.

However, the obligation to maintain the Works Insurance Policy will not apply to the following (both of which relate to practical completion):

- if the Works are divided into Sections, any Section after the date of issue of its Section Completion Statement; and
- if partial possession is taken by the Employer under clause 2.30 (Partial Possession by Employer: Contractor's consent), the Relevant Part, from the Relevant Date.

Clause 6.8: Related definitions

> 6.8 *In these Conditions the following phrases shall have the following meanings...*

Clause 6.8 contains a number of definitions which primarily relate to matters under section 6, and include:

- All Risks Insurance;
- Excepted Risks;
- Joint Names Policy;

- Pool Re Cover;
- Specified Perils; and
- Terrorism Cover.

Where relevant and appropriate, reference to these defined terms has been made throughout Part 1 (and particularly this section 6), rather than considering them further here, in isolation.

An exception will be made for the definition of '*Joint Names Policy*', which applies to several provisions:

> 6.8 *Joint Names Policy: a policy of insurance which includes the Employer and the Contractor as composite insured and under which the insurers have no right of recourse against any person named as an insured, or, pursuant to clause 6.9, recognised as an insured thereunder.*

In other words, under a Joint Names Policy, both the Employer and Contractor are named as composite insured.

The insurers will have no right of recourse against any person named as an insured, namely the Employer and the Contractor.

Furthermore, in accordance with clause 6.9 (Sub-contractors – Specified Perils cover under Works Insurance Policies), the insurers will have no right of recourse against any person *recognised* as an insured, namely the Contractor's sub-contractors.

Clause 6.9: Sub-contractors – Specified Perils cover under Works Insurance Policies

> 6.9.1 *The Contractor, where Insurance Option A applies, and the Employer, where Insurance Option B or C applies, shall ensure that the Works Insurance Policy either:*
>
> 6.9.1.1 *provides for recognition of each sub-contractor as an insured under the policy; or*
>
> 6.9.1.2 *includes a waiver by the insurers of any right of subrogation which they may have against any such sub-contractor*
>
> *in respect of loss or damage by the Specified Perils to the Works or relevant Section, work executed and Site Materials and that this recognition or waiver continues up to and including the date of issue of any statement or other document which states that in relation to the Works, the sub-contractor's works are practically complete or, if earlier, the date of termination of the sub-contractor's employment. Where there are Sections, the recognition or waiver for a sub-contractor in relation to a Section shall cease upon the issue of such statement or other document for the sub-contractor's work in that Section.*

Clause 6.9 relates to the extension of Specified Perils cover under the Works Insurance Policy to the Contractor's sub-contractors.

The obligation will be on the Party responsible for taking out the Works Insurance Policy, namely:

- the Contractor, where Insurance Option A applies; and
- the Employer, where Insurance Option B or Insurance Option C applies.

The relevant Party must ensure that the Works Insurance Policy either:

- recognises each of the Contractor's sub-contractors as an insured party under the Works Insurance Policy, or alternatively,
- includes a waiver by the insurers of any right of subrogation which they may have against any such sub-contractor. This is to prevent the insurers from bringing a claim against those sub-contractors, where they have paid out to the insured parties (such as the Employer or the Contractor), and who may otherwise then have sought to recover their losses from those sub-contractors.

This is in respect of loss or damage by any of the Specified Perils to:

- the Works, or (where the Works are divided into Sections) the relevant Section;
- work executed; and
- Site Materials.

This recognition or waiver must remain in place until:

- the date of issue of any document confirming practical completion of the sub-contractor's works (in relation to the Works under the Contract); or
- where the Works are divided into Sections, the date of issue of any document confirming practical completion of the sub-contractor's works in relation to each Section; or
- if earlier, the date of termination of the sub-contractor's employment under their sub-contract with the Contractor.

The '*Specified Perils*' are defined in clause 6.8 as follows:

6.8 Specified Perils: fire, lightning, explosion, storm, flood, escape of water from any water tank, apparatus or pipe, earthquake, aircraft and other aerial devices or articles dropped therefrom, riot and civil commotion, but excluding Excepted Risks.

6.9.2 Clause 6.9.1 applies also in respect of any Works Insurance Policy taken out in default under clause 6.12.2.

Clause 6.9.2 provides that clause 6.9.1 also applies in circumstances where the Party who was meant to take out the Works Insurance Policy fails to provide evidence of it, allowing the other Party to take out the Works Insurance Policy instead, in accordance with clause 6.12.2 (Evidence of insurance), considered below.

Clause 6.10: Terrorism Cover – policy extensions and premiums

> 6.10.1 *To the extent that the Works Insurance Policy excludes (or would otherwise exclude) loss or damage caused by terrorism, the Contractor, where Insurance Option A applies, or the Employer, where Insurance Option B or C applies, shall unless otherwise agreed effect and maintain, either as an extension to the Works Insurance Policy or as a separate Joint Names Policy, in the same amount and for the same period as the Works Insurance Policy, such Terrorism Cover as is specified in or by the Contract Particulars, subject to clauses 6.10.4 and 6.11.*

Clause 6.10 relates to the requirement to take out and maintain insurance for loss or damage caused by terrorism, or '*Terrorism Cover*', which is defined in clause 6.8 as follows:

> 6.8 *Terrorism Cover: Pool Re Cover or other insurance against loss or damage to work executed and Site Materials (and/or, for the purposes of clause 6.11.1, to an Existing Structure and/or its contents) caused by or resulting from terrorism.*

In turn, '*Pool Re Cover*' is also defined in clause 6.8 as follows:

> 6.8 *Pool Re Cover: such insurance against loss or damage to work executed and Site Materials caused by or resulting from terrorism as is from time to time generally available from insurers which are members of the Pool Reinsurance Company Limited scheme or of any similar successor scheme.*

Again, the obligation will on the Party responsible for taking out the Works Insurance Policy, namely:

- the Contractor, where Insurance Option A applies; and
- the Employer, where Insurance Option B or Insurance Option C applies.

To the extent that the Works Insurance Policy *excludes* loss or damage caused by terrorism, then (unless the Parties otherwise agree) the relevant Party must take out and maintain Terrorism Cover either:

- as an extension to the Works Insurance Policy; or
- as a separate Joint Names Policy.

In both cases, the Terrorism Cover must be taken out in the same amount, and for the same period, as the Works Insurance Policy, and as specified against the reference to clause 6.10 and Schedule 3 in the Contract Particulars.

This is subject to the exceptions set out in clause 6.10.4 and clause 6.11 (Terrorism Cover – non-availability – Employer's options), both considered below.

6.10.2 *Where Insurance Option A applies and the Contractor is required to take out and maintain Pool Re Cover, the cost of that cover and its renewal shall be deemed to be included in the Contract Sum save that, if at any renewal of the cover there is a variation in the rate on which the premium is based, the Contract Sum shall be adjusted by the net amount of the difference between the premium paid by the Contractor and the premium that would have been paid but for the change in rate.*

Clause 6.10.2 relates to the cost of taking out and maintaining Terrorism Cover (and in particular, Pool Re Cover).

It is relevant where both:

- Insurance Option A (New Buildings – All Risks Insurance of the Works by the Contractor) applies; and
- the Contractor is required to take out and maintain Pool Re Cover.

In such circumstances, the cost of both that cover, and its renewal, will be considered included in the Contract Sum.

The only exception is where upon any renewal, there is a variation in the rate on which the insurance premium is based. In these circumstances, the Contract Sum will be adjusted by the net amount of the difference between (i) the insurance premium paid by the Contractor, and (ii) the insurance premium which the Contractor *would* have paid, but for the change in rate.

6.10.3 *Where Insurance Option A applies and Terrorism Cover other than Pool Re Cover is specified as required, the cost of such other cover and of its renewal shall be added to the Contract Sum.*

Clause 6.10.3 also relates to the cost of taking out and maintaining Terrorism Cover (but in this case, *not* Pool Re Cover).

It is relevant where both:

- Insurance Option A applies; and
- Terrorism Cover *other* than Pool Re Cover is specified against the reference to clause 6.10 and Schedule 3 in the Contract Particulars.

In such circumstances, the cost of both that other cover, and its renewal, will be added to the Contract Sum.

> 6.10.4 *Where Insurance Option A applies and the Employer is a Local or Public Authority, if at any renewal of the Terrorism Cover (of any type) there is an increase in the rate on which the premium is based, the Employer may instruct the Contractor not to renew the Terrorism Cover. If it so instructs, clause 6.13.5.3 shall apply with effect from the renewal date.*

Clause 6.10.4 entitles the Employer in certain circumstances to instruct the Contractor *not* to renew the Terrorism Cover.

It is relevant where:

- Insurance Option A applies;
- the Employer is a Local or Public Authority; and
- at any renewal of the Terrorism Cover (whether Pool Re Cover or otherwise), there is an increase in the rate on which the premium is based.

If the Employer issues this instruction, clause 6.13.5.3 (Loss or damage – insurance claims and reinstatement) will apply from the renewal date, to the effect that any reinstatement work will be treated as a Change, to the extent that it is no longer recoverable under the Terrorism Cover.

Resource 206 page 533	Instruction: Employer to Contractor	An instruction from the Employer not to renew Terrorism Cover.

Clause 6.11: Terrorism Cover – non-availability – Employer's options

> 6.11.1 *If the insurers named in any Joint Names Policy notify either Party that, with effect from a specified date (the 'cessation date'), Terrorism Cover will cease and will no longer be available or will only continue to be available with a reduction in the scope or level of such cover, the recipient shall immediately notify the other Party.*

Clause 6.11 provides for the circumstances where Terrorism Cover will either:

- cease, and will no longer be available at all, or
- will only continue to be available in a more limited way, with a reduction in the scope or level of cover.

If the insurers of any Joint Names Policy notify one of the Parties of either of these circumstances, which will take effect from a certain date (defined as the '*cessation date*'), that Party must immediately notify the other.

Resource 207 page 533	Notice: Contractor or Employer to the other Party	A notice of the non-availability of Terrorism Cover.

6.11.2 The Employer, after receipt of such notification but before the cessation date, shall give notice to the Contractor either:

6.11.2.1 that, notwithstanding the cessation or reduction in scope or level of Terrorism Cover, the Employer requires that the Works continue to be carried out; or

6.11.2.2 that on the date stated in the Employer's notice (which shall be a date after the date of the insurers' notification but no later than the cessation date) the Contractor's employment under this Contract shall terminate.

Clause 6.11.2 requires the Employer to make a decision, following receipt of the notice under clause 6.11.1 (whether from the insurers or from the Contractor).

Before the cessation date, the Employer must notify the Contractor either:

- in accordance with clause 6.11.2.1, that the Employer requires the Works to continue to be carried out, even though the Terrorism Cover will cease, or will be reduced in its scope or level; or
- in accordance with clause 6.11.2.2, that the Contractor's employment under the Contract will terminate, on the date stated in the Employer's notice (which must be between (i) the date of the insurers' notification, and (ii) the cessation date).

Resource 208 page 534	Notice: Employer to Contractor	A notice from the Employer to the Contactor, following the non-availability of Terrorism Cover.

6.11.3 Where Insurance Option A applies and the Employer gives notice under clause 6.11.2.1 requiring continuation of the Works, the Employer may instruct the Contractor to effect and maintain any alternative or additional form of Terrorism Cover then reasonably obtainable by the Contractor; the net additional cost to the Contractor of any such cover and its renewal shall be added to the Contract Sum.

Clause 6.11.3 entitles the Employer in certain circumstances to instruct the Contractor to place and maintain any '*reasonably obtainable*' alternative or additional form of Terrorism Cover.

It is relevant where:

- Insurance Option A applies; and
- the Employer has given a notice to the Contractor under clause 6.11.2.1, requiring the Contractor to continue carrying out the Works.

If the Employer issues such an instruction, the net additional cost to the Contractor of both the cover, and its renewal, will be added to the Contract Sum.

Resource 209 page 534	Instruction: Employer to Contractor	An instruction from the Employer for the Contractor to effect and maintain an alternative or additional form of Terrorism Cover.

6.11.4 *If the Employer gives notice of termination under clause 6.11.2.2, then upon and from such termination the provisions of clause 8.12 shall apply.*

Clause 6.11.4 provides for the consequences of the termination of the Contractor's employment under the Contract in accordance with clause 6.11.2.2, by directing the Parties to clause 8.12 (Consequences of Termination under clauses 8.9 to 8.11, etc.).

6.11.5 *If the Employer does not give notice of termination under clause 6.11.2.2, but work executed and/or Site Materials thereafter suffer physical loss or damage caused by terrorism, clauses 6.13 and 6.14 shall as appropriate apply.*

Clause 6.11.5 provides for where:

- the Employer does *not* give a notice of termination to the Contractor, under clause 6.11.2.2; but
- work executed or Site Materials subsequently suffer physical loss or damage, as a result of terrorism.

In these circumstances, the following clauses (considered below) apply, as appropriate:

- clause 6.13 (Loss or damage – insurance claims and reinstatement), which relates to insurance claims under the Works Insurance Policy (or any policies covering Existing Structures or their contents that are placed by the Employer), and reinstatement work to repair the damage caused; and

- clause 6.14 (Loss or damage to Existing Structures – right of termination), which entitles the Employer to terminate the Contractor's employment under the Contract, in the event of material loss or damage to any of the Existing Structures.

Clause 6.12: Evidence of insurance

6.12.1 *Where a Party is required by this Contract to effect and maintain an insurance policy or cover under any of clauses 6.4, 6.5, 6.7 and 6.10, or is responsible for ensuring that it is effected and maintained, that Party shall at the request of the other Party supply such documentary evidence as the other Party may reasonably require that the policy or cover has been effected and remains in force.*

Clause 6.12 relates to the provision of documentary evidence by the Party responsible for effecting and maintaining insurance in accordance with the following provisions:

- clause 6.4 (Contractor's insurance of its liability), which is the responsibility of the Contractor;
- clause 6.5 (Contractor's insurance of liability of Employer), which is also the responsibility of the Contractor;
- clause 6.7 (Insurance Options and period), which is the responsibility of either the Contractor (where Insurance Option A applies) or the Employer (where Insurance Option B or Insurance Option C applies); and
- clause 6.10 (Terrorism Cover – policy extensions and premiums), which is the responsibility of either the Contractor (where Insurance Option A applies) or the Employer (where Insurance Option B or Insurance Option C applies).

The Party responsible for effecting and maintaining the relevant insurance policy must, upon request from the other Party, supply any documentary evidence as that other Party may reasonably require, to provide reassurance that the policy or cover is in place.

In practice, this will usually take the form of a copy of the relevant policy documents, or a letter from the relevant insurance broker.

Resource 210 page 535	Request: Contractor or Employer to the other Party	A request for evidence of insurance.
Resource 211 page 535	Reply: Contractor or Employer to the other Party	The provision of evidence of insurance.

6.12.2 *If a Party required to provide such documentary evidence fails to provide it within 7 days of a request being made, the other Party may assume that there has been a failure to insure, and may insure against any risk, liability or expense to which it may be exposed as a consequence, but shall not be obliged to do so. If the other Party insures, the defaulting Party shall be liable for the costs that the other Party incurs in taking out and maintaining that insurance. Any costs payable to the Contractor shall be added to the Contract Sum; any costs payable to the Employer may be deducted from any sums due or to become due to the Contractor or shall be recoverable from the Contractor as a debt.*

Clause 6.12.2 sets out the consequences if the Party responsible for effecting and maintaining the relevant insurance policy listed in clause 6.12.1 *fails* to provide the necessary documentary evidence to the other Party, within 7 days of a request being made.

In these circumstances:

- The other Party is entitled to then assume that there has been a failure to insure (on the basis that no evidence to the contrary was provided).
- To protect itself, the other Party *may* therefore insure against any consequent risk, liability or expense.
- If it does so, then the Party responsible for effecting and maintaining the relevant insurance policy will be liable for the associated costs in taking out and maintaining the replacement insurance:
  - If the Employer was responsible, then the costs payable to the Contractor will be added to the Contract Sum.
  - If the Contractor was responsible, then the costs payable to the Employer either (i) may be deducted from any sums due, or to become due, to the Contractor, or (ii) may be recovered from the Contractor as a debt.

Resource 212 page 536	Correspondence: Contractor or Employer to the other Party	Confirmation of a failure to provide evidence of insurance.

Clause 6.13: Loss or damage – insurance claims and reinstatement

6.13.1 *If during the carrying out of the Works any loss or damage affecting any executed work or Site Materials is occasioned by any of the risks covered by the Works Insurance Policy or an Excepted Risk or there is any loss of or damage of any kind to any of the Existing Structures or their contents, the Contractor shall forthwith upon it occurring or becoming apparent give notice to the Employer of its nature, location and extent.*

Clause 6.13.1 relates to:

- loss or damage affecting any executed work or Site Materials, caused by (i) any of the risks covered by the Works Insurance Policy, or (ii) an Excepted Risk; or
- loss or damage to any of the Existing Structures, or their contents.

If any of these circumstances arise during the carrying out of the Works, the Contractor must immediately notify the Employer of the nature, location, and extent of the loss or damage.

Resource 213 page 537	Notice: Contractor to Employer	A notice from the Contractor to the Employer of loss or damage affecting executed work, Site Materials, or the Existing Structures.

6.13.2 Subject to clauses 6.13.5.1 and 6.13.6, the occurrence of such loss or damage to executed work or Site Materials shall be disregarded in calculating any amounts payable to the Contractor under this Contract.

Clause 6.13.2 provides that the occurrence of such loss or damage described in clause 6.13.1 to either (i) executed work, or (ii) Site Materials, will be disregarded, in calculating any amounts payable to the Contractor. In other words, the relevant executed work or Site Materials should be considered as though the loss or damage did *not* occur.

This is subject to clause 6.13.5.1 and clause 6.13.6 (considered below).

6.13.3 The Contractor, for itself and for all its sub-contractors recognised as an insured under the Works Insurance Policy, shall authorise the insurers to pay to the Employer all monies from such insurance, and from any policies covering Existing Structures or their contents that are effected by the Employer.

Clause 6.13.3 relates to payments from insurers to the Employer.

It requires the Contractor, for both itself and for its sub-contractors who are recognised as insured under the Works Insurance Policy, to authorise the insurers to pay the Employer under both (i) the Works Insurance Policy, and (ii) any policies covering Existing Structures (or their contents), which are placed by the Employer.

Resource 214 page 537	Authorisation: Contractor to insurers	An authorisation from the Contractor to the insurers to pay insurance monies to the Employer.

6.13.4 *Where loss or damage affecting executed work or Site Materials is occasioned by any risk covered by the Works Insurance Policy, and subject to clause 6.14 where relevant, the Contractor shall after any inspection required by the insurers under the Works Insurance Policy and with due diligence restore the damaged work, replace or repair any lost or damaged Site Materials, remove and dispose of any debris (collectively 'reinstatement work') and proceed with the carrying out and completion of the Works.*

Clause 6.13.4 relates to the carrying out of '*reinstatement work*' by the Contractor, following loss or damage affecting any executed work or Site Materials, which has been caused by any risk covered by the Works Insurance Policy.

Reinstatement work includes:

- restoring the damaged work;
- replacing or repairing any lost or damaged Site Materials; and
- removing and disposing of any debris.

After any inspection required by the insurers under the Works Insurance Policy, the Contractor must then:

- carry out the reinstatement work; and
- proceed with the carrying out and completion of the Works.

These obligations are subject to clause 6.14 (Loss or damage to Existing Structures – right of termination), which entitles the Employer to terminate the Contractor's employment under the Contract, in the event of material loss or damage to any of the Existing Structures.

Recommendation: Record keeping of reinstatement work

In accordance with clause 6.13.6, where Insurance Option B or paragraph C.2 of Insurance Option C applies, or where loss or damage is caused by an Excepted Risk, this reinstatement work shall be treated as a Change.

The Contractor and Employer are therefore encouraged to keep records of each element of the reinstatement work required.

The Contractor and Employer are further encouraged to check with the insurers under the Works Insurance Policy as soon as possible as to whether they will require an inspection.

6.13.5 Where Insurance Option A applies:

6.13.5.1 the Employer shall pay all monies from such insurance to the Contractor by instalments under separate reinstatement work statements issued by the Employer at the same dates as those for Payment Notices under clause 4.7.5 but without deduction of Retention and less only the amounts referred to in clause 6.13.5.2;

6.13.5.2 the Employer may retain from those monies any amounts properly incurred by the Employer and notified by it to insurers in respect of professional fees up to the aggregate amount of the percentage cover for those fees or (if less) the amount paid by insurers in respect of those fees;

6.13.5.3 in respect of reinstatement work, the Contractor shall not be entitled to any payment other than amounts received under the Works Insurance Policy except where there has been a cessation of or reduction in Terrorism Cover under clause 6.10.4 or 6.11 and loss or damage is then caused by or results from terrorism, in which case the reinstatement work shall, to the extent that its cost is no longer recoverable under the policy, be treated as a Change and under clause 4.12.2.4, 4.13.2.4 or 6.13.5.1 included in Payment Notices. In neither case shall there be any reduction in any amount payable by reason of any act or neglect of the Contractor or of any sub-contractor which may have contributed to the physical loss or damage;

6.13.5.4 for the purposes of clause 2.26.1 (but not otherwise under these Conditions except as referred to in clause 6.13.5.3) reinstatement work shall be treated as a Change.

Clause 6.13.5 is relevant where Insurance Option A (New Buildings – All Risks Insurance of the Works by the Contractor) applies.

Clause 6.13.5.1 provides that the Employer must pay to the Contractor all money received under the Works Insurance Policy (apart from in respect of professional fees, considered in clause 6.13.5.2). This is to be paid in instalments, under '*reinstatement work statements*', which the Employer must issue alongside Payment Notices under clause 4.7.5 (Interim Payments – Contractor's Interim Payment Applications, due dates and Payment Notices), but in respect of which no deduction of Retention will be made.

Clause 6.13.5.2 provides that the only amounts which may be retained by the Employer are in respect of professional fees. These fees must be properly incurred by the Employer, and notified by the Employer to the insurers under the Works Insurance Policy. They may be retained up to either:

- the aggregate amount of the percentage cover for those fees, stated against the reference to clause 6.7 and Schedule 3 in the Contract Particulars; or, if less,
- the amount paid by the insurers under the Works Insurance Policy for those fees.

Clause 6.13.5.3 relates to reinstatement work, and provides that the Contractor will *only* be entitled to payment of amounts received under the Works Insurance Policy. The exception applies in the following circumstances:

- where Terrorism Cover has either ceased or reduced under clause 6.10.4 (Terrorism Cover – policy extensions and premiums) or clause 6.11 (Terrorism Cover – non-availability – Employer's options); and
- loss and damage is then caused by terrorism.

In this case, reinstatement work will be treated as a Change (to the extent that the cost of the reinstatement work cannot be recovered under the Works Insurance Policy), and will be included in Payment Notices.

Irrespective of whether this exception applies, clause 6.13.5.3 further provides that there will be no reduction in the amount payable to the Contractor as a result of any act or neglect of the Contractor (or any sub-contractor), which may have contributed to the physical loss or damage.

Clause 6.13.5.4 provides that reinstatement work will be treated as a Change, for the purposes of clause 2.26.1 (Relevant Events), but not otherwise (except as outlined in clause 6.13.5.3 above). In other words, reinstatement work constitutes a Relevant Event, which may entitle the Contractor to an extension of time. However, it will not constitute a Relevant Matter, and so the Contractor will have no entitlement to reimbursement of loss and/or expense.

Resource 215 page 538	Reinstatement work statements: Employer to Contractor	The provision by the Employer to the Contractor of a reinstatement work statement.

6.13.6 *Where Insurance Option B or paragraph C.2 of Insurance Option C applies or where loss or damage is caused by an Excepted Risk, reinstatement work shall be treated as a Change.*

Clause 6.13.6 is relevant where either:

- Insurance Option B (New Buildings – All Risks Insurance of the Works by the Employer) applies; or
- paragraph C.2 (The Works – Joint Names Policy for All Risks) of Insurance Option C (Joint Names Insurance by the Employer of Existing Structures and Works in or Extensions to them) applies; or
- loss or damage is caused by an '*Excepted Risk*' (as defined in clause 6.8 (Related definitions)).

In such circumstances, clause 6.13.6 provides that any reinstatement work will be treated as a Change.

Clause 6.14: Loss or damage to Existing Structures – right of termination

6.14 *If there is material loss of or damage to any of the Existing Structures, the Employer shall be under no obligation to reinstate those structures, but either Party may, if it is just and equitable, terminate the Contractor's employment under this Contract by notice given to the other in accordance with clause 1.7.4 within 28 days of the occurrence of that loss or damage. If such notice is given, then:*

6.14.1 *unless within 7 days of receiving the notice (or such longer period as may be agreed) the Party to which it is given invokes a dispute resolution procedure of this Contract to determine whether the termination is just and equitable, it shall be deemed to be so;*

6.14.2 *upon the giving of that notice or, where a dispute resolution procedure is invoked within that period, upon any final upholding of the notice, the provisions of clause 8.12 shall apply.*

Clause 6.14 provides for the circumstance where there is '*material*' loss or damage to any of the Existing Structures.

In this event, the Employer will not be obliged to reinstate (or repair) those structures.

Instead, either Party may terminate the Contractor's employment under the Contract, by giving a notice to the other Party.

This notice must be given:

- in accordance with clause 1.7.4 (Notices and other communications); and
- within 28 days of the occurrence of the loss or damage.

However, clause 6.14 only permits the termination of the Contractor's employment under the Contract where it is '*just and equitable*' to do so. This is very broadly drafted, and open to interpretation.

Clause 6.14 therefore affords the Party to whom the notice is given the opportunity to refer the matter to one of the dispute resolution procedures under the Contract, within 7 days of receiving the notice (or a longer period, if the Parties agree), in order to determine whether the termination was, in fact, '*just and equitable*'. These dispute resolution procedures are included in:

- clause 9.1 (Notification and negotiation of disputes);
- clause 9.2 (Mediation);
- Article 8 (Adjudication) and clause 9.3 (Adjudication);
- Article 9 (Arbitration) and clause 9.4 to clause 9.9 (Arbitration); and
- Article 10 (Legal proceedings).

If that Party does not refer the matter to one of these dispute resolution procedures within the relevant period, then the termination will be deemed to be '*just and equitable*'.

Clause 6.14.2 provides for consequences of termination, by directing the Parties to clause 8.12 (Consequences of Termination under clauses 8.9 to 8.11, etc.). This is in the event that either (i) the Contractor's employment is terminated under the Contract following the giving of the notice, or (ii) where the other Party challenged the notice in accordance with clause 6.14.1, it was finally upheld (on the basis that the termination was considered to be '*just and equitable*').

Resource 216 page 539	Notice: Contractor or Employer to the other Party	A notice of termination following material loss or damage to any of the Existing Structures.
Resource 217 page 540	Correspondence: Contractor or Employer to the other Party	The invocation of a dispute resolution procedure to determine whether the termination under clause 6.14 is just and equitable.

Professional Indemnity Insurance

Overview

This sub-section relates to the Contractor's obligation to take out and maintain a Professional Indemnity insurance policy, and provides for the circumstance where that insurance ceases to be available at commercially reasonable rates or terms.

Very briefly, Professional Indemnity insurance provides cover against liability relating to professional negligence by the Contractor. In the context of the Contract, and clause 2.17 (Design Work – liabilities and limitation) in particular, this cover typically applies in respect of negligent design.

Key provisions

- clause 6.15 (Obligation to insure)
- clause 6.16 (Increased cost and non-availability)

Clause 6.15: Obligation to insure

6.15 The Contractor shall:

6.15.1 forthwith after this Contract has been entered into, take out (unless it has already done so) a Professional Indemnity insurance policy with limits of indemnity of the types and in amounts not less than those stated in the Contract Particulars;

6.15.2 thereafter, provided it is available at commercially reasonable rates and terms, maintain such insurance until the expiry of the period stated in the Contract Particulars from the date of practical completion of the Works; and

6.15.3 as and when reasonably requested to do so by the Employer, produce for inspection documentary evidence that such insurance has been effected and/or is being maintained.

DOI: 10.1201/9781003485919-53

Clause 6.15 imposes three obligations on the Contractor in respect of Professional Indemnity insurance. The Contractor must:

- immediately after the date of the Contract (and unless it has already done so), take out a Professional Indemnity insurance policy, with the limits of indemnity of the types and in amounts at least equal to those stated against the reference to clause 6.15 in the Contract Particulars;
- maintain that Professional Indemnity insurance (typically on an annual renewal basis) until the expiry of the period stated against the reference to clause 6.15 in the Contract Particulars, from the date of practical completion of the Works, provided the insurance is available at both '*commercially reasonable*' rates and terms; and
- upon request, provide the Employer with documentary evidence that the Professional Indemnity insurance policy has been placed, or is being maintained.

Resource 218 page 540	Request: Employer to Contractor	A request from the Employer to the Contractor for evidence of professional indemnity insurance.
Resource 219 page 541	Reply: Contractor to Employer	The provision by the Contractor to the Employer of evidence of professional indemnity insurance.

Market practice: 'Commercially reasonable' rates and terms

Clause 6.15.2 is often amended to the effect that any increased or additional premium as a result of the Contractor's claims record will be deemed to be within the meaning of '*commercially reasonable rates and terms*'. In other words, the Contractor will still be obliged to maintain Professional Indemnity insurance, even though that increased or additional premium may have otherwise fallen outside the meaning of '*commercially reasonable rates and terms*'.

Clause 6.16: Increased cost and non-availability

6.16 *If the insurance referred to in clause 6.15 ceases to be available at commercially reasonable rates or terms, the Contractor shall immediately give notice to the Employer so that the Contractor and the Employer can discuss the means of best protecting their respective positions in the absence of such insurance.*

Clause 6.16 provides for the circumstance where the Contractor's Professional Indemnity insurance is no longer available at either '*commercially reasonable*' rates or terms.

In such a case, the Contractor must immediately notify the Employer, so that the Parties can discuss how to best protect their respective positions, in the absence of such Professional Indemnity insurance.

Resource 220 page 541	Notice: Contractor to Employer	A notice from the Contractor to the Employer of the increased cost or non-availability of professional indemnity insurance.
Resource 221 page 542	Tracker: Contractor and Employer	A tracker template for insurances required under the Contract, and their renewal.

Joint Fire Code – compliance

Overview

This sub-section relates to the '*Joint Code of Practice on the Protection from Fire of Construction Sites and Buildings Undergoing Renovation*', published by Construction Industry Publications Ltd and the Fire Protection Association, and defined as the '*Joint Fire Code*'.

The objective of the Joint Fire Code is to prevent fires on construction sites, which in most cases can be achieved by '*designing out risks, taking simple precautions, and by adopting safe working practices*'.[20]

The Parties can decide whether or not the Joint Fire Code is to apply (against the reference to clause 6.17 in the Contract Particulars). However, given its subject matter, the Parties are encouraged to select it to apply. Indeed, compliance is often required by insurers.

Key provisions

- clause 6.17 (Application of clauses)
- clause 6.18 (Compliance with Joint Fire Code)
- clause 6.19 (Breach of Joint Fire Code – Remedial Measures)
- clause 6.20 (Joint Fire Code – amendments/revisions)

Clause 6.17: Application of clauses

6.17 *Clauses 6.18 to 6.20 apply where the Contract Particulars state that the Joint Fire Code applies.*

Clause 6.17 provides that the remaining clauses of this sub-section (namely clause 6.18 to clause 6.20) apply where stated against the reference to clause 6.17 in the Contract Particulars.

In accordance with the definition of the '*Joint Fire Code*' in clause 1.1 (Definitions), the relevant edition of the Joint Fire Code is that current at the Base Date (subject to any amendments or revisions, considered in clause 6.18 and clause 6.20).

[20] Joint Fire Code (10th edition, August 2022, incorporating amendment 1, January 2023), at page 7

DOI: 10.1201/9781003485919-54

Clause 6.18: Compliance with Joint Fire Code

6.18 *The Parties shall comply with the Joint Fire Code and any amendments or revisions to it; the Employer shall ensure such compliance by all Employer's Persons and the Contractor shall ensure such compliance by all Contractor's Persons.*

Clause 6.18 requires both Parties to comply with the Joint Fire Code, along with any amendments or revisions to it.

Furthermore, the Employer and Contractor are both required to ensure that all Employer's Persons and all Contractor's Persons respectively comply with the Joint Fire Code.

Recommendation: Amendments and revisions to the Joint Fire Code

The Parties are encouraged to provide copies of the Joint Fire Code to all Employer's Persons and Contractor's Persons, and to account for the Joint Fire Code in any relevant agreements (for example, sub-contracts).

The Parties are also encouraged to check periodically for any amendments or revisions to the Joint Fire Code, to ensure compliance with clause 6.18.

Clause 6.19: Breach of Joint Fire Code – Remedial Measures

6.19.1 *If a breach of the Joint Fire Code occurs and the insurers under the Works Insurance Policy specify by notice to the Employer or the Contractor the remedial measures they require (the 'Remedial Measures'), the Party receiving the notice shall copy it to the other and the Contractor shall ensure that the Remedial Measures are carried out.*

Clause 6.19 provides for the circumstance where there is a breach of the Joint Fire Code, and the insurers under the Works Insurance Policy then notify either the Employer or the Contractor the remedial measures they require (defined as the '*Remedial Measures*'), in which case:

- the Party receiving the notice from the insurers under the Works under the Works Insurance Policy must provide a copy to the other Party; and
- the Contractor must ensure that the Remedial Measures are carried out.

Resource 222 page 543	Correspondence from insurers: Employer/Contractor	The provision of a notice from the insurers of the Remedial Measures required, as a result of a breach of the Joint Fire Code.
Resource 223 page 544	Reply: Contractor to Employer	A reply from the Contractor to the Employer following receipt of the Remedial Measures required by the insurers, as a result of a breach of the Joint Fire Code.

6.19.2 *If the Contractor, within 7 days of receipt of a notice specifying Remedial Measures, does not begin to carry out or thereafter fails without reasonable cause regularly and diligently to proceed with the Remedial Measures, then the Employer may employ and pay other persons to carry out those Remedial Measures. The Contractor shall be liable for all additional costs incurred by the Employer in connection with such employment and an appropriate deduction shall be made from the Contract Sum.*

Clause 6.19.2 provides for the circumstance where the Contractor, despite having received a notice specifying Remedial Measures (whether from the Employer or directly from the insurers under the Works Insurance Policy) either:

- does not begin to carry out the Remedial Measures; or
- fails, without reasonable grounds, to '*regularly and diligently*' proceed with the Remedial Measures,

within 7 days from receipt of that notice.

Clause 6.19.2 then entitles the Employer to employ and pay others to carry out those Remedial Measures, in which case:

- the Contractor will be liable for all additional costs incurred by the Employer in connection with the employment of those other persons, and
- an appropriate deduction will be made from the Contract Sum, in respect of those additional costs.

Resource 224 page 544	Notice: Employer to Contractor	Confirmation from the Employer to the Contractor that the Employer will employ and pay others to carry out the Remedial Measures, following the Contractor's failure to proceed with the Remedial Measures.
Resource 225 page 545	Notice: Employer to Contractor	Confirmation from the Employer to the Contractor that the Employer has employed and paid others to carry out the Remedial Measures, following the Contractor's failure to proceed with the Remedial Measures.

Clause 6.20: Joint Fire Code – amendments/revisions

6.20 *Where the Joint Fire Code is, under a Joint Names Policy, applicable to the Works and amendments or revisions are made to it after the Base Date, any cost of compliance by the Contractor with amendments or revisions made after that date shall be borne as stated in the Contract Particulars. If the cost is to be borne by the Employer, it shall be added to the Contract Sum.*

Clause 6.20 relates to the allocation of the commercial risk of compliance by the Contractor with any amendments or revisions made to the Joint Fire Code after the Base Date, in circumstances where the Joint Fire Code is applicable to the Works under a Joint Names Policy.

The Party responsible for the cost of the Contractor's compliance with any such amendments or revisions will be identified against the reference to clause 6.20 in the Contract Particulars.

If the cost is to be borne by the Employer, that cost will be added to the Contract Sum. (If the cost is to be borne by the Contractor, then there will be no need to adjust the Contract Sum.)

Section 7: Assignment, Performance Bonds and Guarantees, Third Party Rights and Collateral Warranties

Section overview

Section 7 includes some of the key provisions in the Contract relating to the following third parties:

- those to whom the Employer or Contractor may transfer any of their rights under the Contract, described as '*assignees*';
- banks and sureties, who may provide a performance bond to the Employer, as security for the Contractor's performance of the Contract;
- the Contractor's parent company, who may provide a parent company guarantee in favour of the Employer, again to guarantee the Contractor's performance of the Contract;
- Purchasers and Tenants, who may benefit from P&T Rights granted by the Contractor and its sub-contractors, whether by way of third party rights or collateral warranties;
- Funders, who may benefit from Funder Rights granted by the Contractor and its sub-contractors, again, whether by way of third party rights or collateral warranties; and
- the Contractor's sub-contractors, who may also grant third party rights or a collateral warranty in favour of the Employer.

Section 7 therefore covers the following three areas:

- **Assignment**: the ability of the Parties to transfer their rights under the Contract to a third party;
- **Performance bonds and guarantees**: performance security from the Contractor, in the form of a performance bond (from a bank or surety) and parent company guarantees (from the Contractor's parent company); and
- **Third party rights and collateral warranties**: third party rights and collateral warranties from the Contractor and its sub-contractors, in favour of Purchasers, Tenants, Funders, and (in the case of sub-contractors) the Employer.

Sub-sections

- Assignment (clause 7.1 – clause 7.2)
- Performance Bonds and Guarantees (clause 7.3)
- Clauses 7.7 to 7.11 – Preliminary (clause 7.4 – clause 7.6)
- Third Party Rights from Contractor (clause 7.7 – clause 7.8)
- Collateral Warranties from Contractor (clause 7.9 – clause 7.10)
- Third Party Rights and Collateral Warranties from Sub-Contractors (clause 7.11)

DOI: 10.1201/9781003485919-56

Assignment

Overview

Assignment describes the transfer of a Party's rights under the Contract (sometimes described as the 'benefit') to a third party.

This sub-section relates to the assignment by the Parties (and just the Employer, in the case of clause 7.2 (Rights of enforcement)) of any of their rights under the Contract to a third party.

Key provisions

- clause 7.1 (General)
- clause 7.2 (Rights of enforcement)

Clause 7.1: General

7.1 *Subject to clause 7.2, neither the Employer nor the Contractor shall without the consent of the other assign this Contract or any rights thereunder.*

Clause 7.1 simply provides that neither Party may assign the Contract, or any of their respective rights under the Contract, without the consent of the other Party.

This is subject to clause 7.2 (Assignment: Rights of enforcement) below.

Assignment is to be distinguished from novation, which describes the transfer of one party's rights *and* obligations under a contract (sometimes described as the 'benefit' and the 'burden') to a third party, and effectively creates a new contract between the two new parties.

Novation is common in the context of design consultants, which may have been appointed by the Employer during the pre-construction phase, and the appointments of which are then novated from the Employer to the Contractor at the same time as the Parties enter into the Contract.

DOI: 10.1201/9781003485919-57

Resource 226 page 545	Request for consent: Contractor or Employer to the other Party	A request for consent from the other Party to assign the Contract or any rights thereunder.
Resource 227 page 546	Consent given: Contractor or Employer to the other Party	Consent given to assign the Contract or any rights thereunder.
Resource 228 page 546	Consent withheld: Contractor or Employer to the other Party	Consent withheld to assign the Contract or any rights thereunder.

Market practice: Contractor's consent

It is common in practice for clause 7.1 to be amended to permit the Employer to assign the Contract without the Contractor's consent in certain circumstances, for example, to a Funder, or to a group or affiliated company.

Clause 7.2: Rights of enforcement

7.2 *Where clause 7.2 is stated in the Contract Particulars to apply, then, in the event of transfer by the Employer of its freehold or leasehold interest in, or of a grant by the Employer of a leasehold interest in, the whole of the premises comprising the Works or (if the Contract Particulars so state) any Section, the Employer may at any time after practical completion of the Works or of the relevant Section grant or assign to any such transferee or lessee the right to bring proceedings in the name of the Employer (whether by arbitration or litigation, whichever applies under this Contract) to enforce any of the terms of this Contract made for the benefit of the Employer. The assignee shall be estopped from disputing any enforceable agreements reached between the Employer and the Contractor which arise out of and relate to this Contract (whether or not they are or appear to be a derogation from the right assigned) and which are made prior to the date of any grant or assignment.*

Clause 7.2 (where it is stated against the reference to clause 7.2 in the Contract Particulars to apply) provides a mechanism for certain third parties to bring a claim under the Contract in the name of the Employer.

It is relevant in the event of either (i) a transfer by the Employer of its freehold or leasehold in, or (ii) a grant by the Employer of a leasehold interest in, the whole of the premises comprising the Works or (if stated against the reference to clause 7.2 in the Contract Particulars) any Section.

In such circumstances, after practical completion of the Works or relevant Section, the Employer may at any time grant or assign to any transferee or lessee (now described as the '*assignee*') the right to bring proceedings in the Employer's name, to enforce any of the terms of the Contract made for the Employer's benefit.

These proceedings will be either:

- arbitration, in accordance with Article 9 (Arbitration) and clause 9.4 to clause 9.9; or
- litigation, in accordance with Article 10 (Legal proceedings).

However, the assignee will be '*estopped*' (or prevented) from disputing any enforceable agreements made between the Parties which relate to the Contract (for example, a settlement agreement, or a deed of variation to the Contract), if they were made before the grant or assignment.

This is the case even if those agreements '*derogate*' (or depart) from the right assigned to the third party, or they appear to do so. In other words, if they limit or affect the Employer's ability (and therefore the assignee's ability) to bring proceedings in relation to the right assigned.

Performance Bonds and Guarantees

Overview

This sub-section relates to the provision of performance security to the Employer, in the form of performance bonds and parent company guarantees.

A performance bond or guarantee is a contract between the Employer and a guarantor, whereby the guarantor agrees to pay the Employer damages for losses it has sustained, as a result of the breach of the Contract by the Contractor. The Guarantor's liability under the performance bond is usually limited to a percentage of the Contract Sum, and the performance bond will typically expire at a date linked to practical completion (for example, upon the issue of the Practical Completion Statement).

A parent company guarantee is a contract between the Employer and the Contractor's parent company, whereby the parent company guarantees the performance of the Contractor under the Contract. The Guarantor's liability to the Employer is usually co-extensive with that of the Contractor.

Key provisions

- clause 7.3 (Performance Bonds and Guarantees)

Clause 7.3: Performance Bonds and Guarantees

7.3 The Contractor shall on the execution of this Contract provide to the Employer whichever of the following the Contract Particulars state as being required:

7.3.1 a performance bond or guarantee of the Contractor's due performance of the Contract from a bank or other surety approved by the Employer in an amount equal to the percentage of the Contract Sum and for the period stated in the Contract Particulars;

7.3.2 a guarantee by the Contractor's parent company identified in the Contract Particulars;

any such bond or guarantee, unless otherwise agreed by the Employer, being substantially in the form of the document identified by the Contract Particulars.

DOI: 10.1201/9781003485919-58

Upon the execution of the Contract, clause 7.3 requires the Contractor to provide the Employer with a performance bond or guarantee, and/or a parent company guarantee (each as stated against the reference to clause 7.3.1 and clause 7.3.2 in the Contract Particulars).

The performance bond or guarantee must meet the following requirements:

- the guarantor under the performance bond must be either a bank or other surety approved by the Employer (which should ideally be established before the execution of the Contract);
- the performance bond must be in the amount equal to the percentage of the Contract Sum stated in the Contract Particulars (against the reference to clause 7.3.1); and
- the performance bond must be for the period stated in the Contract Particulars (also against the reference to clause 7.3.1).

The performance bond and the parent company guarantee must also be '*substantially*' in the form identified in the Contract Particulars (against the reference to clause 7.3.1 and clause 7.3.2 respectively), unless the Employer agrees otherwise.

Resource 229 page 547	Request: Contractor to Employer	A request from the Contractor for the Employer's consent to provide a performance bond from a particular surety.
Resource 230 page 547	Approval given: Employer to Contractor	Approval given from the Employer for the Contractor to provide a performance bond from a particular surety.
Resource 231 page 548	Approval withheld: Employer to Contractor	Approval withheld from the Employer for the Contractor to provide a performance bond from a particular surety.
Resource 232 page 548	Request: Contractor to Employer	A request by the Contractor to the Employer to amend the form of performance bond or parent company guarantee.
Resource 233 page 549	Agreement: Employer to Contractor	An agreement by the Employer for the Contractor to change the proposed form of performance bond or parent company guarantee.
Resource 234 page 549	No agreement: Employer to Contractor	A refusal by the Employer for the Contractor to change the proposed form of performance bond or parent company guarantee.
Resource 235 page 550	Tracker: Contractor	A tracker template for recording performance bonds and parent company guarantees, across multiple building contracts.

Market practice: ABI Model Form of Guarantee Bond

It is common for the form of performance bond to be based on the Model Form of Guarantee Bond published by the Association of British Insurers (ABI), drafted specifically for the construction industry, and adopted by the National Joint Consultative Committee for Building.

Market practice: Condition precedent to payment

Clause 7.3 is often amended to the effect that the provision of the performance bond and/or parent company guarantee is a condition precedent to the Employer's obligation to make any payment to the Contractor.

Clauses 7.7 to 7.11 – Preliminary

Overview

This sub-section includes a number of provisions to assist with the interpretation and administration of the three following sub-sections:

- Third Party Rights from Contractor (clause 7.7 and clause 7.8);
- Collateral Warranties from Contractor (clause 7.9 and clause 7.10); and
- Third Party Rights and Collateral Warranties from Sub-Contractors (clause 7.11).

These sub-sections relate to rights in favour of the following:

- Purchasers and Tenants (defined as '*P&T Rights*');
- Funders (defined as '*Funder Rights*'); and
- the Employer (defined as '*Employer Rights*').

These rights are granted by way of third party rights or collateral warranties, which provide alternative means by which a third party may enforce the terms of a contract, even though they are not a party to it:

- third party rights allow the third party to enforce the terms of a contract, by way of the Contracts (Rights of Third Parties) Act 1999; and
- collateral warranties create a direct contractual link between the parties to the 'primary' contract, and the third party.

Key provisions

- clause 7.4 (Rights Particulars)
- clause 7.5 (Notices)
- clause 7.6 (Execution of Collateral Warranties)

DOI: 10.1201/9781003485919-59

Clause 7.4: Rights Particulars

7.4 *The requirements for the grant of P&T Rights and Funder Rights by the Contractor and sub-contractors and any requirement for the grant of Employer Rights by any sub-contractors ('Rights Particulars') are set out in the document(s) identified in the Contract Particulars against the reference to clause 7.4. As respects those requirements:*

7.4.1 *such rights are conferred only on persons sufficiently identified (by name, class or description) in the Rights Particulars;*

7.4.2 *if in relation to an identified beneficiary the Rights Particulars fail to specify the method by which such rights are to be conferred, the Contractor in relation to rights to be granted by it may elect to do so either as third party rights or by collateral warranty;*

7.4.3 *unless otherwise stated in the Rights Particulars, the term 'the Consultants' shall in all third party rights and/or collateral warranties to be granted mean the Employer's Agent (including any replacements) and any other consultants providing design services to the Employer in connection with the Works.*

Clause 7.4 relates to the grant of the following:

- P&T Rights, by the Contractor and sub-contractors;
- Funder Rights, by the Contractor and sub-contractors; and
- Employer Rights, by sub-contractors.

The requirements for the grant of these rights are defined as the '*Rights Particulars*', and are set out in the document identified against the reference to clause 7.4 in the Contract Particulars.

Clause 7.4 clarifies the following in respect of the Rights Particulars:

- **Sufficient identification:** the relevant rights will be conferred only on persons who have been '*sufficiently identified*', whether by their name, class, or description, in the Rights Particulars. This reflects section 1 of the Contracts (Rights of Third Parties) Act 1999. It is therefore important to ensure that the corresponding identifying details are clear. To assist with this, the JCT has published a Model Form for the Rights Particulars.[21]
- **Third party rights or collateral warranties:** if the Rights Particulars do not specify the method by which such rights are to be conferred on an identified beneficiary, the Contractor has a choice: it may grant those rights either as third party rights or by collateral warranty.
- **Consultants:** unless stated otherwise, reference within third party rights or collateral warranties to '*the Consultants*' mean both (i) the Employer's Agent (and any replacement), and (ii) any other consultants providing design services to the

[21] Available at: https://www.jctltd.co.uk/docs/DB-2024-Rights-Particulars.pdf

Employer in connection with the Works. Consultants are referenced in paragraph 1.3 of the P&T Rights, contained in Part 1 of Schedule 5 (Third Party Rights for Purchasers and Tenants), and paragraph 1.1 of the Funder Rights, contained in Part 2 of Schedule 5 (Third Party Rights for a Funder). These are both 'net contribution' provisions, which limit the Contractor's liability to the proportion of either the Purchaser, Tenant, or Funder's losses which it would be '*just and equitable to require the Contractor to pay having regard to the extent of the Contractor's responsibility for the same*'.

Clause 7.5: Notices

> 7.5 *Each notice to the Contractor referred to in clauses 7.7 to 7.11 shall be given in accordance with clause 1.7.4.*

Clause 7.5 simply provides that any notice from the Employer to the Contractor must be sent in accordance with clause 1.7.4 (Notices and other communications), in respect of the following clauses:

- clause 7.7 (Rights for Purchasers and Tenants)
- clause 7.8 (Rights for a Funder)
- clause 7.9 (Contractor's Warranties – Purchasers and Tenants)
- clause 7.10 (Contractor's Warranty – Funder)
- clause 7.11 (Third Party Rights and Collateral Warranties from Sub-Contractors)

Clause 7.6: Execution of Collateral Warranties

> 7.6 *Where this Contract is executed as a deed, any collateral warranty to be entered into by the Contractor pursuant to clause 7.9 or 7.10 shall be executed as a deed. Where this Contract is executed under hand, any such warranty may be executed under hand.*

Clause 7.6 relates to the execution of collateral warranties in accordance with clause 7.9 (Contractor's Warranties – Purchasers and Tenants) and clause 7.10 (Contractor's Warranty – Funder):

- where the Contract is executed as a deed, those collateral warranties *must* also be executed as deeds.
- where the Contract is executed under hand, those collateral warranties *may* also be executed under hand.

Further guidance on the execution of deeds is included in the Attestation section above.

Third Party Rights from Contractor

Overview

This sub-section relates to P&T Rights for Purchasers and Tenants, and Funder Rights for Funders.

These rights are conferred by the Contractor on Purchasers, Tenants, and Funders, as third party rights (rather than as collateral warranties, which are considered in the next sub-section).

The P&T Rights and the Funder Rights permit Purchasers, Tenants, and Funders to enforce the Contract for their benefit, even though they are not a party to it, by way of the Contracts (Rights of Third Parties) Act 1999.

Key provisions

- clause 7.7 (Rights for Purchasers and Tenants)
- clause 7.8 (Rights for a Funder)

Clause 7.7: Rights for Purchasers and Tenants

7.7.1 *Where the Rights Particulars state that the Contractor shall confer P&T Rights on a Purchaser or Tenant as third party rights, those rights shall vest in that Purchaser or Tenant on the date of receipt by the Contractor of the Employer's notice to that effect, stating the name of the Purchaser or Tenant and the nature of its interest in the Works, and (where paragraph 7.2, Part 1 of Schedule 5 (Third Party Rights for Purchasers and Tenants) is to apply) an email address for that Purchaser or Tenant.*

Clause 7.7 relates to P&T Rights, which are conferred by the Contractor on Purchasers and Tenants identified in the Rights Particulars, as third party rights (rather than by way of collateral warranties, which are considered in clause 7.9 (Contractor's Warranties – Purchasers and Tenants)).

Clause 7.7.1 provides that P&T Rights will '*vest*' in (or be granted to) a Purchaser or Tenant on the date the Contractor receives the Employer's notice to that effect, which will include the following details:

- the name of the Purchaser or Tenant;
- the nature of their interest in the Works; and

DOI: 10.1201/9781003485919-60

- an email address for the Purchaser or Tenant (where paragraph 7.2 of Part 1 of Schedule 5 (Third Party Rights for Purchasers and Tenants) applies, which relates to the service of notices by email).

Resource 236 page 550	Notice: Employer to Contractor	A notice from the Employer to the Contractor of the vesting of P&T Rights as third party rights in a Purchaser or Tenant.

7.7.2 Where P&T Rights have vested in any Purchaser or Tenant, the Employer and the Contractor shall not be entitled without the consent of that Purchaser or Tenant to amend or vary the express provisions of this clause 7.7 or of Part 1 of Schedule 5 (Third Party Rights for Purchasers and Tenants) but, subject thereto, the rights of the Employer and/or the Contractor:

7.7.2.1 to terminate the Contractor's employment under this Contract (whether under section 8 or otherwise), or to agree to rescind this Contract;

7.7.2.2 to agree to amend or otherwise vary or to waive any terms of this Contract;

7.7.2.3 to agree to settle any dispute or other matter arising out of or in connection with this Contract, in each case in or on such terms as they shall in their absolute discretion think fit,

shall not be subject to the consent of any Purchaser or Tenant.

Clause 7.7.2 applies once P&T Rights have '*vested*' in a Purchaser or Tenant in accordance with clause 7.7.1.

It sets out a number of matters for which the Employer and Contractor require the Purchaser or Tenant's consent, and a number of matters for which they do not.

The Employer and Contractor require the Purchaser or Tenant's consent if they wish to amend or vary either:

- the express provisions of clause 7.7; or
- the express provisions of Part 1 of Schedule 5 (Third Party Rights for Purchasers and Tenants).

However, the Employer and Contractor do not require the Purchaser or Tenant's consent if they wish to do any of the following:

- terminate the Contractor's employment under the Contract (either under section 8 (Termination), or otherwise), or agree to rescind (or cancel or 'set aside') the Contract;
- agree to (i) amend or otherwise vary, or (ii) waive, any terms of the Contract; or
- agree to settle any dispute arising out of or in connection with the Contract, on any terms.

Resource 237 page 551	Request: Contractor to Purchaser or Tenant	A request from the Contractor to a Purchaser or Tenant to amend or vary clause 7.7 or Part 1 of Schedule 5, relating to P&T Rights.

Clause 7.8: Rights for a Funder

7.8.1 *Where the Rights Particulars state that the Contractor shall confer Funder Rights on a Funder as third party rights, those rights shall vest in the Funder on the date of receipt by the Contractor of the Employer's notice to that effect, which (where paragraph 11.2, Part 2 of Schedule 5 (Third Party Rights for a Funder) is to apply) shall state an email address for the Funder.*

Clause 7.8 relates to Funder Rights, which are conferred by the Contractor on Funders identified in the Rights Particulars, as third party rights (rather than by way of collateral warranties, which are considered in clause 7.10 (Contractor's Warranty – Funder)).

Clause 7.8.1 provides that Funder Rights will '*vest*' in (or be granted to) a Funder on the date the Contractor receives the Employer's notice to that effect, which will include an email address for the Funder (where paragraph 11.2 of Part 2 of Schedule 5 (Third Party Rights for a Funder) applies, which relates to the service of notices by email).

7.8.2 *Where Funder Rights have been vested in the Funder pursuant to clause 7.8.1:*

7.8.2.1 *no amendment or variation shall be made to the express terms of this clause 7.8, to Part 2 of Schedule 5 (Third Party Rights for a Funder) or to the relevant Rights Particulars without the prior written consent of the Funder; and*

7.8.2.2 *neither the Employer nor the Contractor shall agree to rescind this Contract, and the rights of the Contractor to terminate its employment under this Contract or to treat this Contract as repudiated shall in all respects be subject to the provisions of paragraph 6 of Part 2 of Schedule 5*

but, subject thereto, unless and until the Funder gives notice under paragraph 5 or paragraph 6.4 of Part 2 of Schedule 5, the Contractor shall remain free without the consent of the Funder to agree with the Employer to amend or otherwise vary or to waive any term of this Contract and to settle any dispute or other matter arising out of or in connection with this Contract, in each case in such terms as they think fit, without any requirement that the Contractor obtain the consent of the Funder.

Clause 7.8.2 applies once Funder Rights have '*vested*' in a Funder in accordance with clause 7.8.1.

It sets out a number of matters for which the Employer and Contractor require the Funder's consent, and a number of matters for which they do not.

The Employer and Contractor require the Funder's prior written consent if they wish to amend or vary either:

- the express provisions of clause 7.8;
- the express provisions of Part 2 of Schedule 5 (Third Party Rights for a Funder); or
- the relevant Rights Particulars.

The Employer and the Contractor are not permitted to rescind (or cancel or 'set aside') the Contract.

Furthermore, the rights of the Contractor to either (i) terminate its employment under the Contract, or (ii) treat the Contract as repudiated, will be subject to the provisions of paragraph 6 of Part 2 of Schedule 5 (Third Party Rights for a Funder). Paragraph 6 provides 'step-in' rights to the Funder, whereby the Funder may, in certain circumstances, 'step in' to take the place of the Employer under the Contract.

However, apart from these exceptions, and unless and until the Funder exercises their 'step-in' rights under either paragraph 5 or paragraph 6.4 of Part 2 of Schedule 5, the Contractor does not require the Funder's consent if it wishes to do any of the following:

- agree with the Employer to (i) amend or otherwise vary, or (ii) waive, any terms of the Contract, as the Parties see fit; or
- agree to settle any dispute arising out of or in connection with the Contract, on any terms as the Parties see fit.

Resource 238 page 552	Notice: Employer to Contractor	A notice from the Employer to the Contractor of the vesting of Funder Rights as third party rights in a Funder.
Resource 239 page 552	Request: Contractor to Funder	A request from the Contractor to the Funder to amend or vary clause 7.8 or Part 2 of Schedule 5, relating to Funder Rights.

Collateral Warranties from Contractor

Overview

Similar to the sub-section above, this sub-section also relates to P&T Rights for Purchasers and Tenants, and Funder Rights for Funders.

However, in this sub-section, rights are conferred by the Contractor on Purchasers, Tenants, and Funders, by way of collateral warranties (rather than by way of third party rights).

Collateral warranties are separate contracts, but which have a connection with (or are 'collateral' to) the 'primary' Contract.

In the case of collateral warranties between the Contractor and Purchasers, Tenants, and Funders, the Contractor will warrant to those other parties (usually described as 'beneficiaries') that it has complied with the Contract (between the Contractor and the Employer).

This creates a direct contractual link for those beneficiaries to bring a claim against the Contractor under the collateral warranty, if the Contractor does not comply with the Contract.

Key provisions

- clause 7.9 (Contractor's Warranties – Purchasers and Tenants)
- clause 7.10 (Contractor's Warranty – Funder)

Clause 7.9: Contractor's Warranties – Purchasers and Tenants

7.9 *Where the Rights Particulars state that the Contractor shall confer P&T Rights on a Purchaser or Tenant by way of collateral warranty, the Employer may by notice to the Contractor, identifying the Purchaser or Tenant and its interest in the Works, require that the Contractor within 14 days from receipt of that notice enter into a Collateral Warranty with such Purchaser or Tenant in the form CWa/P&T, completed in accordance with the relevant Rights Particulars.*

DOI: 10.1201/9781003485919-61

Clause 7.9 relates to P&T Rights, which are conferred by the Contractor on Purchasers and Tenants identified in the Rights Particulars, by way of collateral warranties (rather than as third party rights, which are considered above in clause 7.7 (Third Party Rights from Contractor: Rights for Purchasers and Tenants)).

The Employer may notify the Contractor, identifying both (i) the Purchaser or Tenant, and (ii) their interest in the Works, that it requires the Contractor to enter into a Collateral Warranty with that Purchaser or Tenant, within 14 days from receipt of that notice.

The form of Collateral Warranty is that published by the JCT (reference CWa/P&T), and it must be completed in accordance with the relevant Rights Particulars.

Resource 240 page 553	Notice: Employer to Contractor	A notice from the Employer to the Contractor to enter into a Collateral Warranty with a Purchaser or Tenant.

Clause 7.10: Contractor's Warranty – Funder

7.10 *Where the Rights Particulars state that the Contractor shall confer Funder Rights on a Funder by way of collateral warranty, the Employer may by notice to the Contractor require that the Contractor within 14 days from receipt of the Employer's notice enter into a Collateral Warranty with the Funder in the form CWa/F, completed in accordance with the relevant Rights Particulars.*

Clause 7.10 relates to Funder Rights, which are conferred by the Contractor on Funders identified in the Rights Particulars, by way of collateral warranties (rather than as third party rights, which are considered above in clause 7.8 (Third Party Rights from Contractor: Rights for a Funder)).

The Employer may notify the Contractor that it requires the Contractor to enter into a Collateral Warranty with the Funder, within 14 days from receipt of that notice.

The form of Collateral Warranty is that published by the JCT (reference CWa/F), and it must be completed in accordance with the relevant Rights Particulars.

Recommendation: P&T Rights and Funder Rights

The Contractor is encouraged to maintain a tracker of collateral warranties requested and provided, as their provision is often a precondition of practical completion, and sometimes payment.

Resource 241 page 553	Notice: Employer to Contractor	A notice from the Employer to the Contractor to enter into a Collateral Warranty with a Funder.
Resource 242 page 554	Tracker: Contractor or Employer	A tracker template to record P&T Rights and Funder Rights, by way of third party rights or collateral warranty.

Market practice: Condition precedent to payment

Clause 7.9 and clause 7.10 are often amended to the effect that the provision of the collateral warranties is a condition precedent to the Employer's obligation to make payment to the Contractor.

Third Party Rights and Collateral Warranties from Sub-Contractors

Overview

Clause 7.7 to clause 7.10 deal with third party rights and collateral warranties from the Contractor.

This sub-section relates to the grant of third party rights or the provision of collateral warranties from the Contractor's sub-contractors, in favour of Purchasers, Tenants, Funders, and also the Employer.

These third party rights and collateral warranties allow those beneficiaries to enforce the terms of the sub-contract between the Contractor and the sub-contractor, even though they are not a party to it. This may be helpful, for example, in the event that the Contractor becomes '*Insolvent*', as defined in clause 8.1 (Meaning of insolvency).

Key provisions

- clause 7.11 (Third Party Rights and Collateral Warranties from Sub-Contractors)

Clause 7.11: Third Party Rights and Collateral Warranties from Sub-Contractors

7.11 *Where the Rights Particulars state that a sub-contractor shall confer third party rights on a Purchaser, Tenant or Funder and/or the Employer or execute and deliver a Collateral Warranty in favour of such person:*

Clause 7.11 relates to sub-contractors, and applies where the Rights Particulars state that they must either (i) confer third party rights on, or (ii) execute and deliver a Collateral Warranty in favour of, a Purchaser, Tenant, Funder, and/or the Employer.

DOI: 10.1201/9781003485919-62

7.11.1 *the Contractor shall comply with the Contract Documents as to the obtaining of such rights or warranties including:*

7.11.1.1 *on receipt of notice from the Employer identifying in each case the sub-contractor, type of right or warranty and beneficiary, promptly giving notice under clause 2.26.3 or, where appropriate, 2.26.4 of the JCT Design and Build Sub-Contract Conditions or other equivalent sub-contract condition to each sub-contractor identified in the Employer's notice; and*

7.11.1.2 *in the case of each Collateral Warranty specified in the Employer's notice and within 21 days of receipt of that notice, taking such steps as are required to obtain each warranty, promptly forwarding the executed document to the Employer or as the Employer may direct and, where Collateral Warranty SCWa/F is required, having itself also executed and delivered the document;*

Clause 7.11.1 sets out the Contractor's obligations in this regard:

- **Third party rights:** following receipt of a notice from the Employer to the Contractor, which identifies (i) the sub-contractors, (ii) the type of right or warranty, and (iii) the beneficiary, the Contractor must promptly give notice to the sub-contractor under (i) clause 2.26.3 or, where appropriate, clause 2.26.4 of the JCT Design & Build Sub-Contract Conditions, or (ii) other equivalent sub-contract condition, to each sub-contractor identified in the notice.
- **Collateral warranties:** if the Employer specifies that Collateral Warranties are required, the Contractor must obtain and forward the executed Collateral Warranties to the Employer (or as the Employer may otherwise direct) within 21 days from receipt of the Employer's notice. Where the JCT Sub-Contractor Collateral Warranty for a Funder (reference SCWa/F) is required, the Contractor must also execute and deliver the document.

Recommendation: Form of Sub-Contract

The Contractor should ensure that each of its relevant sub-contracts complies with the requirements of clause 7.11, to ensure that the sub-contractors' obligations to confer third party rights or collateral warranties in favour of Purchasers, Tenants, Funders, and the Employer align with the Contractor's corresponding obligations under clause 7.11.

The JCT Design and Build Sub-Contract Agreement (reference DBSub/A 2024) and Design and Build Sub-Contract Conditions (reference DBSub/A 2024) already provide for this.

Resource 243 page 554	Notice: Employer to Contractor	A notice from the Employer to the Contractor requesting third party rights or collateral warranties from a sub-contractor.
Resource 244 page 555	Correspondence: Contractor to Employer	Confirmation or provision by the Contractor to the Employer of third party rights or a collateral warranty from a sub-contractor.

7.11.2 *any amendment to the form of any third party rights or collateral warranty proposed by a sub-contractor shall require approval by both the Contractor and the Employer;*

Clause 7.11.2 provides that if the sub-contractor wishes to amend the form of any third party rights or collateral warranty, it must obtain the approval of both the Contractor and the Employer.

Resource 245 page 555	Approval given/ withheld: Contractor or Employer to sub-contractor	Approval given/withheld to the sub-contractor to amend the form of third party rights or collateral warranty.

7.11.3 *in the case of vested third party rights, the Contractor shall not without the consent of each beneficiary in which those rights have been vested:*

7.11.3.1 *agree any amendment or variation to the express terms of clause 2.26, clause 2.27 or Schedule 5 (Third Party Rights) of the JCT Design and Build Sub-Contract Conditions or other equivalent conditions of the sub-contract; or*

7.11.3.2 *where such beneficiary is the Employer or a Funder, agree to rescind the sub-contract.*

Clause 7.11.3 relates to third party rights which have already vested in a particular beneficiary.

In such circumstances, the Contractor requires the beneficiary's consent to:

- agree any amendment or variation to the following express terms of the JCT Design and Build Sub-Contract Conditions (reference JCT DBSub/C 2024):
  - clause 2.26 (Third Party Rights and Collateral Warranties etc.: Particulars and notices);
  - clause 2.27 (Third Party Rights and Collateral Warranties etc.: Rights of termination, variation etc.; and
  - Schedule 5 (Third Party Rights),

  or other equivalent conditions of the sub-contract; or
- agree to rescind (or cancel or 'set aside') the sub-contract, in circumstances where the beneficiary is either the Employer or a Funder.

Resource 246 page 556	Request: Contractor to beneficiary	A request from the Contractor to a beneficiary to amend, vary, or rescind the sub-contract.
Resource 247 page 557	Consent given/ withheld: Employer to Contractor	Consent given/withheld by the Employer to the Contractor to amend the sub-contract.
Resource 248 page 557	Tracker: Contractor and Employer	A tracker template to record third party rights and collateral warranties from sub-contractors.

Section 8: Termination

Section overview

Whilst it will hopefully never be required, section 8 relates to the termination of the Contractor's employment under the Contract.

It provides several circumstances in which:

- the Employer may terminate the Contractor's employment;
- the Contractor may terminate its employment; and
- either party may terminate the Contractor's employment.

It also sets out the consequences of any such termination. (As stated, it is the Contractor's *employment* under the Contract that is terminated, rather than the Contract itself, in order to provide for these consequences.)

If any of the relevant circumstances arise, these provisions should be considered and administered very carefully, as termination is known to be fertile ground for disputes, particularly in the context of a potential breakdown in the relationship between the Parties.

Sub-sections

- General (clause 8.1 – clause 8.3)
- Termination by Employer (clause 8.4 – clause 8.8)
- Termination by Contractor (clause 8.9 – clause 8.10)
- Termination by either Party and regulation 73(1)(a) of the PC Regulations (clause 8.11)
- Consequences of Termination under clauses 8.9 to 8.11, etc. (clause 8.12)
- Termination Payment – final date, notices and amount (clause 8.13)

DOI: 10.1201/9781003485919-64

General

Overview

This sub-section sets out a number of preliminary matters to assist with the interpretation and administration of the remainder of section 8, including:

- the definition of '*Insolvency*', which will be particularly relevant in respect of clause 8.5 (Termination by Employer: Insolvency of Contractor) and clause 8.10 (Termination by Contractor: Insolvency of Employer);
- notices required under section 8, including their use, effect, and how they should be served;
- that certain rights under section 8 will not affect any other rights and remedies of the Parties; and
- the possibility that the Contractor's employment under the Contract may be reinstated, if it has been terminated.

Key provisions

- clause 8.1 (Meaning of insolvency)
- clause 8.2 (Notices under section 8)
- clause 8.3 (Other rights, reinstatement)

DOI: 10.1201/9781003485919-65

Clause 8.1: Meaning of insolvency

8.1 *For the purposes of these Conditions:*

8.1.1 *a company becomes Insolvent:*

8.1.1.1 *when it enters administration within the meaning of Schedule B1 to the Insolvency Act 1986;*

8.1.1.2 *on the appointment of an administrative receiver or a receiver or manager of its property under Chapter I of Part III of that Act, or the appointment of a receiver under Chapter II of that Part;*

8.1.1.3 *on the passing of a resolution for voluntary winding-up without a declaration of solvency under section 89 of that Act; or*

8.1.1.4 *on the making of a winding-up order under Part IV or V of that Act.*

8.1.2 *a partnership becomes Insolvent:*

8.1.2.1 *on the making of a winding-up order against it under any provision of the Insolvency Act 1986 as applied by an order under section 420 of that Act; or*

8.1.2.2 *when sequestration is awarded on the estate of the partnership under section 22 of the Bankruptcy (Scotland) Act 2016 or the partnership grants a trust deed for its creditors.*

8.1.3 *an individual becomes Insolvent:*

8.1.3.1 *on the making of a bankruptcy order against them under Part IX of the Insolvency Act 1986; or*

8.1.3.2 *on the sequestration of their estate under the Bankruptcy (Scotland) Act 2016 or when they grant a trust deed for their creditors.*

8.1.4 *a person also becomes Insolvent if:*

8.1.4.1 *it enters into an arrangement, compromise or composition in satisfaction of its debts (excluding a scheme of arrangement as a solvent company for the purposes of amalgamation or reconstruction);*

8.1.4.2 *(in the case of a partnership) each partner is the subject of an individual arrangement or any other event or proceedings referred to in this clause 8.1;*

8.1.4.3 *(in the case of a company) a moratorium pursuant to Part A1 of the Insolvency Act 1986 comes into force with respect to it; or*

8.1.4.4 *(in the case of a company) an order is made sanctioning a compromise or arrangement pursuant to Part 26A of the Companies Act 2006 with respect to it.*

Each of clauses 8.1.1 to 8.1.4 also includes any analogous arrangement, event or proceedings in any other jurisdiction.

Clause 8.1 provides a detailed definition of the term '*Insolvent*'.

This definition will be particularly relevant for two key provisions considered below, namely:

- clause 8.5 (Termination by Employer: Insolvency of Contractor); and
- clause 8.10 (Termination by Contractor: Insolvency of Employer).

It is categorised as follows:

- clause 8.1.1 relates to companies (along with clause 8.1.4.3 and clause 8.1.4.4);
- clause 8.1.2 relates to partnerships (along with clause 8.1.4.2);
- clause 8.1.3 relates to individuals; and
- clause 8.1.4 relates to '*persons*' which, in accordance with clause 1.4.4 (Headings, references to persons, legislation etc.), includes individuals, firms, partnerships, companies, and any other body corporate.

The classification of the Contractor and the Employer will therefore determine which of the sub-clauses within clause 8.1 will apply to each of them.

The final paragraph of the clause further expands the definition to include any equivalent or similar '*arrangement, event or proceedings*' in any other jurisdiction.

Recommendation: Specialist advice

Given the complexity of this area, and its underlying legislation, the Parties are encouraged to take advice from a specialist insolvency practitioner, when determining whether a Party is Insolvent under clause 8.1, or for any other matters relating to insolvency.

Clause 8.2: Notices under section 8

8.2.1 Notice of termination of the Contractor's employment shall not be given unreasonably or vexatiously.

8.2.2 Such termination shall take effect on receipt of the relevant notice.

8.2.3 Each notice referred to in this section, except for the notices referred to in clause 8.13, shall be given in accordance with clause 1.7.4.

Clause 8.2 sets out three separate matters, all of which relate to notices under section 8:

- Clause 8.2.1 requires that any notice of termination of the Contractor's employment under the Contract, whether given by the Employer or the Contractor, must not be given unreasonably, or vexatiously (in other words, in a manner intended to cause annoyance to the other Party). These are sound principles for the giving of any notice under the Contract, and align with the principle in Article 3 (Collaborative working), but it is noteworthy that it is expressly stated in respect of termination, likely due to the heightened possibility of a breakdown in the relationship between the Parties.

- Clause 8.2.2 simply provides that the termination of the Contractor's employment under the Contract will take effect when the relevant notice is received by either the Contractor or the Employer (depending on which provision below applies).
- Clause 8.2.3 requires that, with the exception of the notices referred to in clause 8.13 (Termination Payment – final date, notices and amount), *all* notices in section 8 must be given in accordance with clause 1.7.4 (Notices and other communications). This applies to many different notices throughout section 8, across several provisions, but no further reminder is given.

Clause 8.3: Other rights, reinstatement

8.3.1 *The provisions of clauses 8.4 to 8.7 are without prejudice to any other rights and remedies of the Employer. The provisions of clauses 8.9 and 8.10, and (in the case of termination under either of those clauses) the provisions of clause 8.12, are without prejudice to any other rights and remedies of the Contractor.*

As considered further below, clause 8.4 to clause 8.7 (Termination by Employer) provide certain rights and remedies to the Employer. They entitle the Employer to terminate the Contractor's employment under the Contract in certain circumstances, and provide for the consequences of such termination.

Similarly, clause 8.9 and clause 8.10 (Termination by Contractor), and (as it relates to clause 8.9 and clause 8.10) clause 8.12 (Consequences of Termination under clauses 8.9 to 8.11, etc.) provide certain rights and remedies to the Contractor. They entitle the Contractor to terminate its employment under the Contract in certain circumstances, and provide for the consequences of such termination.

Clause 8.3.1 provides that these contractual rights will not limit or adversely affect any other rights and remedies available to the Parties in law.

A key example is the ability of either Party to terminate the Contractor's employment under the Contract for repudiatory breach. In other words, a breach of the Contract so significant that it goes to the heart of the agreement between the Parties (even though it is not expressly identified in the clauses above), and which gives the innocent Party the right to treat the Contract as having been brought to an end, along with the right to bring a claim for damages against the other Party.

8.3.2 *Irrespective of the grounds of termination, the Contractor's employment may at any time be reinstated if and on such terms as the Parties agree.*

Clause 8.3.2 provides that, no matter the grounds of termination (and whether by the Employer or the Contractor), the Contractor's employment under the Contract may be reinstated, on whichever terms as the Parties may agree.

Resource 249 page 558	Correspondence: Contractor or Employer to the other Party	Confirmation of the reinstatement of the Contractor's employment under the Contract.

Termination by Employer

Overview

This sub-section sets out a number of circumstances where the Employer may terminate the Contractor's employment under the Contract, and the consequences of such a termination.

These are circumstances where the Contractor is in some way at fault, whether by way of a breach of the Contract, becoming Insolvent, or failing to comply with legislation relating to bribery and corruption.

Key provisions

- clause 8.4 (Default by Contractor)
- clause 8.5 (Insolvency of Contractor)
- clause 8.6 (Corruption and regulation 73(1)(b) of the PC Regulations)
- clause 8.7 (Consequences of termination under clauses 8.4 to 8.6)
- clause 8.8 (Employer's decision not to complete the Works)

Clause 8.4: Default by Contractor

8.4.1 *If, before practical completion of the Works, the Contractor:*

8.4.1.1 *without reasonable cause wholly or substantially suspends the carrying out of the Works; or*

8.4.1.2 *fails to proceed regularly and diligently with the performance of its obligations under this Contract; or*

8.4.1.3 *refuses or neglects to comply with a notice or instruction from the Employer requiring it to remove any work, materials or goods not in accordance with this Contract and by such refusal or neglect the Works are materially affected; or*

8.4.1.4 *fails to comply with clause 3.3 or 7.1; or*

8.4.1.5 *fails to comply with clause 3.16,*

the Employer may give to the Contractor a notice specifying the default or defaults (a 'specified' default or defaults).

DOI: 10.1201/9781003485919-66

Clause 8.4.1 sets out the following circumstances occurring before practical completion of the Works, described as '*specified*' defaults, which may entitle the Employer to terminate the Contractor's employment under the Contract:

- The Contractor suspends the carrying out of the Works, whether entirely or substantially, without having reasonable grounds for doing so (such as the Contractor's entitlement to suspend for non-payment, under clause 4.11 (Contractor's right of suspension)).
- The Contractor fails to proceed '*regularly and diligently*' (in other words, properly) with the performance of its obligations under the Contract. This is very broadly drafted, and in the context of a complex project, with interrelating obligations on the Parties, potentially difficult to demonstrate.
- The Contractor either refuses or neglects to comply with a notice or instruction from the Employer requiring the Contractor to remove work, materials or goods which are not in accordance with the Contract, and as a consequence, the Works are '*materially affected*'.
- The Contractor fails to comply with clause 3.3 (Consent to sub-contracting), for example, by failing to obtain the Employer's consent to sub-contract any part of the design or the carrying out of the Works.
- The Contractor fails to comply with clause 7.1 (Assignment: General), for example, by attempting to assign the Contract (or any of the Contractor's rights under the Contract), without first obtaining the Employer's consent.
- The Contractor fails to comply with clause 3.16 (CDM Regulations and Part 2A of the Building Regulations), including (where applicable) in its capacity as (i) Principal Designer or Principal Contractor for the purposes of the CDM Regulations, and (ii) Principal Designer or Principal Contractor for the purposes of the Building Regulations.

In any of these circumstances, clause 8.4.1 entitles the Employer to notify the Contractor of the relevant specified defaults.

Resource 250 page 558	Notice: Employer to Contractor	A notice from the Employer to the Contractor of a specified default.
Resource 251 page 559	Reply: Contractor to Employer	A reply from the Contractor to the Employer's notice of a specified default.

Market practice: Further termination events

Clause 8.4.1 is often expanded to increase the number of potential specified defaults, for example:

- a breach of a specific provision of the Contract, such as a failure by the Contractor to provide a performance bond or parent company guarantee in accordance with clause 7.3 (Performance Bonds and Guarantees); or
- a more general provision, such as a 'material' breach of the Contract by the Contractor.

8.4.2 *If the Contractor continues a specified default for 14 days from receipt of the notice under clause 8.4.1, the Employer may on, or within 21 days from, the expiry of that 14 day period by a further notice to the Contractor terminate the Contractor's employment under this Contract.*

Clause 8.4.2 then sets out when the Employer may terminate the Contractor's employment under the Contract, namely where:

- the Employer has notified the Contractor of a specified default, in accordance with clause 8.4.1;
- however, the Contractor has continued with this specified default for 14 days from receipt of the Employer's notice.

In such circumstances, the Employer may then give another notice to the Contractor, this time terminating the Contractor's employment under the Contract. If the Employer wishes to give this further notice, it must do so within 21 days from the expiry of the initial 14-day period (which begins from receipt by the Contractor of the Employer's notice under clause 8.4.1).

Resource 252 page 560	Notice: Employer to Contractor	A notice of termination from the Employer to the Contractor, following the continuation of a specified default.

8.4.3 *If the Employer does not give the further notice referred to in clause 8.4.2 (whether as a result of the ending of any specified default or otherwise) but the Contractor repeats a specified default (whether previously repeated or not), then, upon or within a reasonable time after such repetition, the Employer may by notice to the Contractor terminate that employment.*

Clause 8.4.3 gives the Employer a further opportunity to terminate the Contractor's employment under the Contract, in the case of repeated specified defaults, namely where:

- the Employer has notified the Contractor of a specified default, in accordance with clause 8.4.1 (considered above);
- however, the Employer does not give the Contractor a further notice in accordance with clause 8.4.2 (for any reason, including where the specified default has ended); and
- the Contractor then repeats a specified default, whether or not that specified default was previously repeated.

In such circumstances, the Employer may then give another notice to the Contractor, this time terminating the Contractor's employment under the Contract.

If the Employer wishes to give this further notice, it must do so within '*a reasonable time*' after the repetition of the specified default.

Resource 253 page 560	Notice: Employer to Contractor	A notice of termination from the Employer to the Contractor, following the repetition of a specified default.

Clause 8.5: Insolvency of Contractor

8.5.1 If the Contractor is Insolvent, the Employer may at any time by notice to the Contractor terminate the Contractor's employment under this Contract.

Clause 8.5.1 entitles the Employer to terminate the Contractor's employment under the Contract, in the event that the Contractor becomes '*Insolvent*', as defined by clause 8.1 (Meaning of insolvency).

In such circumstances, the Employer may give a notice to the Contractor terminating the Contractor's employment under the Contract. This notice may be given at any time following the Contractor becoming Insolvent.

Resource 254 page 561	Notice: Employer to Contractor	A notice of termination from the Employer to the Contractor, following the Contractor becoming Insolvent.

8.5.2 *The Contractor shall immediately notify the Employer if it makes any proposal, gives notice of any meeting or becomes the subject of any proceedings or appointment relating to any of the matters referred to in clause 8.1.*

Clause 8.5.2 requires the Contractor to notify the Employer immediately of certain circumstances relating to the Contractor's insolvency.

In particular, the Contractor must notify the Employer if it:

(i) makes any proposal,
(ii) gives notice of any meeting, or
(iii) becomes the subject of any proceedings or appointment,

relating to any of the insolvency matters referred to in clause 8.1 (Meaning of insolvency).

Resource 255 page 561	Notice: Contractor to Employer	A notice from the Contractor to the Employer relating to the Contractor's insolvency.

8.5.3 *As from the date the Contractor becomes Insolvent, whether or not the Employer has given such notice of termination:*

8.5.3.1 *clauses 8.7.3 to 8.7.6 and (if relevant) clause 8.8 shall apply as if such notice had been given;*

8.5.3.2 *the Contractor's obligations under Article 1 and these Conditions to carry out and complete the Works shall be suspended; and*

8.5.3.3 *the Employer may take reasonable measures to ensure that the site, the Works and Site Materials are adequately protected and that such Site Materials are retained on site; the Contractor shall allow and shall not hinder or delay the taking of those measures.*

Clause 8.5.3 sets out the following three consequences following the Contractor becoming Insolvent (whether or not the Employer has notified the Contractor of the termination of its employment under the Contract, in accordance with clause 8.5.1):

- Clause 8.7.3 to clause 8.7.6 (Termination by Employer: Consequences of termination under clauses 8.4 to 8.6) and, if relevant, clause 8.8 (Employer's decision not to complete the Works), will apply, as if the Employer *had* notified the Contractor of the termination of its employment under the Contract, in accordance with clause 8.5.1. (These provisions are considered further below.)

- The Contractor's obligations under both (i) Article 1 (Contractor's obligations), namely to complete the design for the Works and carry out and complete the construction of the Works, in accordance with the Contract Documents, and (ii) the Conditions more generally, will be suspended.
- The Employer will be entitled to take '*reasonable measures*' to ensure that each of (i) the site, (ii) the Works, and (iii) any Site Materials, are adequately protected, and that any such Site Materials are kept on site. Furthermore, the Contractor must not obstruct or delay the Employer in this regard.

Recommendation: Record keeping for protection measures

In accordance with clause 8.5.3.3, the Employer may take '*reasonable measures*' to ensure that the site, the Works, and Site Materials are adequately protected, and that such Site Materials are kept on site.

The Employer should therefore keep records of those reasonable measures and their associated costs, as these will form part of the calculation of the amount due on termination, in accordance with clause 8.7.6.1 (Consequences of termination under clauses 8.4 to 8.6).

Clause 8.6: Corruption and regulation 73(1)(b) of the PC Regulations

8.6 *The Employer shall be entitled by notice to the Contractor to terminate the Contractor's employment under this or any other contract with the Employer if, in relation to this or any other such contract, the Contractor or any person employed by it or acting on its behalf shall have committed an offence under the Bribery Act 2010, or, where the Employer is a Local or Public Authority, shall have given any fee or reward the receipt of which is an offence under sub-section (2) of section 117 of the Local Government Act 1972, or, where this Contract is one to which regulation 73(1) of the PC Regulations applies, the circumstances set out in regulation 73(1)(b) of the PC Regulations apply.*

Clause 8.6 entitles the Employer to terminate the Contractor's employment under the Contract in the event of bribery or corruption by either (i) the Contractor, or (ii) any person employed by the Contractor or acting on its behalf, in respect of any of the following:

- If they commit an offence under the Bribery Act 2010.
- Where the Employer is a Local or Public Authority, if they give a '*fee or reward*', the receipt of which constitutes an offence under sub-section (2) of section 117

of the Local Government Act 1972. Sub-section (2) states: '*An officer of a local authority shall not, under colour of his office or employment, accept any fee or reward whatsoever other than his proper remuneration.*'

- Where the Contract is one to which regulation 73(1) of the PC Regulations applies, if the circumstances set out in regulation 73(1)(b) of the PC Regulations apply. Regulation 73(1) states: '*Contracting authorities* [Employers] *shall ensure that every public contract which they award contains provisions enabling the contracting authority* [Employers] *to terminate the contract where: ... (b) the contractor has, at the time of contract award, been in one of the situations referred to in regulation 57(1), including as a result of the application of regulation 57(2), and should therefore have been excluded from the procurement procedure.*' The situations referred to in regulation 57(1) relate to offences concerning bribery and corruption, and are mandatory grounds for excluding the Contractor from the procurement procedure.

In such circumstances, the Employer may give a notice to the Contractor terminating the Contractor's employment under either (i) the Contract, or (ii) any other contract between the Contractor and the Employer.

Resource 256 page 562	Notice: Employer to Contractor	A notice of termination from the Employer to the Contractor, relating to corruption and regulation 73(1)(b) of the PC Regulations.

Clause 8.7: Consequences of termination under clauses 8.4 to 8.6

> *8.7 If the Contractor's employment is terminated under clause 8.4, 8.5 or 8.6:*

Clause 8.7 sets out a number of consequences, if the Contractor's employment under the Contract is terminated under either:

- clause 8.4 (Default by Contractor);
- clause 8.5 (Insolvency of Contractor); or
- clause 8.6 (Corruption and regulation 73(1)(b) of the PC Regulations).

> *8.7.1 the Employer may employ and pay other persons to carry out and complete the Works and to make good any defects of the kind referred to in clause 2.35, or may do so itself, and the Employer and such other persons may enter upon and take possession of the site and the Works and (subject to obtaining any necessary third party consents) may use all temporary buildings, plant, tools, equipment and Site Materials for those purposes;*

Clause 8.7.1 gives the Employer broad powers in respect of the carrying out and completion of the Works. In particular, it entitles the Employer to:

- employ and pay others to (i) carry out and complete the Works, and (ii) to make good any defects of the kind referred to in clause 2.35 (Schedules of defects and instructions);
- undertake the work above itself, rather than employing and paying others to do so;
- along with any others referred to above, enter and take possession of the site and the Works; and
- use all temporary buildings, plant, tools, equipment, and Site Materials, provided that the Employer has obtained any necessary consents from the relevant third parties.

Recommendation: Record keeping for completing the Works and making good defects

In accordance with clause 8.7.1, if the Contractor's employment is terminated under clause 8.4, clause 8.5, or clause 8.6, the Employer may employ and pay other persons to carry out and complete the Works, and to make good any defects of the kind referred to in clause 2.35 (Schedules of defects and instructions).

The Employer should therefore keep records of these works and the associated costs, as these will form part of the calculation of the amount due on termination, in accordance with clause 8.7.6.1 (Consequences of termination under clauses 8.4 to 8.6).

8.7.2 the Contractor shall:

8.7.2.1 when required in writing by the Employer to do so (but not before), remove or procure the removal from the Works of any temporary buildings, plant, tools, equipment, goods and materials belonging to the Contractor or Contractor's Persons;

8.7.2.2 provide the Employer with copies of all Contractor's Design Documents then prepared, whether or not previously provided;

8.7.2.3 if so required by the Employer within 14 days of the date of termination, assign (so far as assignable and so far as it may lawfully be required to do so) to the Employer, without charge, the benefit of any agreement for the supply of materials or goods and/or for the execution of any work for the purposes of this Contract;

Clause 8.7.2 imposes a number of obligations on the Contractor, to reflect the fact that it will no longer be involved in carrying out and completing the Works, and to assist the Employer with the transition. In particular, the Contractor must:

- if required by the Employer (but not before the Employer makes the written request), remove any temporary buildings, plant, tools, equipment, goods, and materials belonging to the Contractor (or any Contractor's Persons) from the Works, in accordance with clause 8.7.2.1;
- give the Employer copies of all Contractor's Design Documents prepared by or on behalf of the Contractor, whether or not these have been previously provided, in accordance with clause 8.7.2.2; and
- if required by the Employer within 14 days of the date of termination of the Contractor's employment under the Contract, assign to the Employer (so far as it is possible to do so) the benefit of any agreement for either (i) the supply of materials or goods, or (ii) the execution of any work, for the purposes of the Contract, in accordance with clause 8.7.2.3. (However, footnote [58] of the Contract advises that clause 8.7.2.3 may not be effectual, in the case of the Contractor's insolvency.)

Resource 257 page 562	Requirement: Employer to Contractor	A requirement by the Employer for the Contractor to remove from the Works temporary buildings, plant, tools, equipment, goods and materials belonging to the Contractor.
Resource 258 page 563	Correspondence: Contractor to Employer	The provision by the Contractor to the Employer of all Contractor's Design Documents then prepared.
Resource 259 page 563	Requirement: Employer to Contractor	A requirement by the Employer for the Contractor to assign agreements for the supply of materials or goods, and/or for the execution of any work, for the purposes of the Contract.

8.7.3 no further sum shall become due to the Contractor under this Contract other than any amount that may become due to it under clause 8.7.6 or 8.8.3 and the Employer need not pay any sum that has already become due either:

8.7.3.1 insofar as the Employer has given or gives a Pay Less Notice under clause 4.9.5; or

8.7.3.2 if the Contractor, after the last date upon which such notice could have been given by the Employer in respect of that sum, has become insolvent within the meaning of clauses 8.1.1 to 8.1.3;

Clause 8.7.3 places a number of restrictions on the payment of further sums to the Contractor, to reflect the fact that the Contractor is responsible for the termination of its employment under the Contract.

It provides that the only further sums which may become due to the Contractor under the Contract will be in accordance with either:

- clause 8.7.6 (Consequences of termination under clauses 8.4 to 8.6); or
- clause 8.8.3 (Employer's decision not to complete the Works).

(Both clauses are considered further below.)

Furthermore, clause 8.7.3 provides that the Employer is not obliged to pay any sum which has *already* become due if either:

- the Employer gives a Pay Less Notice, in accordance with clause 4.9.5 (Interim and final payments – final date and amount); or
- the Contractor has become insolvent (as defined by clause 8.1.1 to clause 8.1.3 (Meaning of insolvency)) after the last date when the Employer could have given a Pay Less Notice in respect of that sum.

8.7.4 upon completion of the Works and the making good of defects in them (or of instructions otherwise, as referred to in clause 2.35) ('completion'), the Employer shall forthwith notify the Contractor of the date of completion and such completion shall be deemed for the purposes of this clause 8.7 to have taken place on the date so notified;

Clause 8.7.4 requires the Employer to immediately notify the Contractor of the date of '*completion*', which is defined as the completion of both (i) the Works, and (ii) the making good of defects in them (or of any instructions otherwise, as referred to in clause 2.35 (Schedules of defects and instructions)).

It further states that '*completion*' will be deemed to have taken place on the date notified by the Employer, for the purposes of clause 8.7 (and clause 8.7.5 in particular, considered below).

Resource 260 page 564	Notice: Employer to Contractor	A notice from the Employer to the Contractor of the completion of the Works and the making good of defects in them.

8.7.5 the due date for the Termination Payment under clause 8.7.6 shall be 2 months after the date of completion as referred to in clause 8.7.4;

Clause 8.7.5 provides the due date for the '*Termination Payment*', which for these purposes, is defined in clause 8.7.6 below.

It states that the due date will be 2 months after the date of '*completion*', as defined by clause 8.7.4 above.

8.7.6 the amount due on termination shall be calculated in accordance with the following amounts:

8.7.6.1 the amount of expenses properly incurred by the Employer, including those incurred pursuant to clause 8.7.1 and, where applicable, clause 8.5.3.3, and of any direct loss and/or damage caused to the Employer and for which the Contractor is liable, whether arising as a result of the termination or otherwise;

8.7.6.2 the amount of payments made to the Contractor; and

8.7.6.3 the total amount which would have been payable for the Works in accordance with this Contract,

and if the sum of the amounts referred to in clauses 8.7.6.1 and 8.7.6.2 exceeds the amount referred to in clause 8.7.6.3, the difference shall be an amount payable by the Contractor to the Employer or, if that sum is less, by the Employer to the Contractor (the 'Termination Payment').

Clause 8.7.6 sets out the definition and method of calculating the '*Termination Payment*' (in this case, for the purposes of clause 8.7).

The Termination Payment will be the amount due to either the Contractor or the Employer upon the termination of the Contractor's employment under the Contract. It is calculated as follows:

Calculation (a): clause 8.7.6.1 and clause 8.7.6.2

Firstly, add together the following:

- in accordance with clause 8.7.6.1, the amount of (i) expenses properly incurred by the Employer, including those incurred in accordance with clause 8.7.1 and, where applicable, clause 8.5.3.3 (Insolvency of Contractor), and (ii) any direct loss and/or damage caused to the Employer, for which the Contractor is liable, and whether that loss and/or damage has arisen as a result of the termination, or otherwise; and
- in accordance with clause 8.7.6.2, the amount of payments previously made to the Contractor.

Calculation (b): clause 8.7.6.3

Secondly, in accordance with clause 8.7.6.3, calculate the total amount which would have been payable for the Works, in accordance with the Contract.

Calculation (c)

Finally, calculate the difference between the sum in Calculation (a) and the amount in Calculation (b).

This is the Termination Payment.

If the sum in Calculation (a) is *greater* than the amount in Calculation (b), the difference is the amount payable by the Contractor to the Employer.

If the sum in Calculation (a) is *less* than the amount in Calculation (b), the difference is the amount payable by the Employer to the Contractor.

Clause 8.8: Employer's decision not to complete the Works

> 8.8.1 *If within the period of 6 months from the date of termination of the Contractor's employment the Employer decides not to have the Works carried out and completed, it shall forthwith notify the Contractor. If no such notification is given but within that 6 month period the Employer does not commence to make arrangements for such carrying out and completion, for the purposes of this clause 8.8 the Employer shall on and after the expiry of that 6 month period be treated as having decided not to have the Works completed.*

Clause 8.8 provides for the circumstances where the Contractor's employment under the Contract has been terminated, and the Employer decides that it does *not* wish to have the Works carried out and completed:

- If the Employer makes that decision within 6 months from the date of termination of the Contractor's employment under the Contract, it must immediately notify the Contractor.
- Alternatively, for the purposes of clause 8.8 (and clause 8.8.2 in particular), if the Employer does *not* notify the Contractor, but within that same 6-month period the Employer does not begin to make arrangements for the carrying out and completion of the Works, the Employer will be treated as having decided not to have the Works completed, upon the expiry of that 6-month period.

Resource 261 page 564	Notice: Employer to Contractor	A notice from the Employer to the Contractor of the Employer's decision not to complete the Works.

8.8.2 The due date for the Termination Payment under clause 8.8.3 shall be 2 months after the last day of the 6 month period referred to in clause 8.8.1 whether or not the Employer has given a notification under clause 8.8.1.

Clause 8.8.2 provides the due date for the '*Termination Payment*', which for these purposes, is defined in clause 8.8.3 below.

It states that the due date will be 2 months after the last day of the 6-month period referred to in clause 8.8.1, and whether or not the Employer has notified the Contractor under clause 8.8.1 of its decision not to have the Works carried out and completed.

8.8.3 The amount due on termination shall be calculated in accordance with the following amounts:

8.8.3.1 the total value of work properly executed at the date of termination or date on which the Contractor became Insolvent, ascertained in accordance with these Conditions as if that employment had not been terminated, together with any amounts due to the Contractor under these Conditions not included in such total value; and

8.8.3.2 the aggregate amount of any expenses properly incurred by the Employer and of any direct loss and/or damage caused to the Employer and for which the Contractor is liable, whether arising as a result of the termination or otherwise,

and after taking into account amounts previously paid to the Contractor under this Contract, if the amount referred to in clause 8.8.3.2 exceeds the amount referred to in clause 8.8.3.1, the difference shall be an amount payable by the Contractor to the Employer or, if the clause 8.8.3.2 amount is less, by the Employer to the Contractor (the 'Termination Payment').

Clause 8.8.3 provides the definition and method of calculating the '*Termination Payment*' (in this case, for the purposes of clause 8.8).

The Termination Payment will be the amount due to either the Contractor or the Employer upon the termination of the Contractor's employment under the Contract. It is calculated as follows:

Calculation (a): clause 8.8.3.1

Firstly, calculate the total value of (i) work properly executed, ascertained in accordance with the Conditions as if the Contractor's employment under the Contract had not been terminated, at either the date of termination, or the date on which the

Contractor became Insolvent, and (ii) any amounts due to the Contractor under the Conditions, which were not included in (i).

These are sums due to the Contractor.

Calculation (b): clause 8.8.3.2

Secondly, calculate the aggregate amount of (i) expenses properly incurred by the Employer, and (ii) any direct loss and/or damage caused to the Employer, for which the Contractor is liable, and whether that loss and/or damage has arisen as a result of the termination, or otherwise.

These are sums due to the Employer.

Calculation (c)

If the amount in Calculation (b) is *greater* than in Calculation (a), taking into account the amounts previously paid to the Contractor, the difference will be the amount payable by the Contractor to the Employer.

If the amount in Calculation (b) is *less* than in Calculation (a), again, taking into account the amounts previously paid to the Contractor, the difference will be the amount payable by the Employer to the Contractor.

This is the Termination Payment.

Termination by Contractor

Overview

This sub-section sets out a number of circumstances where the Contractor may terminate its employment under the Contract, and the consequences of such a termination.

These are circumstances where the Employer is in some way at fault, whether by way of a breach of the Contract, or becoming Insolvent.

Key provisions

- clause 8.9 (Default by Employer)
- clause 8.10 (Insolvency of Employer)

Clause 8.9: Default by Employer

8.9.1 If the Employer:

8.9.1.1 does not pay by the final date for payment the amount due to the Contractor in accordance with clause 4.9 and/or any VAT properly chargeable on that amount; or

8.9.1.2 fails to comply with clause 7.1; or

8.9.1.3 fails to comply with clause 3.16,

the Contractor may give to the Employer a notice specifying the default or defaults (a 'specified' default or defaults).

Clause 8.9.1 sets out the following circumstances, again described as '*specified*' defaults, which may entitle the Contractor to terminate its employment under the Contract:

- The Employer fails to pay to the Contractor by the final date for payment either (i) any amount due in accordance with clause 4.9 (Interim and final payments – final date and amount), or (ii) any VAT properly chargeable on that amount due.
- The Employer fails to comply with clause 7.1 (Assignment: General), for example, by attempting to assign the Contract (or any of the Employer's rights under the Contract), without first obtaining the Contractor's consent.

DOI: 10.1201/9781003485919-67

- The Employer fails to comply with clause 3.16 (CDM Regulations and Part 2A of the Building Regulations), including (where applicable) in its capacity as (i) Principal Designer (or Principal Contractor) for the purposes of the CDM Regulations, and (ii) Principal Designer (or Principal Contractor) for the purposes of the Building Regulations.

In any of these circumstances, clause 8.9.1 entitles the Contractor to notify the Employer of the relevant specified defaults.

Resource 262 page 565	Notice: Contractor to Employer	A notice from the Contractor to the Employer of a specified default.

8.9.2 *If after the Date of Possession (or after any deferred Date of Possession pursuant to clause 2.4) but before practical completion of the Works the carrying out of the whole or substantially the whole of the uncompleted Works is suspended for a continuous period of the length stated in the Contract Particulars by reason of any impediment, prevention or default, whether by act or omission, by the Employer or any Employer's Person, then, unless it is caused by the negligence or default of the Contractor or any Contractor's Person, the Contractor may give to the Employer a notice specifying the event or events (a 'specified' suspension event or events).*

Clause 8.9.2 provides the following further circumstances, described as '*specified*' suspension events, which may also entitle the Contractor to terminate its employment under the Contract:

- the carrying out of the whole (or substantially the whole) of the uncompleted Works is suspended for a continuous period, of the length stated against the reference to clause 8.9.2 in the Contract Particulars (for example, 2 months);
- that suspension take place between (i) the Date of Possession (or any deferred Date of Possession in accordance with clause 2.4 (Deferment of Possession)), and (ii) practical completion of the Works;
- that suspension is caused by reason of any '*impediment, prevention or default*', and whether by act or omission (in other words, a failure to do something), by either the Employer, or any Employer's Person; and
- that suspension is *not* caused by reason of the '*negligence or default*' of either the Contractor, or any Contractor's Person.

In such circumstances, clause 8.9.2 entitles the Contractor to notify the Employer of the relevant specified suspension events.

Resource 263 page 565	Notice: Contractor to Employer	A notice from the Contractor to the Employer of a specified suspension event.
Resource 264 page 566	Reply: Employer to Contractor	A reply from the Employer to the Contractor's notice of a specified default or a specified suspension event.

8.9.3 If a specified default or a specified suspension event continues for 14 days from the receipt of notice under clause 8.9.1 or 8.9.2, the Contractor may on, or within 21 days from, the expiry of that 14 day period by a further notice to the Employer terminate the Contractor's employment under this Contract.

Clause 8.9.3 sets out when the Contractor may terminate its employment under the Contract, namely where:

- the Contractor has notified the Employer of either (i) a specified default, in accordance with clause 8.9.1, or (ii) a specified suspension event, in accordance with clause 8.9.2;
- however, this specified default or specified suspension event has continued for 14 days from receipt of the Contractor's notice.

In such circumstances, the Contractor may give another notice to the Employer, this time terminating the Contractor's employment under the Contract.

If the Contractor wishes to give this further notice, it must do so within 21 days from the expiry of the initial 14-day period (which begins from receipt by the Employer of the Contractor's notice under clause 8.9.1 or clause 8.9.2).

Resource 265 page 567	Notice: Contractor to Employer	A notice of termination from the Contractor to the Employer, following the continuation of a specified default or a specified suspension event.

> 8.9.4 *If the Contractor for any reason does not give the further notice referred to in clause 8.9.3, but (whether previously repeated or not):*
>
> 8.9.4.1 *the Employer repeats a specified default; or*
>
> 8.9.4.2 *a specified suspension event is repeated for any period, such that the regular progress of the Works is or is likely to be materially affected thereby,*
>
> *then, upon or within a reasonable time after such repetition, the Contractor may by notice to the Employer terminate the Contractor's employment under this Contract.*

Clause 8.9.4 gives the Contractor a further opportunity to terminate its employment under the Contract, in the case of either repeated specified defaults or repeated specified suspension events (whether or not either of them were previously repeated), namely where:

- the Contractor has notified the Employer of either (i) a specified default, in accordance with clause 8.9.1, or (ii) a specified suspension event, in accordance with clause 8.9.2;
- however, the Contractor does not give the Employer the further notice in accordance with clause 8.9.3, for any reason;
- then either (i) the Employer repeats a specified default, or (ii) a specified suspension event is repeated, for any period, which will (or will likely) materially affect the regular progress of the works.

In such circumstances, the Contractor may give another notice to the Employer, this time terminating the Contractor's employment under the Contract.

If the Contractor wishes to give this further notice, it must do so within '*a reasonable time*' after the repetition of the specified default or specified suspension event.

The Employer should take care in this regard, following a recent decision by the Court of Appeal,[22] where it held that the Contractor's right to terminate under clause 8.9.4 extended to circumstances where the further notice in clause 8.9.3 had not been given, because the Contractor's right to give it had not arisen, on the basis that the Employer had remedied the previous specified default within the relevant time period.

In other words, repeated specified defaults (such as non-payment), even if the Employer remedies them within the relevant time period, may give the Contractor a right to terminate its employment under the Contract, in accordance with clause 8.9.4.

Resource 266 page 568	Notice: Contractor to Employer	A notice of termination from the Contractor to the Employer, following the repetition of a specified default or a specified suspension event.

[22] *Providence Building Services Limited v Hexagon Housing Association Limited* [2024] EWCA Civ 962

Clause 8.10: Insolvency of Employer

8.10.1 *If the Employer is Insolvent, the Contractor may by notice to the Employer terminate the Contractor's employment under this Contract;*

Clause 8.10.1 entitles the Contractor to terminate its employment under the Contract, in the event that the Employer becomes '*Insolvent*', as defined by clause 8.1 (Meaning of insolvency).

In such circumstances, the Contractor may give a notice to the Employer terminating the Contractor's employment under the Contract. This notice may be given at any time following the Employer becoming Insolvent.

Resource 267 page 569	Notice: Contractor to Employer	A notice of termination from the Contractor to the Employer, following the Employer becoming Insolvent.

8.10.2 *the Employer shall immediately notify the Contractor if it makes any proposal, gives notice of any meeting or becomes the subject of any proceedings or appointment relating to any of the matters referred to in clause 8.1;*

Clause 8.10.2 requires the Employer to notify the Contractor immediately of certain circumstances relating to the Employer's insolvency.

In particular, the Employer must notify the Contractor if it:

(i) makes any proposal,
(ii) gives notice of any meeting, or
(iii) becomes the subject of any proceedings or appointment,

relating to any of the insolvency matters referred to in clause 8.1 (Meaning of insolvency).

Resource 268 page 569	Notice: Employer to Contractor	A notice from the Employer to the Contractor relating to the Employer's insolvency.

8.10.3 as from the date the Employer becomes Insolvent, the Contractor's obligations under Article 1 and these Conditions to carry out and complete the Works shall be suspended.

Clause 8.10.3 provides that, from the date of the Employer becoming Insolvent, the Contractor's obligations to carry out and complete the Works under both (i) Article 1 (Contractor's obligations), and (ii) the Conditions, will be suspended.

Termination by either Party and regulation 73(1)(a) of the PC Regulations

Overview

This sub-section sets out a number of circumstances where either the Employer or the Contractor may terminate the Contractor's employment under the Contract.

These are primarily circumstances where neither the Employer nor the Contractor is at fault, but rather relate to 'neutral' events.

Key provisions

- clause 8.11 (Termination by either Party and regulation 73(1)(a) of the PC Regulations)

Clause 8.11: Termination by either Party and regulation 73(1)(a) of the PC Regulations

8.11.1 If, before practical completion of the Works, the carrying out of the whole or substantially the whole of the uncompleted Works is suspended for the relevant continuous period of the length stated in the Contract Particulars by reason of one or more of the following events:

8.11.1.1 force majeure;

8.11.1.2 Employer's instructions under clause 2.13, 3.9 or 3.10 issued as a result of the negligence or default of any Statutory Provider;

8.11.1.3 loss or damage to the Works occasioned by any risk covered by the Works Insurance Policy or by an Excepted Risk;

8.11.1.4 civil commotion or the use or threat of terrorism and/or the activities of the relevant authorities in dealing with such event or threat;

8.11.1.5 an epidemic:

DOI: 10.1201/9781003485919-68

8.11.1.5.1 *first occurring after the Base Date which affects the execution of the Works; or*

8.11.1.5.2 *first occurring before the Base Date whose effects change after the Base Date and any such change affects the execution of the Works*

by limiting the availability or use of labour or the availability of persons engaged in providing services for the Works where such labour and/or persons are necessary for the proper carrying out of the Works, or preventing the Contractor from, or delaying the Contractor in, securing such goods or materials or such services as are necessary for the proper carrying out of the Works;

8.11.1.6 *the occurrence after the Base Date of any of the following which affects the execution of the Works:*

8.11.1.6.1 *the passing into law of any statute, statutory instrument or other subordinate legislation, regulation or bye-law, whether to make a new law or change or repeal an existing law;*

8.11.1.6.2 *the exercise of any statutory power, except to the extent caused or contributed to by any default, whether by act or omission, of the Contractor or any Contractor's Person; or*

8.11.1.6.3 *the publication of any guidance,*

by the United Kingdom Government or any of the devolved administrations, or any Local or Public Authority or any equivalent authority governed by public law in any of the devolved administrations, or in the case of guidance by the Construction Leadership Council or its successor; or

8.11.1.7 *delay in receipt of any permission or approval for the purposes of Development Control Requirements necessary for the Works to be carried out or proceed, which delay the Contractor has taken all practicable steps to avoid or reduce,*

then either Party, subject to clause 8.11.2, may upon the expiry of that relevant period of suspension give notice to the other that, unless the suspension ceases within 7 days after the date of receipt of that notice, it may terminate the Contractor's employment under this Contract. Failing such cessation within that 7 day period, it may then by further notice terminate that employment.

Clause 8.11.1 sets out several circumstances (which are events beyond the reasonable control of the Parties) where either the Employer or the Contractor may terminate the Contractor's employment under the Contract.

They apply where, before practical completion of the Works, the carrying out of the whole (or substantially the whole) of the uncompleted Works is suspended, for the relevant continuous period of the length stated against the reference to clause 8.11.1.1 to clause 8.1.1.7 in the Contract Particulars.

The circumstances are as follows:

- force majeure;
- Employer's instructions issued as a result of the '*negligence or default*' of any Statutory Provider, under either (i) clause 2.13 (Notice of discrepancies etc.), (ii) clause 3.9 (Instructions requiring Changes), or (iii) clause 3.10 (Postponement of work);
- loss or damage to the Works caused by (i) any risk covered by the Works Insurance Policy, or (ii) an '*Excepted Risk*' (as defined in clause 6.8 (Insurance of the Works and Existing Structures: Related definitions));
- civil commotion, or the use or threat of terrorism, and/or the activities of the relevant authorities in dealing with them;
- epidemics either (i) first occurring *after* the Base Date, which affects the execution of the Works, or (ii) first occurring *before* the Base Date, but whose effects change *after* the Base Date, and any such change affects the execution of the Works, by:
  - limiting (i) the availability or use of labour, or (ii) the availability of persons engaged in providing services for the Works, where such labour or persons are necessary to properly carry out the Works; or
  - either (i) preventing the Contractor from, or (ii) or delaying the Contractor in, securing goods, materials, or services as are necessary to properly carry out the Works;
- the occurrence after the Base Date of any of the following, which affects the execution of the Works:
  - the passing into law of legislation, regulations, or bye-laws (whether to make a new law, or to change or repeal an existing law);
  - the exercise of any statutory power, except to the extent that the exercise of the power has been caused or contributed to by any default (whether by act or omission) of the Contractor, or any Contractor's Person; or
  - the publication of guidance,

by (i) the United Kingdom Government (or any devolved administration), or (ii) any Local or Public Authority (or any equivalent authority governed by public law in any devolved administration), or (iii) in the case of guidance, by the Construction Leadership Council, or its successor;

- the occurrence after the Base Date of the following, which affects the execution of the Works:
  - the passing into law of legislation, regulations, or bye-laws (whether to make a new law, or to change or repeal an existing law);
  - the exercise of any statutory power, except to the extent that the exercise of the power has been caused or contributed to by any default (whether by act or omission) of the Contractor, or any Contractor's Person; or
  - the publication of guidance,

by (i) the United Kingdom Government (or any devolved administration), or (ii) any Local or Public Authority (or any equivalent authority governed by public law in any devolved administration), or (iii) in the case of guidance, by the Construction Leadership Council, or its successor;

- delay in receipt of any permission or approval necessary for the Works to be carried out or to proceed, for the purposes of Development Control Requirements (thought to relate primarily to planning), provided that the Contractor has taken '*all practicable steps*' to avoid or reduce the delay.

In any of these circumstances, and upon the expiry of the period of suspension stated in the Contract Particulars (against the reference to clause 8.11.1.1 to clause 8.11.1.7), either the Employer or the Contractor may give a notice to the other Party, stating that, unless that suspension ceases within 7 days after the date of receipt of the notice, then it may terminate the Contractor's employment under the Contract.

(In the Contractor's case, this is subject to clause 8.11.2 below.)

If that suspension has *not* ceased within that 7-day period, then that Party may give another notice to the other Party, this time terminating the Contractor's employment under the Contract.

Resource 269 page 570	Notice: Contractor or Employer to the other Party	A notice of a potential termination of the Contractor's employment under the Contract, following a suspension of the carrying out of the uncompleted Works for a continuous period.
Resource 270 page 572	Notice: Contractor or Employer to the other Party	A notice of termination of the Contractor's employment under the Contract, following a suspension of the carrying out of the uncompleted Works for a continuous period.

8.11.2 The Contractor shall not be entitled to give notice under clause 8.11.1 in respect of the matter referred to in clause 8.11.1.3 where the loss or damage to the Works was caused by the negligence or default of the Contractor or any Contractor's Person.

Clause 8.11.2 prohibits the Contractor from giving a notice under clause 8.11.1 above, in respect of clause 8.11.1.3 (which relates to loss or damage to the Works caused by (i) any risk covered by the Works Insurance Policy, or (ii) an Excepted Risk), in circumstances where that loss or damage was caused by the '*negligence or default*' of the Contractor (or of any Contractor's Person).

8.11.3 Where this Contract is one to which regulation 73(1) of the PC Regulations applies the Employer shall be entitled by notice to the Contractor to terminate the Contractor's employment under this Contract where the grounds set out in regulation 73(1)(a) of the PC Regulations apply.

Clause 8.11.3 is relevant where regulation 73(1) of the PC Regulations applies to the Contract. (The '*PC Regulations*' are defined in clause 1.1 (Definitions) as the Public Contracts Regulations 2015.)

If so, it entitles the Employer to give a notice to the Contractor, terminating the Contractor's employment under the Contract, in circumstances where the grounds set out in regulation 73(1)(a) of the PC Regulations apply.

Regulation 73(1) states (emphasis added): '*Contracting authorities* [Employers] *shall ensure that every public contract which they award contains provisions enabling the contracting authority* [Employers] *to terminate the contract where: (a)* ***the contract has been subject to a substantial modification which would have required a new procurement procedure*** *in accordance with regulation 72(9)...*'

Resource 271 page 572	Notice: Employer to Contractor	A notice of termination from the Employer to the Contractor, relating to regulation 73(1) of the PC Regulations.

Consequences of Termination under clauses 8.9 to 8.11, etc.

Overview

This sub-section sets out a number of consequences, if the Contractor's employment under the Contract is terminated under either:

- clause 8.9 (Termination by Contractor: Default by Employer);
- clause 8.10 (Termination by Contractor: Insolvency of Employer);
- clause 8.11 (Termination by either Party and regulation 73(1)(a) of the PC Regulations);
- clause 6.11.2.2 (Terrorism Cover – non-availability – Employer's options), in the event that Terrorism Cover will cease and will no longer be available, or will only continue to be available with a reduction in the scope or level of such cover; or
- clause 6.14 (Loss or damage to Existing Structures – right of termination), in the event of material loss or damage to any of the Existing Structures.

Key provisions

- clause 8.12 (Consequences of Termination under clauses 8.9 to 8.11, etc.)

Clause 8.12: Consequences of Termination under clauses 8.9 to 8.11, etc.

8.12 If the Contractor's employment is terminated under any of clauses 8.9 to 8.11 or under clause 6.11.2.2 or 6.14:

As stated above, clause 8.12 applies where the Contractor's employment under the Contract is terminated under either clause 8.9, clause 8.10, clause 8.11, clause 6.11.2.2, or clause 6.14.

8.12.1 no further sums shall become due to the Contractor otherwise than in accordance with this clause 8.12;

DOI: 10.1201/9781003485919-69

Clause 8.12.1 provides that *only* sums in accordance with clause 8.12 (and in particular, clause 8.12.5) will become due to the Contractor, but not otherwise.

> *8.12.2 the Contractor shall:*
>
> *8.12.2.1 with all reasonable dispatch remove or procure the removal from the site of any temporary buildings, plant, tools and equipment belonging to the Contractor and Contractor's Persons and, subject to the provisions of clause 8.12.5, all goods and materials (including Site Materials); and*
>
> *8.12.2.2 provide to the Employer copies of the documents referred to in clause 2.37 then prepared;*

Clause 8.12.2 imposes a number of obligations on the Contractor, to reflect the fact that it will no longer be involved in carrying out and completing the Works, and to assist the Employer with the transition. In particular, the Contractor must:

- remove any temporary buildings, plant, tools, and equipment belonging to the Contractor (or any Contractor's Persons) from the site (reasonably quickly);
- subject to clause 8.12.5 (considered below), remove all goods and materials, including Site Materials, from the site (again, reasonably quickly); and
- give the Employer copies of the documents referred to in clause 2.37 (As-built Drawings) which the Contractor has already prepared.

Resource 272 page 573	Correspondence: Contractor to Employer	The provision of documents relating to clause 2.37 (As-built Drawings) from the Contractor to the Employer.

> *8.12.3 the Contractor shall not later than 2 months after the date of termination provide the Employer with all documents necessary for calculating the amounts referred to in clause 8.12.5. Not later than 3 months after receipt of those documents, the Employer shall prepare and send to the Contractor an account which shall set out the amounts referred to in clause 8.12.5;*

Clause 8.12.3 relates to the preparation of the account to calculate the Termination Payment, in accordance with clause 8.12.5 below:

- Firstly, the Contractor must provide the Employer with '*all documents necessary*' to calculate the amounts referred to in clause 8.12.5. This must be done within 2 months after the date of the termination of the Contractor's employment under the Contract, as referred to in clause 8.12.1.
- The Employer must then prepare an account, setting out the amounts referred to in clause 8.12.5, and send it to the Contractor. This must be done within 3 months after the Employer has received the documents referred to above.

Resource 273 page 573	Correspondence: Contractor to Employer	The provision of all documents necessary for calculating the amounts referred to in clause 8.12.5, from the Contractor to the Employer.
Resource 274 page 574	Account: Employer to Contractor	The provision of an account of the amounts referred to in clause 8.12.5, from the Employer to the Contractor.

8.12.4 the due date for the Termination Payment under clause 8.12.5 shall be the last date for issue of the Employer's account under clause 8.12.3;

Clause 8.12.4 provides the due date for the '*Termination Payment*', which for these purposes, is defined in clause 8.12.5 below.

It states that the due date will be the last date for issue of the Employer's account of the amounts referred to in clause 8.12.5, in accordance with clause 8.12.3 above.

8.12.5 the amount due on termination from the Employer to the Contractor or (if a negative amount) from the Contractor to the Employer shall be the aggregate of:

8.12.5.1 the total value of work properly executed at, and of any design work properly carried out before, the date of termination of the Contractor's employment, ascertained in accordance with these Conditions as if the employment had not been terminated, together with any other amounts due to the Contractor under these Conditions;

8.12.5.2 any sums ascertained in respect of direct loss and/or expense under clause 4.20 (whether ascertained before or after the date of termination);

8.12.5.3 the reasonable cost of removal under clause 8.12.2;

8.12.5.4 the cost of materials or goods (including Site Materials) properly ordered for the Works for which the Contractor then has paid or is legally bound to pay;

8.12.5.5 (only where the Contractor's employment is terminated either under clause 8.9 or 8.10, or under clause 8.11.1.3 if the loss or damage to the Works was caused by the negligence or default of the Employer or any Employer's Person) any direct loss and/or damage caused to the Contractor by the termination,

less amounts previously paid to the Contractor under this Contract, but without deduction of any Retention (the 'Termination Payment'). Payment by the Employer for any such materials and goods as are referred to in clause 8.12.5.4 shall be subject to those materials and goods thereupon becoming the Employer's property.

Clause 8.12.5 provides the definition and method of calculating the '*Termination Payment*' (in this case, for the purposes of clause 8.12).

The Termination Payment will be the amount due to either the Contractor or the Employer upon the termination of the Contractor's employment under the Contract. It is calculated as follows:

Calculation (a)

Calculate the total of the following:

- in accordance with clause 8.12.5.1, the total value of (i) work properly executed at, and (ii) any design work properly carried out before, the date of termination of the Contractor's employment, both ascertained in accordance with the Conditions, as if the Contractor's employment under the Contract had not been terminated, and (iii) any other amounts due to the Contractor under the Conditions;

- in accordance with clause 8.12.5.2, any sums ascertained in respect of direct loss and/or expense under clause 4.20 (Loss and Expense: Notification and ascertainment), whether those sums were ascertained before or after the date of termination;
- in accordance with clause 8.12.5.3, the reasonable cost of removal by the Contractor of any temporary buildings, plant, tools, equipment, and (where applicable) goods and materials, including Site Materials, under clause 8.12.2;
- in accordance with clause 8.12.5.4, the cost of materials or goods (including Site Materials) which have been properly ordered for the Works, for which the Contractor has either (i) paid already, or (ii) is legally bound to pay (for example, under a supply contract); and
- in accordance with clause 8.12.5.5, any direct loss and/or damage caused to the Contractor by the termination of its employment under the Contract. This only applies where the Contractor's employment was terminated under either:
  - clause 8.9 (Termination by Contractor: Default by Employer);
  - clause 8.10 (Termination by Contractor: Insolvency of Employer); or
  - clause 8.11.1.3 (Termination by either Party and regulation 73(1)(a) of the PC Regulations), if the loss or damage to the Works was caused by the '*negligence or default*' of the Employer, or any Employer's Person.

Calculation (b)

Calculate the total amounts previously paid to the Contractor under the Contract, but *without* deducting any of the Retention.

Calculation (c)

Subtract the amount in Calculation (b) from the amount in Calculation (a).

This is the Termination Payment.

If it is a *positive* amount, it is the amount due from the Employer to the Contractor.

If it is a *negative* amount, it is the amount due from the Contractor to the Employer.

The final sentence of clause 8.12.5 provides that any payment by the Employer to the Contractor for any materials or goods (including Site Materials) referred to in clause 8.12.5.4 above will be subject to those materials and goods then becoming the Employer's property.

Termination Payment – final date, notices, and amount

Overview

This sub-section contains the mechanism for the payment of the Termination Payment to either the Contractor or the Employer, whether that Termination Payment is calculated in accordance with either:

- clause 8.7 (Termination by Employer: Consequences of termination under clauses 8.4 to 8.6);
- clause 8.8 (Termination by Employer: Employer's decision not to complete the Works); or
- clause 8.12 (Consequences of Termination under clauses 8.9 to 8.11, etc.).

It was introduced to expressly provide for both the payment and notice requirements of the Housing Grants, Construction and Regeneration Act 1996.

Key provisions

- clause 8.13 (Termination Payment – final date, notices and amount)

Clause 8.13: Termination Payment – final date, notices and amount

8.13.1 Subject to clause 8.13.5, the final date for payment of the Termination Payment shall be 14 days from its due date, as fixed in accordance with clause 8.7.5, 8.8.2 or 8.12.4.

Clause 8.13.1 provides the final date for payment of the Termination Payment.

Subject to clause 8.13.5 (considered below), the final date for payment of the Termination Payment will be 14 days from the due date, whether that due date has been fixed in accordance with either:

- clause 8.7.5 (Termination by Employer: Consequences of termination under clauses 8.4 to 8.6);

DOI: 10.1201/9781003485919-70

- clause 8.8.2 (Termination by Employer: Employer's decision not to complete the Works); or
- clause 8.12.4 (Consequences of Termination under clauses 8.9 to 8.11, etc.).

8.13.2 *Not later than 5 days after the due date for the Termination Payment the Employer shall give a notice to the Contractor, stating the sum that it considers to be or have been due at the due date, calculated in accordance with clause 8.7.6, 8.8.3 or 8.12.5, and the basis on which that sum has been calculated (the 'Employer's Termination Payment Notice'). Such notice shall show the difference or amount referred to in clause 8.7.6, 8.8.3 or 8.12.5 as a balance due to the Contractor from the Employer or vice versa.*

Clause 8.13.2 requires the Employer to give an '*Employer's Termination Payment Notice*' to the Contractor.

Similar to other Payment Notices, the Employer's Termination Payment Notice must be given not later than 5 days after the due date for the Termination Date.

It must state the sum that the Employer considers to be (or have been) due at the due date, calculated in accordance with either:

- clause 8.7.6 (Termination by Employer: Consequences of termination under clauses 8.4 to 8.6);
- clause 8.8.3 (Termination by Employer: Employer's decision not to complete the Works); or
- clause 8.12.5 (Consequences of Termination under clauses 8.9 to 8.11, etc.).

It must also:

- state the basis on which that sum has been calculated; and
- show the difference or amount referred to in either clause 8.7.6, clause 8.8.3, or clause 8.12.5, as a balance due either (i) to the Contractor from the Employer, or (ii) to the Employer from the Contractor.

Resource 275 page 575	Template: Employer's Termination Payment Notice	A template Employer's Termination Payment Notice, from the Employer to the Contractor.

8.13.3 *If the Party by which the Termination Payment is stated to be payable ('the payer') intends to pay less than the stated balance, it shall not later than 5 days before the final date for payment give the other Party a Pay Less Notice in accordance with clause 4.10.1.*

If the Party by which the Termination Payment is stated in either (i) the Employer's Termination Payment Notice, or (ii) the '*Contractor's Termination Payment Notice*' (defined in clause 8.13.5 below), to be payable (defined as '*the payer*') intends to pay less than the stated balance, it must give the other Party a Pay Less Notice in accordance with clause 4.10.1 (Pay Less Notices and other general provisions).

This Pay Less Notice must be given not later than 5 days before the final date for payment.

8.13.4 Where a Pay Less Notice is given under clause 8.13.3, the payment to be made on or before the final date for payment shall not be less than the amount stated in it as due.

If either Party gives a Pay Less Notice to the other Party under clause 8.13.3 above, it must not make a payment of less than the amount stated as due in that Pay Less Notice, on or before the final date for payment. (In other words, it must pay at least the amount stated as due in that Pay Less Notice.)

8.13.5 If an Employer's Termination Payment Notice is not given in accordance with clause 8.13.2:

8.13.5.1 the Contractor may at any time after the 5 day period referred to in clause 8.13.2 give a notice to the Employer, stating the sum that the Contractor considers to have become due under clause 8.7.6, 8.8.3 or 8.12.5 at the due date and the basis on which that sum has been calculated (the 'Contractor's Termination Payment Notice') and, subject to any Pay Less Notice given under clause 8.13.5.3, the Termination Payment shall be the sum stated as due in the Contractor's Termination Payment Notice;

8.13.5.2 if the Contractor gives a Contractor's Termination Payment Notice, the final date for payment of the sum specified in it shall for all purposes be regarded as postponed by the same number of days as the number of days after expiry of the 5 day period referred to in clause 8.13.2 that the Contractor's Termination Payment Notice is given;

8.13.5.3 following the Contractor's Termination Payment Notice the Employer may not later than 5 days before the final date for payment give a Pay Less Notice in accordance with clause 4.10.1 and, if it gives such notice, the provisions of clause 8.13.4 shall correspondingly apply.

Clause 8.13.5 provides for the circumstance where the Employer does *not* give the Employer's Termination Payment Notice in accordance with clause 8.13.2.

It states that the Contractor may give a '*Contractor's Termination Payment Notice*' to the Employer, as follows:

- This may be given at any time after the 5-day period referred to in clause 8.13.2 above.
- Again, it must state (i) the sum that the Contractor considers to have become due under either clause 8.7.6, clause 8.8.3, or clause 8.12.5, at the due date, and (ii) the basis on which that sum has been calculated.

Subject to any Pay Less Notice given under clause 8.13.5.3, the Termination Payment will be the sum stated as due in the Contractor's Termination Payment Notice.

If the Contractor does give a Contractor's Termination Payment Notice, the final date for payment will be regarded as postponed by the same number of days as the number of days after expiry of the 5-day period referred to in clause 8.13.2 that the Contractor's Termination Payment Notice is given. (For example, if the Contractor gives the Contractor's Termination Payment Notice 10 days after the expiry of the 5-day period referred to in clause 8.13.2, the final date for payment will be postponed by a corresponding 10 days.)

Once the Contractor gives the Contractor's Termination Payment Notice, the Employer may give a Pay Less Notice in accordance with clause 4.10.1 (Pay Less Notices and other general provisions). Again, this Pay Less Notice must be given not later than 5 days before the final date for payment.

If the Employer does give this Pay Less Notice, clause 8.13.4 above will apply, to the effect that the Employer must pay at least the sum stated as due in that Pay Less Notice.

Resource 276 page 577	Template: Contractor's Termination Payment Notice	A template Contractor's Termination Payment Notice, from the Contractor to the Employer.

8.13.6 If the payer fails to pay the Termination Payment, or any part of it, by the final date for its payment, the payer shall, in addition to any unpaid amount that should properly have been paid, pay the other Party simple interest on that amount at the Interest Rate for the period from the final date for payment until payment is made. Acceptance of a payment of any such interest shall not in any circumstances be construed as a waiver of any right to proper payment of the principal amount due. Any such unpaid amount and any interest under this clause 8.13.6 shall be recoverable as a debt.

Similar to clause 4.9.6 (Interim and final payments – final date and amount), clause 8.13.6 provides that if the payer fails to pay the Termination Payment (whether in whole or in part) by the final date for payment, then it will also be liable to pay the other Party simple interest on that amount.

This interest will be paid at the '*Interest Rate*' (as defined in clause 1.1 (Definitions)), for the period from (i) the final date for payment (when payment *should* have been made), to (ii) the date on which payment is actually made.

Clause 8.13.6 further provides both that:

- acceptance of a payment of *interest* will not constitute a waiver of the right to proper payment of the *principal* amount due; and
- any such unpaid principal amount and interest under clause 8.13.6 will be recoverable by the other Party as a debt.

8.13.7 In relation to the requirements in this clause 8.13 for the giving of Employer's Termination Payment Notices, and Pay Less Notices, it is immaterial that the amount then considered to be due may be zero.

Similar to clause 4.10.3 (Pay Less Notices and other general provisions) in respect of other Payment Notices and Pay Less Notices, clause 8.13.7 provides that it is immaterial if the amount considered to be due in either (i) any Employer's Termination Payment Notice, or (ii) any Pay Less Notice given under clause 8.13, is zero.

Section 9: Settlement of Disputes

Section overview

Section 9 sets out some the key forms of dispute resolution under the Contract.

It distinguishes between an internal form of dispute resolution, namely negotiation, and external forms, including mediation, adjudication, and Arbitration.

If Article 9 (Arbitration) does not apply, then (unless the Parties subsequently agree to refer a dispute to arbitration) litigation will be the final means of dispute resolution.

Sub-sections

- Notification and negotiation of disputes (clause 9.1)
- Mediation (clause 9.2)
- Adjudication (clause 9.3)
- Arbitration (clause 9.4 – clause 9.9)

DOI: 10.1201/9781003485919-72

Notification and negotiation of disputes

Overview

This sub-section relates to the negotiation of disputes by the Parties, in the hope that they can be resolved without having to refer them to the other external (and typically more expensive and time-consuming) methods of dispute resolution, namely mediation, adjudication, arbitration, and litigation.

Key provisions

- clause 9.1 (Notification and negotiation of disputes)

Clause 9.1: Notification and negotiation of disputes

9.1 With a view to avoidance or early resolution of disputes or differences (subject to Article 8), each Party shall promptly notify the other of any matter that appears likely to give rise to a dispute or difference. The senior executives nominated in the Contract Particulars (or if either is not available, a colleague of similar standing) shall meet as soon as practicable for direct, good faith negotiations to resolve the matter.

Clause 9.1 was previously included in (the optional) Supplemental Provision 10 of the 2016 edition of the Contract, but it is now mandatory.

It aims to resolve disputes at an early stage, or better still, avoid them altogether.

The Parties are first required to notify each other of any matter which seems likely to result in a dispute.

Following this notification, the senior executives nominated in the Contract Particulars must meet as soon as they can to negotiate in good faith, and hopefully thereby avoid or resolve the dispute.

DOI: 10.1201/9781003485919-73

Recommendation: Promoting ongoing dialogue and negotiations

Given the Parties' competing commercial interests, disputes and differences under the Contract will sometimes be unavoidable, so it is to be hoped that they can be resolved by way of negotiation.

The Parties are therefore encouraged to consider practical ways to help ensure that these negotiations are successful, as this will likely prove considerably quicker and less expensive than the other external methods of dispute resolution. The Parties may wish to consider the following:

- The senior executives from each Party should be introduced before the Parties enter into the Contract (a time which often represents the high-water mark of good relations between the Parties).
- Rather than wait for a dispute or difference to arise (and therefore potentially place great pressure on the subsequent negotiations), the senior executives may wish to meet regularly throughout the course of the project, to discuss the progress of the Works. This should help the senior executives gain a better shared understanding of the project, and its potential difficulties.
- Depending on the nature of the dispute or difference, it may be helpful for the senior executives to jointly appoint an independent expert or specialist, to help improve their shared understanding.
- When negotiations take place, each senior executive should take an attendance note of their meeting, or better still, agree a joint attendance note, of the discussions and actions required.

Taken together, these measures should help foster an environment of trust and collaboration, where good faith negotiations and dialogue can take place, even in the face of multiple and complex disputes.

Resource 277 page 579	Notice: Contractor or Employer to the other Party	A notice of a matter which appears likely to give rise to a dispute or difference.

Mediation

Overview

This sub-section contains only one provision, clause 9.1, which relates to mediation.

Mediation is one of the external methods of dispute resolution provided for in the Contract.

In summary, the mediator, who will be an independent third party chosen by the Parties, will aim to facilitate an agreement.

Mediation will typically be much quicker and less expensive than arbitration or litigation. It is usually confidential, and can be conducted in a more flexible way. The mediator does not make a binding decision, however, so if agreement is reached in principle, it should promptly be recorded in a written settlement agreement.

Key provisions

- clause 9.2 (Mediation)

Clause 9.2: Mediation

9.2 *Subject to Article 8, if a dispute or difference arises under this Contract which cannot be resolved by direct negotiations, each Party shall give serious consideration to any request by the other to refer the matter to mediation.*

If negotiations to resolve a dispute (such as those referred to in clause 9.1) are unsuccessful, the next potential step is a mediation.

Unlike adjudication, arbitration, and litigation, which can be unilaterally commenced by one of the Parties, both Parties need to agree to mediate.

Therefore, subject to the rights of the Parties to refer any dispute to adjudication in accordance with Article 8 (Adjudication), clause 9.2 requires the Parties to seriously consider any request by the other to refer the dispute to mediation.

DOI: 10.1201/9781003485919-74

Recommendation: Nomination of a mediator

Whilst not required by the Contract, the Parties may wish to consider nominating an independent third party as a potential mediator at the outset of the project (similar to the naming of an Adjudicator against the reference to clause 9.3.1 (Adjudication) in the Contract Particulars). This could be done on an informal basis, but it may have the effect of instilling greater confidence in both Parties in the mediation process, should it be required.

Resource 278 page 580	Request: Contractor or Employer to the other Party	A request to mediate a difference or dispute between the Parties.

Adjudication

Overview

This sub-section also contains only one provision, clause 9.2, which relates to adjudication.

Article 8 (Adjudication) gives the Parties a contractual right to refer any dispute or difference arising under the Contract to adjudication, in addition to the Parties' statutory right under the Housing Grants, Construction and Regeneration Act 1996 (often referred to simply as 'the Construction Act').

In summary, the Party who wishes to refer the dispute to adjudication (known as the 'Referring Party') must issue a notice of their intention to refer the dispute to adjudication on the other Party (known as the 'Responding Party'), following which either the Adjudicator named in the Contract Particulars (against the reference to clause 9.3.1) will be appointed, or the nominating body named in the Contract Particulars (also against the reference to clause 9.3.1) will nominate the Adjudicator.

Very briefly, the Parties will typically exchange written submissions, following which the Adjudicator will issue their decision.

Adjudication is typically less expensive than arbitration or litigation (although the costs can still be considerable), less formal, and much quicker: section 108(2)(c) of the Housing Grants, Construction and Regeneration Act 1996 provides for the Adjudicator's decision to be made within 28 days from receipt of the Referring Party's referral notice (although this period can be extended).

Adjudication is therefore the most common form of external dispute resolution in the industry.

However, the Adjudicator's decision is only temporarily binding on the Parties, unless and until the dispute is finally determined by arbitration, litigation, or agreement.

Key provisions

- clause 9.3 (Adjudication)

DOI: 10.1201/9781003485919-75

Clause 9.3: Adjudication

9.3 *If a dispute or difference arises under this Contract which either Party wishes to refer to adjudication, the Scheme shall apply, subject to the following:*

9.3.1 *for the purposes of the Scheme the Adjudicator shall be the person (if any) and the nominating body shall be that stated in the Contract Particulars;*

9.3.2 *where the dispute or difference is or includes a dispute or difference relating to clause 3.13.3 and as to whether an instruction issued thereunder is reasonable in all the circumstances:*

9.3.2.1 *the Adjudicator to decide such dispute or difference shall (where practicable) be an individual with appropriate expertise and experience in the specialist area or discipline relevant to the instruction or issue in dispute;*

9.3.2.2 *if the Adjudicator does not have the appropriate expertise and experience, the Adjudicator shall appoint an independent expert with such expertise and experience to advise and report in writing on whether or not the instruction under clause 3.13.3 is reasonable in all the circumstances.*

Clause 9.3 provides that, in respect of any adjudication, Part 1 of The Scheme for Construction Contracts (England and Wales) Regulations 1998 (defined as '*the Scheme*') will apply.

The Scheme supplements the Housing Grants, Construction and Regeneration Act 1996, and provides a statutory framework for adjudication.

Clause 9.3 states that the application of the Scheme will be subject to the following:

- **the Adjudicator and the nominating body:** the Adjudicator will be the person (if any) identified in the Contract Particulars (against the reference to clause 9.3), and similarly, the nominating body will be that identified in the Contract Particulars (again, against the reference to clause 9.3); and
- **Work not in accordance with the Contract:** where the dispute includes a difference relating to clause 3.13.3 (Work not in accordance with the Contract), and as to whether an instruction issued in accordance with that clause is '*reasonable in all the circumstances*':
  - where practicable, the Adjudicator must be appropriately qualified and experienced in the specialist discipline relevant to the instruction or dispute; and
  - if they are not appropriately qualified and experienced, the Adjudicator must appoint a suitably qualified and experienced independent expert to advise them, and to provide a written report on whether or not the relevant instruction was reasonable in all the circumstances.

Recommendation: Residential occupiers

Section 106 of the Housing Grants, Construction and Regeneration Act 1996 provides that Part 2 of that Act, which relates to adjudication, does *not* apply to construction contracts with a '*residential occupier*'.

In section 106(2), it defines a construction contract with a residential occupier as '*a construction contract which principally relates to operations on a dwelling which one of the parties to the contract occupies, or intends to occupy*' as their residence.

If the Employer is a residential occupier in this regard, the Parties should seek advice in respect of both clause 9.3 and Article 8 (Adjudication).

Market practice: Multiple disputes

Clause 9.3 is often amended to permit the Adjudicator to adjudicate on more than one related dispute under the Contract at the same time.

Arbitration

Overview

This sub-section relates to arbitration.

Article 9 (Arbitration), where it applies, provides the Parties with an alternative to litigation (referred to in Article 10 (Legal proceedings)). The Parties may also subsequently agree to refer disputes to arbitration, after the date of the Contract.

Section 1 of the Arbitration Act 1996 sets out the following general principles for arbitration:

- the object of arbitration is to obtain the fair resolution of disputes by an impartial tribunal without unnecessary delay or expense;
- the Parties should be free to agree how their disputes are resolved, subject only to such safeguards as are necessary in the public interest; and
- the courts should have limited involvement.

Very briefly, the Arbitrator, who will be an independent third party, will consider the evidence from both Parties, and make their award.

The procedure will be governed by the JCT 2024 edition of the Construction Industry Model Arbitration Rules (CIMAR).

Whilst arbitration is usually more expensive and not as quick as mediation or adjudication, the Arbitrator's award is binding, the process allows for greater flexibility, and the proceedings are typically private.

Key provisions

- clause 9.4 (Conduct of arbitration)
- clause 9.5 (Notice of reference to arbitration)
- clause 9.6 (Powers of Arbitrator)
- clause 9.7 (Effect of award)
- clause 9.8 (Appeal – questions of law)
- clause 9.9 (Arbitration Act 1996)

DOI: 10.1201/9781003485919-76

Clause 9.4: Conduct of arbitration

> 9.4 *Any arbitration pursuant to Article 9 shall be conducted in accordance with the JCT 2024 edition of the Construction Industry Model Arbitration Rules (CIMAR), provided that if any amendments to that edition of the Rules have been issued by the JCT the Parties may, by a joint notice in writing to the Arbitrator, state that they wish the arbitration to be conducted in accordance with the Rules as so amended. References in clause 9.5 to a Rule or Rules are references to such Rule(s) as set out in the JCT 2024 edition of CIMAR.*

Clause 9.4 provides that any arbitration must be conducted in accordance with the JCT 2024 edition of the Construction Industry Model Arbitration Rules (CIMAR).

If any amendments to the JCT 2024 edition are subsequently published, the Parties may notify the Arbitrator that they wish the arbitration to be conducted with the amended Rules.

It further clarifies that references in clause 9.5 to a '*Rule*' or '*Rules*' are references to the JCT 2024 edition of CIMAR.

Clause 9.5: Notice of reference to arbitration

> 9.5.1 *Where pursuant to Article 9 either Party requires a dispute or difference to be referred to arbitration, that Party shall serve on the other Party a notice of arbitration to such effect in accordance with Rule 2.1 identifying the dispute and requiring the other Party to agree to the appointment of an arbitrator. The Arbitrator shall be an individual agreed by the Parties or, failing such agreement within 14 days (or any agreed extension of that period) after the notice of arbitration is served, appointed on the application of either Party in accordance with Rule 2.3 by the person selected in accordance with the Contract Particulars.*
>
> 9.5.2 *Where two or more related arbitral proceedings in respect of the Works fall under separate arbitration agreements, Rules 2.6, 2.7 and 2.8 shall apply.*
>
> 9.5.3 *After the Arbitrator has been appointed either Party may give a further notice of arbitration to the other Party and to the Arbitrator referring any other dispute which falls under Article 9 to be decided in the arbitral proceedings and Rule 3.3 shall apply.*

Clause 9.5 sets out some of the mechanics of the arbitration process:

- clause 9.5.1 provides that, where a Party requires a dispute to be referred to arbitration, it must serve on the other Party a notice of arbitration in accordance with Rule 2.1 of CIMAR, identifying the dispute, and requiring the other Party to agree to the appointment of an arbitrator. The Arbitrator will be an individual agreed by the Parties, or if they fail to reach agreement within 14 days (or any agreed extension) after the notice of arbitration is served, the Arbitrator will be appointed on the application of either Party in accordance with Rule 2.3 of CIMAR, by the person selected against the reference to clause 9.5.1 in the Contract Particulars.
- clause 9.5.2 provides that where multiple related arbitrations in respect of the Works fall under separate arbitration agreements, Rule 2.6, Rule 2.7, and Rule 2.8 of CIMAR will apply.
- clause 9.5.3 provides that after the Arbitrator has been appointed, either Party may give a *further* notice of arbitration to the other Party and to the Arbitrator, referring any other dispute which falls under Article 9 to be decided in the arbitral proceedings, and Rule 3.3 of CIMAR will apply.

Clause 9.6 Powers of Arbitrator

9.6 *Subject to the provisions of Article 9 and clause 1.8, the Arbitrator shall, without prejudice to the generality of their powers, have power to rectify this Contract so that it accurately reflects the true agreement made by the Parties, to direct such measurements and/or valuations as may in their opinion be desirable in order to determine the rights of the Parties and to ascertain and award any sum which ought to have been the subject of or included in any payment and to open up, review and revise any account, opinion, decision, requirement or notice and to determine all matters in dispute which shall be submitted to them in the same manner as if no such account, opinion, decision, requirement or notice had been given.*

Subject to Article 9 (Arbitration) and clause 1.8 (Effect of Final Statement), the Arbitrator has a very broad range of powers. They have the ability to:

- '*rectify*' (or 'correct') the Contract, so that it accurately reflects the true agreement made by the Parties;
- direct such measurements and/or valuations which the Arbitrator considers may be required, in order to determine the rights of the Parties; and
- ascertain and award any sum which should have been the subject of, or included in, any payment, and to open up, review, and revise any account, opinion, decision, requirement, or notice, and to determine all matters in dispute which must be submitted to them in the same manner, as though no such account, opinion, decision, requirement or notice had been given in the first place.

Clause 9.7: Effect of award

9.7 *Subject to clause 9.8 the award of the Arbitrator shall be final and binding on the Parties.*

With the exception set out in clause 9.8, the Arbitrator's award is final and binding on the Parties.

Clause 9.8: Appeal – questions of law

9.8 *The Parties hereby agree pursuant to section 45(2)(a) and section 69(2)(a) of the Arbitration Act 1996 that either Party may (upon notice to the other Party and to the Arbitrator):*

9.8.1 *apply to the courts to determine any question of law arising in the course of the reference; and*

9.8.2 *appeal to the courts on any question of law arising out of an award made in an arbitration under this arbitration agreement.*

The exception to clause 9.7 is set out in clause 9.8, where, in accordance with section 45(2)(a) and section 69(2)(a) of the Arbitration Act 1996, either Party may (having given notice to the other Party and to the Arbitrator):

- **section 45(2)(a)**: apply to the courts to determine any question of law, arising during the course of the arbitration; and
- **section 69(2)(a)**: appeal to the courts on any question of law, arising out of an award made in an arbitration.

Clause 9.9: Arbitration Act 1996

9.9 *The provisions of the Arbitration Act 1996 shall apply to any arbitration under this Contract wherever the same, or any part of it, shall be conducted.*

Clause 9.9 provides that the Arbitration Act 1996 applies to any arbitration under the Contract, wherever it is conducted.

The Parties should also have regard to the Arbitration Act 2025, which introduces a number of reforms to the Arbitration Act 1996.

The Schedules

This section briefly considers the six Schedules to the Conditions, along with the Fluctuations Options.

Given its importance to the design of the Works, the Design Submission Procedure contained in Schedule 1 is considered in greater detail.

Key provisions

- Schedule 1 (Design Submission Procedure)
- Schedule 2 (Supplemental Provisions)
- Schedule 3 (Insurance Options)
- Schedule 4 (Code of Practice)
- Schedule 5 (Third Party Rights)
- Schedule 6 (Forms of Bonds)
- Fluctuations Options

DOI: 10.1201/9781003485919-77

Schedule 1: Design Submission Procedure

Overview

Clause 2.8 (Construction information) requires the Contractor to provide the Employer with copies of the Contractor's Design Documents, in accordance with the Design Submission Procedure. Furthermore, it prohibits the Contractor from commencing any work to which a Contractor's Design Document relates, until the Design Submission Procedure has been complied with.

Schedule 1 contains this Design Submission Procedure, which is designed to help ensure that the Contractor completes the design for the Works in accordance with the Contract.

'*Contractor's Design Documents*' are defined in clause 1.1 (Definitions) as *'the drawings, details and specifications of materials, goods and workmanship and other related documents and information prepared by or for the Contractor in relation to the design of the Works (including such as are contained in the Contractor's Proposals) together, where applicable, with any other design documents or information to be provided by the Contractor under the BIM Protocol'*.

1 *The Contractor shall prepare and submit each of the Contractor's Design Documents to the Employer by the means and in the format stated in the Employer's Requirements or the Contractor's Proposals and in sufficient time to allow any comments of the Employer to be incorporated prior to the relevant Contractor's Design Document being used for procurement and/or in the carrying out of the Works. Where the means and format are not so stated, then, unless and until otherwise agreed with the Employer, the Contractor shall submit to the Employer 2 copies of each of the Contractor's Design Documents.*

DOI: 10.1201/9781003485919-78

Paragraph 1 requires the Contractor to both prepare and submit each Contractor's Design Document to the Employer:

- by the means and format stated in either (i) the Employer's Requirements, or (ii) the Contractor's Proposals; and
- sufficiently early to allow the Employer's comments to be incorporated, before the relevant Contractor's Design Documents are used for either procurement or carrying out the Works.

If the Employer's Requirements or Contractor's Proposals do not state the means and format for the submission of Contractor's Design Documents, then unless otherwise agreed between the Parties, the Contractor must submit two (paper) copies of each Contractor's Design Document to the Employer.

Recommendation: Protocol for submitting Contractor's Design Documents

The Parties are encouraged to agree a protocol for submitting Contractor's Design Documents (and ideally include it in either the Employer's Requirements or the Contractor's Proposals).

This protocol should include both the means and format for the submission of each Contractor's Design Document. For example, by way of a document management platform (to which all relevant parties have access), and with an agreed referencing system.

The Parties could also agree dates for submission of each Contractor's Design Document, perhaps arranged by type or package of work.

Resource 279 page 580	Submission: Contractor to Employer	The submission by the Contractor of Contractor's Design Documents.

2 *Within 14 days from the date of receipt of any Contractor's Design Document, or (if later) 14 days from either the date or expiry of the period for submission of the same stated in the Contract Documents, the Employer shall return one copy of that Contractor's Design Document to the Contractor marked 'A', 'B' or 'C' provided that a document shall be marked 'B' or 'C' only where the Employer considers that it is not in accordance with this Contract.*

Paragraph 2 requires the Employer to return a copy of each Contractor's Design Document to the Contractor, within 14 days from either:

- the date of receipt of that Contractor's Design Document, or if later,
- the date or expiry of the period for submission of that Contractor's Design Document, as stated in the Contract Documents.

The Employer must mark (or label) each Contractor's Design Document with either 'A', 'B', or 'C'.

However, the Employer may only mark a Contractor's Design Document with either 'B' or 'C' where the Employer considers that the Contractor's Design Document is *not* in accordance with the Contract.

Resource 280 page 581	Return: Employer to Contractor	The return by the Employer of Contractor's Design Documents.

3 *If the Employer does not respond to a Contractor's Design Document in the time stated in paragraph 2, it shall be regarded as marked 'A'.*

Paragraph 3 provides for a default position, if the Employer does not return a Contractor's Design Document within the relevant 14-day period set out in paragraph 2: that Contractor's Design Document will be deemed to have been marked 'A'.

Resource 281 page 582	Confirmation: Contractor to Employer	Confirmation by the Contractor to the Employer of Contractor's Design Documents regarded as marked 'A'.

4 *Where the Employer marks a Contractor's Design Document 'B' or 'C', it shall identify by means of a written comment why it considers that the document is not in accordance with this Contract.*

In accordance with paragraph 2, if an Employer marks a Contractor's Design Document 'B' or 'C', it must consider that the Contractor's Design Document is, in some way, *not* in accordance with the Contract.

Paragraph 4 therefore requires the Employer to provide a written explanation.

> 5 *When a Contractor's Design Document is returned by the Employer:*
>
> 5.1 *if it is marked 'A', the Contractor shall carry out the Works in strict accordance with that document;*
>
> 5.2 *if it is marked 'B', the Contractor may carry out the Works in accordance with that document, provided that the Employer's comments are incorporated into it and an amended copy of it is promptly submitted to the Employer; or*
>
> 5.3 *if it is marked 'C', the Contractor shall take due account of the Employer's comments on it and shall either forthwith resubmit it to the Employer in amended form for comment in accordance with paragraph 1 or notify the Employer under paragraph 7.*

Paragraph 5 sets out the consequences for when Contractor's Design Documents are marked either 'A', 'B', or 'C' by the Employer:

- If a Contractor's Design Document is marked 'A', then the Contractor *must* carry out the Works in strict accordance with it.
- If a Contractor's Design Document is marked 'B', then the Contractor *may* carry out the Works in accordance with it, but provided that (i) the Contractor incorporates the Employer's comments into the Contractor's Design Document, and then (ii) the Contractor promptly submits an amended copy of the Contractor's Design Document to the Employer.
- If a Contractor's Design Document is marked 'C', then the Contractor must have due regard to the Employer's comments, and then:
  - if the Contractor agrees with the Employer's comment, resubmit the Contractor's Design Document as soon as possible, in amended form, or
  - if the Contractor does *not* agree with the Employer's comment, notify the Employer under paragraph 7 below.

Resource 282 page 582	Reply: Contractor to Employer	A reply from the Contractor to the Employer to marked Contractor's Design Documents.
Resource 283 page 583	Resubmission: Contractor to Employer	The resubmission by the Contractor of a Contractor's Design Document marked 'B'.
Resource 284 page 583	Resubmission: Contractor to Employer	The resubmission by the Contractor of Contractor's Design Document marked 'C'.

6 *The Contractor shall not carry out any work in accordance with a Contractor's Design Document marked 'C' and the Employer shall not be liable to pay for any work within the Works executed otherwise than in accordance with Contractor's Design Documents marked 'A' or 'B'.*

Paragraph 6 prohibits the Contractor from carrying out any work in accordance with a Contractor's Design Document which has been marked 'C' by the Employer, on the basis that it is not in accordance with the Contract.

It further provides that the Employer will not be liable to pay the Contractor for any work which has been carried out otherwise than in accordance with Contractor's Design Documents marked 'A' or 'B'. In other words, Contractor's Design Documents marked 'C', or not submitted to the Employer in accordance with the Design Submission Procedure.

Recommendation: Carrying out work in accordance with Contractor's Design Documents

Given the potentially significant adverse financial consequences under this paragraph 6 (for example, relating to abortive works), the Contractor is encouraged to introduce internal controls to ensure that works can only ever be carried out in accordance with either:

- Contractor's Design Documents marked 'A', or
- Contractor's Design Documents marked 'B', provided that the Contractor has incorporated the Employer's comments into them.

For example, by labelling the correct versions of each Contractor's Design Document as '*for construction*'.

This is to ensure that none of the Contractor's operatives (or any Contractor's Persons) inadvertently carry out work which is not in accordance with the Contract.

This is particularly the case for larger, more complex projects, where there may be a very great number of Contractor's Design Documents.

7 *If the Contractor disagrees with a comment of the Employer and considers that the Contractor's Design Document in question is in accordance with this Contract, it shall within 7 days of receipt of the comment notify the Employer that it considers that compliance with the comment would give rise to a Change. Such notification shall be accompanied by a statement setting out the Contractor's reasons. Upon receipt of such a notification the Employer shall within 7 days either confirm or withdraw the comment and, where the comment is confirmed, the Contractor shall amend and resubmit the document accordingly.*

Paragraph 7 applies where:

- the Employer has marked a Contractor's Design Document 'C', and has made a written comment as to why the Employer considers that the Contractor's Design Document is not in accordance with the Contract; and
- the Contractor disagrees with the Employer's comment, and considers that the Contractor's Design Document *is* in accordance with the Contract.

In these circumstances, the Contractor must notify the Employer that the Contractor considers that compliance with the Employer's comment would give rise to a Change (as defined under clause 5.1 (Definition of Changes)), and must include its reasons. This notice must be given within 7 days of receipt of the Employer's comment.

Within 7 days of receiving the Contractor's notice, the Employer must then either confirm or withdraw its comment.

Where the Employer confirms its comment, the Contractor is then required to amend and resubmit the Contractor's Design Document.

Resource 285 page 584	Notice: Contractor to Employer	A notice from the Contractor to the Employer of a disagreement with the Employer's comment on a Contractor's Design Document.
Resource 286 page 585	Confirmation: Employer to Contractor	A reply from the Employer to the Contractor's notice of disagreement with the Employer's comment on a Contractor's Design Document.

8 Provided always that:

8.1 confirmation or withdrawal of a comment in accordance with paragraph 7 shall not signify acceptance by the Employer that the relevant Contractor's Design Document or amended document is in accordance with this Contract or that compliance with the Employer's comment would give rise to a Change;

8.2 where in relation to a comment by the Employer the Contractor does not notify the Employer in accordance with paragraph 7, the comment in question shall not be treated as giving rise to a Change; and

8.3 neither compliance with the design submission procedure in this Schedule nor with the Employer's comments shall diminish the Contractor's obligations to ensure that the Contractor's Design Documents and Works are in accordance with this Contract.

Paragraph 8 sets out a number of further clarifications in respect of the Design Submission Procedure.

Paragraph 8.1 provides that, simply because the Employer confirms or withdraws a comment in accordance with paragraph 7, it does *not* mean that the Employer agrees that either:

- the relevant Contractor's Design Document (or amended Contractor's Design Document) is in accordance with the Contract; or
- compliance with the Employer's comment would give rise to a Change (as stated by the Contractor, under a notice given in accordance with paragraph 7).

Paragraph 8.2 provides that if the Contractor fails to notify the Employer in accordance with paragraph 7, then the Employer's comment will *not* be treated as giving rise to a Change.

Paragraph 8.3 provides that neither:

- compliance with the Design Submission Procedure; nor
- compliance with any Employer's comments on any Contractor's Design Documents,

will reduce or limit the Contractor's obligations to ensure that both the Contractor's Design Documents, and the Works, are in accordance with the Contract.

In other words, the Contractor will still be responsible for ensuring that the Contractor's Design Documents and the Works are in accordance with the Contract, even if it can show that it has complied with both the Design Submission Procedure and any Employer's comments.

Resource 287 page 585	Tracker: Contractor and Employer	A tracker template for recording Contractor's Design Documents.

Schedule 2: Supplemental Provisions

Overview

Schedule 2 contains the nine Supplemental Provisions, and is divided into two parts:

Part 1

- Supplemental Provision 1 (Named Sub-Contractors)
- Supplemental Provision 2 (Valuation of Changes – Contractor's estimates)
- Supplemental Provision 3 (Loss and expense – Contractor's estimates)

Part 2

- Supplemental Provision 4 (Acceleration Quotation)
- Supplemental Provision 5 (Health and safety)
- Supplemental Provision 6 (Cost savings and value improvements)
- Supplemental Provision 7 (Performance Indicators and monitoring)
- Supplemental Provision 8 (Transparency)
- Supplemental Provision 9 (The Public Contracts Regulations 2015)

The Supplemental Provisions apply in the following circumstances:

- Supplemental Provisions 1-3 apply only if stated as such in the Contract Particulars (against the reference to the Seventh Recital and Part 1 of Schedule 2).
- Supplemental Provisions 4-7 apply, unless otherwise stated in the Contract Particulars (against the reference to the Seventh Recital and Part 2 of Schedule 2).
- Supplemental Provision 8 applies where the Employer is a Local or Public Authority, or other body mentioned in that provision.
- Supplemental Provision 9 applies where both (i) the Employer is a Local or Public Authority, and (ii) the Contract is subject to the PC Regulations.

Part 1

Supplemental Provision 1: Named Sub-Contractors

Supplemental Provision 1 relates to circumstances where certain work (defined as '*Named Sub-Contract Work*') must be executed by a particular person, as the Contractor's sub-contractor (defined as a '*Named Sub-Contractor*').

DOI: 10.1201/9781003485919-79

For example, the Employer may wish the Contractor to work with a particular sub-contractor, design consultant, or supplier, in relation to the Works.

It requires the Contractor to enter into a sub-contract with any Named Sub-Contractors, and provides for the circumstances where this is not possible. It also provides for the potential termination of a Named Sub-Contractor's employment under the sub-contract, and the consequences.

Importantly, the Contractor will still remain responsible to the Employer for carrying out and completing the Works in accordance with clause 2.1 (Contractor's Obligations: General obligations), irrespective of any such Named Sub-Contractors or any of the other events referred to in Supplemental Provision 1.

Paragraph 1.1.2

Resource 288 page 586	Notice: Contractor to Employer	A notice from the Contractor to the Employer of the Contractor's inability to enter into a sub-contract with a Named Sub-Contractor.
Resource 289 page 587	Instruction: Employer to Contractor	An instruction from the Employer following a notice from the Contractor of the Contractor's inability to enter into a sub-contract with a Named Sub-Contractor.

Paragraph 1.3.1

Resource 290 page 587	Notice: Contractor to Employer	A notice from the Contractor to the Employer of the potential termination of a Named Sub-Contractor's employment under their sub-contract.

Paragraph 1.3.2

Resource 291 page 588	Request for consent: Contractor to Employer	A request from the Contractor for the Employer's consent to terminate a Named Sub-Contractor's employment under their sub-contract.
Resource 292 page 588	Consent given: Employer to Contractor	Consent given by the Employer for the Contractor to terminate a Named Sub-Contractor's employment under their sub-contract.
Resource 293 page 589	Consent withheld: Employer to Contractor	Consent withheld by the Employer for the Contractor to terminate a Named Sub-Contractor's employment under their sub-contract.

Supplemental Provision 2: Valuation of Changes – Contractor's estimates

Supplemental Provision 2 provides a mechanism which allows the Parties to agree *in advance* some of the most significant consequences of instructions requiring Changes, including:

- the value of the Change;
- any resultant extension of time; and
- any resultant reimbursement of loss and/or expense to the Contractor.

This is achieved by the Parties reaching agreement on estimates provided by the Contractor in respect of the relevant instruction requiring a Change, following which those estimates will become binding on the Parties.

If Supplemental Provision 2 applies, if will modify the following provisions:

- section 5 (Changes);
- clause 2.23 to clause 2.26 (Adjustment of Completion Date); and
- clause 4.19 to clause 4.23 (Loss and Expense).

Paragraph 2.2

Resource 294 page 589	Estimates: Contractor to Employer	The provision of estimates by the Contractor to the Employer of the effects of a Change instruction.
Resource 295 page 590	Correspondence: Employer to Contractor	Confirmation from the Employer to the Contractor that estimates are not required.
Resource 296 page 591	Objection: Contractor to Employer	An objection from the Contractor to the Employer to the provision of estimates.

Paragraph 2.5

Resource 297 page 591	Confirmation / withdrawal of instruction: Employer to Contractor	Confirmation or withdrawal by the Employer of an instruction to the Contractor.

Paragraph 2.6

Resource 298 page 592	Tracker: Contractor and Employer	A tracker template for instructions and estimates relating to Supplemental Provision 2.

Supplemental Provision 3: Loss and expense – Contractor's estimates

Supplemental Provision 3 provides an alternative means for the Contractor to be reimbursed for direct loss and/or expense, by modifying the mechanism included in clause 4.19 to clause 4.23 (Loss and Expense).

It allows the Contractor to apply for reimbursement of such direct loss and/ or expense by submitting estimates to the Employer with Interim Payment Applications.

The Employer may request further information and details in support of the Contractor's estimates, and is then required to give notice to the Contractor either that:

- the Employer accepts the Contractor's estimate, in which case the amount is added to the Contract Sum, and no further additions to the Contract Sum will be made during the relevant period in respect of the same matter;
- the Employer wishes to negotiate, failing which the Parties will refer the matter to adjudication; or
- the Employer does not accept the Contractor's estimate, and the mechanism included in clause 4.19 to clause 4.23 will apply in default.

Paragraph 3.2

Resource 299 page 592	Estimate: Contractor to Employer	The provision by the Contractor to the Employer of an estimate of the direct loss and expense incurred by the Contractor.

Paragraph 3.3

Resource 300 page 593	Further estimates: Contractor to Employer	The provision by the Contractor to the Employer of further estimates of the direct loss and expense incurred by the Contractor.

Paragraph 3.4

Resource 301 page 594	Request: Employer to Contractor	A request by the Employer to the Contractor for further information and details relating to the Contractor's estimate of the direct loss and expense incurred by the Contractor.
Resource 302 page 594	Notice: Employer to Contractor	A notice from the Employer to the Contractor relating to the Contractor's estimate of the direct loss and expense incurred by the Contractor.

Paragraph 3.6

Resource 303 page 595	Tracker: Contractor and Employer	A tracker template for estimates of direct loss and expense incurred by the Contractor, relating to Supplemental Provision 3.

Part 2

Supplemental Provision 4: Acceleration Quotation

Supplemental Provision 4 is relevant where the Employer wishes to investigate the possibility of achieving practical completion *before* the Completion Date for the Works (or relevant Section), a process known as 'acceleration'.

It provides a mechanism for both:

- the Contractor to make proposals as to how this could be achieved (defined as an '*Acceleration Quotation*'), along with details of how much time could be saved, the adjustment to the Contract Sum, and any other conditions; and
- the acceptance of the Contractor's Acceleration Quotation by the Employer (defined as a '*Confirmed Acceptance*'), by way of an instruction.

Recommendation: Record keeping for preparing Acceleration Quotations

In accordance with paragraph 4.4.1 of Supplemental Provision 4 (Acceleration Quotation), if an Acceleration Quotation is not accepted, a fair and reasonable amount shall be added to the Contract Sum in respect of the cost of its preparation, provided that it has been prepared on a fair and reasonable basis.

The Contractor is therefore encouraged to keep separate records of the costs associated with preparing Acceleration Quotations (for example, design fees), such that they can be added to the Contract Sum.

Paragraph 4.1.1

Resource 304 page 595	Request: Employer to Contractor	A request by the Employer for the Contractor to provide an Acceleration Quotation.
Resource 305 page 596	Acceleration Quotation: Contractor to Employer	The provision by the Contractor to the Employer of an Acceleration Quotation, or an explanation why it would be impracticable to achieve practical completion earlier than the Completion Date.
Resource 306 page 596	Template: Acceleration Quotation	An Acceleration Quotation template, from the Contractor to the Employer.

Paragraph 4.1.3

Resource 307 page 597	Request: Employer to Contractor	A request by the Employer for the Contractor to issue revised proposals, in respect of an Acceleration Quotation.

Paragraph 4.2.3

Resource 308 page 598	Notice: Employer to Contractor	A notice from the Employer to the Contractor of an adjustment to the periods referred to in Supplemental Provision 4.

Paragraph 4.3.3

Resource 309 page 598	Instruction: Employer to Contractor	A Confirmed Acceptance from the Employer to the Contractor, by way of an instruction.

Supplemental Provision 5: Health and safety

Supplemental Provision 5 seeks to foster a culture and working environment where health and safety is of utmost importance.

It also imposes a number of further obligations on the Contractor, including:

- compliance with codes of practice produced by the Health and Safety Executive;
- the provision of sufficient site-specific health and safety induction training, and regular refresher training (for example, relating to risk assessments and method statements (or 'RAMS');
- ensuring personnel have access to competent health and safety advice, in accordance with regulation 7 of the Management of Health and Safety at Work Regulations 1999; and
- ensuring there is a proper health and safety consultation with all personnel, in accordance with the Health and Safety (Consultation with Employees) Regulations 1996.

Paragraph 5.2.4

Resource 310 page 599	Checklist: Contractor	A checklist of the Contractor's duties in relation to health and safety.

Supplemental Provision 6: Cost savings and value improvements

Supplemental Provision 6 relates to potential changes to the design and specification of the Works, in order to bring commercial or programme benefits to the Employer (often referred to as 'value engineering'), primarily by either:

- reducing the cost of the Works (and associated life cycle and maintenance costs); or
- achieving practical completion earlier than the Completion Date.

It encourages the Contractor to propose such changes, and then for the Parties to negotiate and agree (i) the share in any financial benefit and (ii) any adjustment to the Completion Date.

Paragraph 6.1

Resource 311 page 599	Proposed change: Contractor to Employer	A proposed change to the Works by the Contractor, for the Employer's benefit.

Paragraph 6.3

Resource 312 page 600	Instruction: Employer to Contractor	An instruction by the Employer of a change proposed by the Contractor.

Paragraph 6.4

Resource 313 page 600	Tracker: Contractor and Employer	A tracker template for proposals made by the Contractor, relating to Supplemental Provision 6.

Supplemental Provision 7: Performance Indicators and monitoring

Supplemental Provision 7 provides the Employer with the ability to monitor and evaluate the Contractor's work against any performance indicators (often described as 'key performance indicators' or 'KPIs') identified in the Contract Documents.

Paragraph 7.2

Resource 314 page 601	Provision of information: Contractor to Employer	The provision by the Contractor to the Employer of information for performance indicators and monitoring.

Paragraph 7.3

Resource 315 page 601	Correspondence: Employer to Contractor	Informing the Contractor that targets for performance indicators may not be met.

Supplemental Provision 8: Transparency

Supplemental Provision 8 relates to disclosures under the Freedom of Information Act 2000, in circumstances where the Employer is a Local or Public Authority, or another body subject to that Act.

The Act aims to promote openness and transparency, by providing public access to information held by public authorities (such as the Employer, in this case), whether by requiring public authorities to publish certain information about their activities, or by permitting members of the public to request certain information from public authorities.

Where Supplemental Provision 8 applies, the content of the Contract is *not* confidential, except in relation to any of the disclosure exemptions included in the Act.

Resource 316 page 601	Correspondence: Employer to Contractor	Informing the Contractor of a request made in accordance with the Freedom of Information Act 2000, in relation to the Contract.

Supplemental Provision 9: The Public Contracts Regulations 2015

The PC Regulations were introduced to modernise the rules for the procurement of goods, services and works above certain thresholds by public authorities (in this case, the Employer).[23]

Supplemental Provision 9 therefore applies where both (i) the Employer is a Local or Public Authority, and (ii) the Contract is subject to the PC Regulations.

It imposes a number of obligations on the Contractor and its sub-contractors, in order to comply with the PC Regulations. These include ensuring that sub-contracts provide for the following:

- compliance with regulation 113(2)(c)(i) and (ii), to see that payments are made within 30 days from the date on which the relevant invoice is regarded as valid and undisputed;
- the provision of information required by the sub-contractor to the Contractor under regulation 71 (*Giving information to contracting authorities),* and specifically regulation 71(3), regulation 71(4), and regulation 71(5), and the requirement for any sub-sub-contractor to do the same;
- that the sub-contractor's employment under the sub-contract may be terminated in circumstances where there are grounds for excluding the sub-contractor under regulation 57 (*Mandatory exclusions*); and
- in circumstances where the Employer requires it, in accordance with regulation 71(9) (*Excluding subcontractors*), that the sub-contractor's employment under the sub-contract must be terminated by the Contractor, and a replacement sub-contractor appointed.

[23] Available at: https://www.legislation.gov.uk/uksi/2015/102/memorandum/contents

Schedule 3: Insurance Options

Overview

Schedule 3 relates to clause 6.7 (Insurance of the Works and Existing Structures: Insurance Options and period), and includes the three Insurance Options:

- Insurance Option A (New Buildings – All Risks Insurance of the Works by the Contractor);
- Insurance Option B (New Buildings – All Risks Insurance of the Works by the Employer); and
- Insurance Option C (Joint Names Insurance by the Employer of Existing Structures and Works in or Extensions to them).

They each identify the Party responsible for effecting a Joint Names Policy under the Insurance Option that applies, defined as the '*Works Insurance Policy*'.

The applicable Insurance Option will be as stated against the reference to clause 6.7 and Schedule 3 in the Contract Particulars.

Insurance Option A

(New Buildings – All Risks Insurance of the Works by the Contractor)

Paragraph A.1 (Contractor to effect and maintain a Joint Names Policy) requires the Contractor to effect and maintain a Joint Names Policy for All Risks Insurance.

Resource 317 page 602	Correspondence: Contractor to Employer	The provision by the Contractor to the Employer of the Joint Names Policy.

As an alternative to the Contractor's obligations under paragraph A.1, paragraph A.2 (Use of Contractor's annual policy – as alternative) permits the Contractor to use its annual insurance policy instead, provided that it provides the necessary All Risks Insurance and is a Joint Names Policy.

Resource 318 page 602	Correspondence: Contractor to Employer	Confirmation by the Contractor to the Employer of the use of the Contractor's annual policy, as an alternative to a Joint Names Policy.

DOI: 10.1201/9781003485919-80

Paragraph A.3 (Loss or damage) relates to loss or damage affecting executed work and Site Materials, and to clause 6.13 (Loss or damage – insurance claims and reinstatement).

Insurance Option B

(New Buildings – All Risks Insurance of the Works by the Employer)

Paragraph B.1 (Employer to effect and maintain a Joint Names Policy) requires the Employer to effect and maintain a Joint Names Policy for All Risks Insurance.

Resource 319 page 603	Correspondence: Employer to Contractor	The provision by the Employer to the Contractor of the Joint Names Policy.

Paragraph B.2 (Loss or damage) relates to loss or damage affecting executed work and Site Materials, and to clause 6.13 (Loss or damage – insurance claims and reinstatement).

Insurance Option C

(Joint Names Insurance by the Employer of Existing Structures and Works in or Extensions to them)

Paragraph C.1 (Existing Structures and contents – Joint Names Policy for Specified Perils) requires the Employer to effect and maintain a Joint Names Policy in respect of the Existing Structures (together with the contents of them either owned by the Employer, or for which the Employer is responsible), against loss or damage due to any of the '*Specified Perils*', which are defined in clause 6.8 (Insurance of the Works and Existing Structures: Related definitions).

Resource 320 page 604	Correspondence: Employer to Contractor	The provision by the Employer to the Contractor of the Joint Names Policy for Specified Perils.

Paragraph C.2 (The Works – Joint Names Policy for All Risks) requires the Employer to effect and maintain a Joint Names Policy for All Risks Insurance.

Resource 321 page 604	Correspondence: Employer to Contractor	The provision by the Employer to the Contractor of the Joint Names Policy for All Risks.

Paragraph C.3 (Loss or damage) relates to loss or damage to any executed work, Site Materials, and Existing Structures or their contents, and to both clause 6.13 (Loss or damage – insurance claims and reinstatement) and clause 6.14 (Loss or damage to Existing Structures – right of termination).

Schedule 4: Code of Practice

Schedule 4 contains a Code of Practice, which relates to clause 3.13.3 (Work not in accordance with the Contract).

The Code aims to assist the Employer in the '*fair and reasonable operation*' of the requirements of clause 3.13.3, in relation to instructions under clause 3.12 (Inspection – tests) to open up works for inspection, or to test materials, goods, or work, in order to help the Employer understand the probability or extent of any *further* instances of materials, goods, or work not in accordance with the Contract.

Whilst the Code encourages the Employer and the Contractor to agree the extent and methodology of any such opening up and testing, it also provides fifteen criteria which the Employer must consider when issuing these instructions, including:

- the significance of the non-compliance;
- the potential consequences for the safety of the building;
- the Contractor's record keeping;
- the reason for the non-compliance; and
- potential alternative testing methods.

DOI: 10.1201/9781003485919-81

Schedule 5: Third Party Rights

Overview

Schedule 5 is divided into two parts:

- Part 1: Third Party Rights for Purchasers and Tenants, described as '*P&T Rights*'; and
- Part 2: Third Party Rights for a Funder, described as '*Funder Rights*'.

Part 1: Third Party Rights for Purchasers and Tenants ('P&T Rights')

The P&T Rights relate to clause 7.7 (Rights for Purchasers and Tenants).

They are relevant where the Rights Particulars (identified against the reference to clause 7.4 in the Contract Particulars) require the Contractor to confer them on a Purchaser or Tenant as third party rights (rather than by way of a collateral warranty), and they will vest in the Purchaser or Tenant when the Contractor receives the relevant notice from the Employer, in accordance with clause 7.7.1.

Paragraph 5

Resource 322 page 605	Notice: Contractor to Purchaser or Tenant	A notice from the Contractor to a Purchaser or Tenant, if Professional Indemnity insurance ceases to be available.
Resource 323 page 606	Correspondence: Contractor to Purchaser or Tenant	The provision by the Contractor to a Purchaser or Tenant of documentary evidence of Professional Indemnity insurance.

Part 2: Third Party Rights for a Funder ('Funder Rights')

The Funder Rights relates to clause 7.8 (Rights for a Funder).

They are relevant where the Rights Particulars (again, identified against the reference to clause 7.4 in the Contract Particulars) require the Contractor to confer them on a Funder as third party rights (rather than by way of a collateral warranty),

DOI: 10.1201/9781003485919-82

and they will vest in the Funder when the Contractor receives the relevant notice from the Employer, in accordance with clause 7.8.1.

Paragraph 6.1.1

Resource 324 page 606	Notice: Contractor to Funder	A notice from the Contractor to a Funder, relating to step-in rights, and prior to the Contractor being entitled to terminate its employment under the Contract.

Paragraph 6.1.2

Resource 325 page 607	Notice: Contractor to Funder	A notice from the Contractor to a Funder, where the Contractor has the right to terminate its employment under the Contract.

Paragraph 6.2

Resource 326 page 608	Notice: Contractor to Funder	A notice from the Contractor to a Funder, where the Contractor has the right to treat the Contract as having been repudiated by the Employer.

Paragraph 9

Resource 327 page 609	Notice: Contractor to Funder	A notice from the Contractor to a Funder, if Professional Indemnity insurance ceases to be available.
Resource 328 page 610	Correspondence: Contractor to Funder	The provision by the Contractor to a Funder of documentary evidence of Professional Indemnity insurance.

Schedule 6: Forms of Bonds

Overview

Schedule 6 is divided into three parts, each containing a form of bond:

- Part 1: Advance Payment Bond
- Part 2: Bond in respect of payment for off-site materials and/or goods
- Part 3: Retention Bond

Each form of bond has been agreed between the JCT and the British Bankers' Association (which merged into UK Finance in 2017).

They all provide performance security for the Employer, whereby the relevant Surety will make a payment to the Employer (up to an agreed maximum amount), following a demand made by the Employer.

Part 1: Advance Payment Bond

The advance payment bond relates to clause 4.6 (Advance payment), and is to be made between the Surety and the Employer.

Whether clause 4.6 applies will be stated against the reference to that clause in the Contract Particulars.

If clause 4.6 does apply, then whether an advance payment bond is required will also be stated against the reference to that clause in the Contract Particulars.

It is not applicable in circumstances where the Employer is a Local Authority or other public sector body.

Part 2: Bond in respect of payment for off-site materials and/or goods

The bond in respect of payment for off-site materials and/or goods relates to clause 4.15 (Listed Items), and is to be made between the Surety and the Employer.

It relates to both Listed Items which are:

- uniquely identified, in accordance with clause 4.15.4; and
- *not* uniquely identified, in accordance with clause 4.15.5.

Whether a bond is required in respect of uniquely identified Listed Items will be stated against the reference to clause 4.15.4 in the Contract Particulars.

If there are no uniquely identified Listed Items, the entry against the reference to clause 4.15.5 in the Contract Particulars should be deleted.

DOI: 10.1201/9781003485919-83

Part 3: Retention Bond

The Retention Bond relates to clause 4.17 (Retention Bond), and is to be issued by the Surety in favour of the Employer.

Whether clause 4.17 applies, and therefore whether a Retention Bond is required, will be stated against the reference to that clause in the Contract Particulars.

It is not applicable in circumstances where the Employer is a Local Authority or other public sector body.

JCT Fluctuations Options

Overview

The Contract allows the Parties to adopt one of three JCT Fluctuations Options: A, B, or C, all of which permit further adjustments of the Contract Sum in certain circumstances, to take account of inflation.

Key provisions

- JCT Fluctuations Option A (Contribution, levy and tax fluctuations)
- JCT Fluctuations Option B (Labour and materials cost and tax fluctuations)
- JCT Fluctuations Option C (Formula adjustment)

The three JCT Fluctuations Options are as follows:

- **JCT Fluctuations Option A (Contribution, levy and tax fluctuations)**, which provides for increases or decreases in contributions, levies, and taxes.
- **JCT Fluctuations Option B (Labour and materials cost and tax fluctuations)**, where the Parties have agreed to allow certain cost and tax fluctuations to labour and materials, in accordance with the following paragraphs of that Option:
  - paragraph B.1 (Deemed calculation of Contract Sum – labour rates etc.);
  - paragraph B.2 (Deemed calculation of Contract Sum – labour levies and taxes); and
  - paragraph B.3 (Deemed calculation of Contract Sum – materials, goods, electricity and fuels).
- **JCT Fluctuations Option C (Formula adjustment)**, where the Parties have agreed that fluctuations should be dealt with by adjustment of the Contract Sum under the JCT Formula Rules.

Fluctuations Provision

If any of the JCT Fluctuations Options are selected within the Contract Particulars, they become the Fluctuations Provision, as provided for in clause 1.1 (Definitions) as follows:

> *1.1* ***Fluctuations Provision:*** *the provision (if any) specified by the Contract Particulars (against the reference to clause 4.2, 4.12 and 4.13).*

DOI: 10.1201/9781003485919-84

The Fluctuations Provision is referenced in the following provisions:

- clause 4.2.3 (Contract Sums and Adjustments: Items included in adjustments), relating to the adjustment of the Contract Sum;
- clause 4.12 (Gross Valuation – Alternative A), relating to the calculation of the Gross Valuation in accordance with Alternative A (Stage Payments);
- clause 4.13 (Gross Valuation – Alternative B), relating to the calculation of the Gross Valuation in accordance with Alternative B (Periodic Payments); and
- the corresponding entries against the references to clause 4.2, clause 4.12, and clause 4.13 in the Contract Particulars.

Further reading: JCT Fluctuations Options A, B, and C

The printed text of the 2024 edition of the JCT Fluctuations Options is no longer included in the Contract, and is therefore not considered in any further detail here, but it continues to be available on the JCT website.[24]

[24] Available at: https://www.jctltd.co.uk/docs/Fluctuations_Options_ABC/DB-2024-FluctuationsOptionsABC.pdf

Record keeping: Further resources

As has been emphasised throughout, consistent and accurate record keeping is key to the successful administration of the Contract.

Below are a number of further general templates which the Parties may find useful in this regard.

Resource 329 page 610	Template: Contractor's progress report	A monthly progress report to be prepared by the Contractor.
Resource 330 page 611	Template: Meeting agenda	A meeting agenda template for use by the Contractor and the Employer.
Resource 331 page 612	Template: Meeting minutes	A meeting minutes template for use by the Contractor and the Employer.
Resource 332 page 612	Template: Memorandum	A memorandum template for use by the Contractor and the Employer.
Resource 333 page 613	Template: Site Diary	A site diary template for use by the Contractor.
Resource 334 page 614	Template: Project calendar	A project calendar template for use by the Contractor and the Employer.
Resource 335 page 615	Template: Instruction	An instruction template for use by the Employer

DOI: 10.1201/9781003485919-85

Market practice: Additional provisions

Overview

As part of the negotiation and drafting of the Contract, and in addition to amending the existing provisions, the Parties will often introduce entirely new provisions, to help address matters specific to the project or to the Parties.

A brief description of some of the more common provisions is included below.

Subject	Description
Entire agreement	The Parties agree that the Contract constitutes the entire agreement between them, and supersedes any previous agreements, such as a letter of intent or a pre-construction services agreement.
Novation: consultants, other contractors, etc.	The Contractor agrees to accept a novation of a consultant appointment or other building contract (for example, in respect of demolition and enabling works) previously entered into by the Employer.
Overall liability cap	The Parties agree to an overall cap on the Contractor's liability under the Contract. The JCT has recognised that the Parties may wish to include such an overall cap in this regard, and paragraph 34 of the JCT Design and Build Contract Guide (2024 edition) provides sample drafting for a new clause 2.1.6: '2.1.6 *Subject to the exclusions set out in clauses 2.1.6.1 to [2.1.6.4 / 2.1.6.5], the Contractor's total aggregate liability for any loss, damage, cost or expense arising in respect of the liability of the Contractor under, arising out of or in connection with this Contract whether under the terms of this Contract, for breach of it, in tort, for negligence or breach of statutory duty or otherwise shall be limited to the amount, if any, stated in the Contract Particulars. The limitation referred to in this clause 2.1.6 excludes the following:* *2.1.6.1 any word for or in connection with the provision of a dwelling or dwellings or work in relation to a dwelling or dwellings to which the Defective Premises Act 1972 applies;*

(*Continued*)

DOI: 10.1201/9781003485919-86

Subject	Description
	2.1.6.2 loss, damage, cost or expense arising from fraud, wilful default or deliberate concealment; *2.1.6.3 bribery or corruption offences as referred to in clause 8.6; [or]* *2.1.6.4 any expense, liability, loss, claim or proceedings as referred to in clauses 6.1 or 6.2[; or* *2.1.6.5 any liability for liquidated damages under clause 2.29].'*
Third party agreements	The Employer has entered into various agreements with third parties in relation to the Works, and the Contractor warrants and undertakes that it will carry out the Works so as not to place the Employer in breach of any of these third party agreements.
Deleterious materials	The Contractor agrees not to specify or use any materials considered to be deleterious (or harmful) in carrying out the design and construction of the Works.
Site conditions	One of the Parties (usually the Contractor) accepts liability for the site conditions (possibly including any Existing Structures).
Building warranty / guarantee provider	The Contractor agrees to carry out the Works in accordance with the requirements of any building warranty or guarantee provider.
BREEAM	The Contractor agrees to carry out and complete the design and construction of the Works so as to meet any specified BREEAM requirements. BREEAM (Building Research Establishment Environmental Assessment Methodology) is a sustainability assessment method, relating to the environmental performance of buildings.
Target areas	The Contractor agrees to carry out and complete the design and construction of the Works so as to achieve any target areas (for example, relating to residential apartments or commercial premises).
Defects priority classification	The Contractor agrees to make good any defects in accordance with a priority classification system (for example, 'emergency' defects, 'urgent' defects, and 'routine' defects).
Key personnel	The Contractor agrees to retain a team of specified key staff for the duration of the project.
Reporting requirements	The Contractor agrees to provide a regular (typically monthly) progress report to the Employer.
Programme requirements	The Contractor agrees to provide the Employer with regular programme updates, illustrating the progress of the Works.
Modern slavery	The Parties agree to further obligations in respect of modern slavery, in addition to the applicable Statutory Requirements.

(*Continued*)

Subject	Description
Anti-bribery and anti-corruption	The Parties agree to further obligations in respect of bribery and corruption, in addition to the applicable Statutory Requirements.
Nuisance and trespass	The Contractor agrees to prevent any public or private nuisance or other interference with any nearby landowners, tenants or occupiers, and to ensure that there is no trespass by the Contractor (or any Contractor's Persons) over any nearby property.
Contractor to vacate the site upon termination	The Contractor agrees to vacate the site immediately upon the termination of its employment under the Contract, irrespective of the grounds of termination.

Further reading: The Chancery Lane Project

The Parties may also wish to consider The Chancery Lane Project,[25] which provides template climate-aligned clauses, for use in commercial contracts.

[25] Available at: https://chancerylaneproject.org/clauses/?jurisdiction=109&practice-area=23

Part 2

Resources for Contract Administration

Overview

Part 2 contains 335 resources to assist both the Contractor and the Employer in the consistent and accurate administration of the Contract, including the following:

- notice and other correspondence templates;
- checklists;
- trackers; and
- further templates to assist with record keeping.

It includes templates for notices and other correspondence which are required to be issued under the Contract, along with templates for other correspondence which is not mandatory, but is considered good practice.

In the interests of clarity and accuracy, the resources have been drafted to adhere as closely as possible to the language and phrasing used in the Contract, and where appropriate, the relevant clause references and their descriptions have been included.

Key principles for drafting correspondence and other documents required under the Contract

When drafting any correspondence or other documents required under the Contract (and whether or not adopting any of the resources included in Part 2), the Parties are encouraged to adhere to the following guidelines:

- Review the relevant provisions of the Contract first.
- Establish the objective of the correspondence or document, and what it is intended to achieve.
- Adhere as closely as possible to the language and phrasing used in the Contract.
- Where appropriate, expressly reference the relevant provisions of the Contract, and their descriptions. In some cases, it may be helpful to set out the relevant provisions in full.
- Include all relevant particulars, including dates, durations, sums, document references, etc.
- As far as possible, try to anticipate any further questions the recipient may have, or any further information or documents they may require.
- Issue the correspondence or document at the earliest possible opportunity, and certainly within any applicable timeframes.
- Copy the correspondence or document to all relevant parties (for example, the Contractor, the Employer, and the Employer's Agent).

DOI: 10.1201/9781003485919-88

Drafting assumptions

The resources in Part 2 have been drafted on the basis of a number of assumptions, including the following:

- that the complete drafting of the 2024 edition of the JCT Design and Build Contract applies, without any amendments (for example, by way of a schedule of amendments);
- that both the Contractor and the Employer are private limited companies registered in England and Wales;
- that correspondence from the Contractor is addressed to the Employer's Agent, rather than to the Employer;
- that correspondence is to be sent by email (for example, with references to 'attachments', rather than 'enclosures'); and
- that, in the interests of brevity, the Parties can include their own salutations and expressions of gratitude (or otherwise).

Resource 1

Description	The nomination by the Employer of a replacement Employer's Agent.
Type	Correspondence: Employer to Contractor
Provision references	Article 4 (Employer's Agent)

Reference: Nomination of replacement Employer's Agent (Article 4)

In accordance with Article 4 (Employer's Agent), we write to inform you that we have nominated a replacement Employer's Agent.

[Effective immediately] / [From [*insert date*]], we have nominated [*insert name of new Employer's Agent*] as the Employer's Agent, in place of [*insert name of previous Employer's Agent.*]

Their contact details are as follows: [*insert contact details.*]

We will arrange an introductory meeting as soon as possible.

In the meantime, we confirm that [*insert name of replacement Employer's Agent*] has our full authority to receive and issue applications, consents, instructions, notices, requests, and statements, and to otherwise act for us under any of the Conditions.

Resource 2

Description	A revision by the Employer to the Employer's Agent's authority to act on its behalf.
Type	Notice: Employer to Contractor
Provision references	Article 4 (Employer's Agent)

Reference: Revision to Employer's Agent's authority to act on Employer's behalf (Article 4)

In accordance with Article 4 (Employer's Agent), we write to notify you of a revision to the Employer's Agent's authority to act on our behalf.

Previously, the Employer's Agent had full authority to receive and issue applications, consents, instructions, notices, requests, and statements, and otherwise to act for us under any of the Conditions.

Effective immediately, the Employer's Agent will now have the authority to act for us as follows: [*insert details of the revision to the Employer's Agent's authority.*]

[As a consequence, we will now perform the following obligations directly: [*insert list of obligations which the Employer will now perform directly, instead of the Employer's Agent. For example, issuing instructions.*]]

Resource 3

Description	The replacement of the Employer's nominee or Contractor's nominee, relating to clause 9.1.
Type	Notice: Contractor or Employer to the other Party
Provision references	Contract Particulars, clause 9.1

Reference: Replacement of the [Contractor's] / [Employer's] nominee (clause 9.1)

In relation to clause 9.1 (Notification and negotiation of disputes), and in accordance with the entry against the reference to clause 9.1 in the Contract Particulars, we write to notify you that we have appointed a replacement nominee.

[Effective immediately] / [From [*insert date*]], we have appointed [*insert name of replacement nominee*], a senior executive of the [Contractor] / [Employer], in place of [*insert name of previous nominee*].

Their contact details are as follows: [*insert contact details.*]

Resource 4

Description	A notice of a conflict within the Contract Documents.
Type	Notice: Contractor or Employer to the other Party
Provision references	clause 1.3
Drafting notes	The Parties are encouraged to notify each other of any such conflict as soon as possible. This template is designed for a relatively simple, direct conflict, which can be readily resolved in favour of the Agreement and/or the Conditions. However, other conflicts may require a more comprehensive application of the principles of contract interpretation. For discrepancies or divergences in or between the Employer's Requirements, Contractor's Proposals, Contractor's Design Documents, Changes, or the Statutory Requirements, refer to the relevant resources relating to clause 2.12 (Employer's Requirements – inadequacy), clause 2.13 (Notice of discrepancies etc.), clause 2.14 (Discrepancies in documents), and clause 2.15 (Divergences from Statutory Requirements).

Reference: Notice of a conflict within the Contract Documents (clause 1.3)

We write to notify you of the following conflict within the Contract Documents:

[*Insert details of the conflict between (i) the Agreement and/or Conditions, and (ii) another Contract Document or Framework Agreement.*]

In accordance with clause 1.3 (Agreement etc. to be read as a whole), nothing contained in any other Contract Document or any Framework Agreement, irrespective of their terms, will override or modify the Agreement or the Conditions.

The conflict will therefore be resolved in favour of the Agreement and/or the Conditions, as follows: [*insert description of how the Contract is to be interpreted in favour of the Agreement and/or the Conditions.*]

Resource 5

Description	A communication protocol template to govern the exchange of notices, communications, and documents between the Parties.
Type	Communication Protocol: Employer and Contractor
Provision references	clause 1.7.2
Drafting notes	The communication protocol and the project directory should be regularly reviewed, updated, and distributed. For the purposes of clause 1.7.4, details of the Royal Mail's Signed For 1st class and Special Delivery Guaranteed services are available at: https://www.royalmail.com/sending/uk

Communication protocol

Revision: [*insert revision*]

Updated: [*insert date of last update*]

Distribution: [*insert list of parties to whom the communication protocol is to be distributed.*]

Objective

In accordance with clause 1.7.2 (Notices and other communications), but subject to clause 1.7.4, notices, other communications, and any documents to be supplied may or (where so required) shall be sent or transmitted by the means (electronic or otherwise) and in such format as the Parties have agreed, or may from time to time agree, in writing for the purposes of the Contract.

The Parties have therefore agreed the following communication protocol for the project:

General correspondence: Email

Subject to clause 1.7.4, all notices, communications, and documents may be sent by email.

All correspondence should be copied to the following email addresses:

- Employer: [*insert email addresses*]
- Employer's Agent: [*insert email addresses*]
- Contractor: [*insert email addresses*]
- Site Manager: [*insert email address*]
- Deputy Site Manager: [*insert email address*]

Design Submission Procedure: Document Management Platform

The project document management platform will be used to give effect to the Design Submission Procedure (contained in Schedule 1).

Please contact [*insert name and email address*] to apply for access to the project document management platform.

Signed For 1st class / Special Delivery Guaranteed post (clause 1.7.4)

In accordance with clause 1.7.4, any notice expressly required by the Contract to be given in accordance with that clause must be delivered by hand, or sent by Signed For 1st class or Special Delivery Guaranteed post.

For this purpose, the Parties intend to use Royal Mail's 'Special Delivery Guaranteed by 1pm' service, the details of which are available at: https://www.royalmail.com/sending/uk/special-delivery-guaranteed-1pm.

[Alternatively, given that clause 1.7.4.2 is stated in the Contract Particulars to apply, any notice expressly required by the Contract to be given in accordance with clause 1.7.4 may be sent by email to the following email addresses stated in the Contract Particulars:

- Employer: [*insert email address*]
- Contractor: [*insert email address*]]

Project directory

The project directory is available to download at the following link: [*insert link to project directory.*]

Resource 6

Description	A notice to the other Party confirming a change of email address.
Type	Notice: Contractor or Employer to the other Party
Provision references	clause 1.7.4.2
Drafting notes	This notice is only applicable where clause 1.7.4.2 is stated in the Contract Particulars to apply. **Timing:** The Parties are encouraged to give this notice as soon as possible after deciding to change their email address, to avoid any uncertainty as to whether a notice under clause 1.7.4 has been correctly sent.

Reference: Notice of change of email address (clause 1.7.4.2)

In accordance with clause 1.7.4.2 (Notices and other communications), which is stated in the Contract Particulars to apply, we write to notify you of the following change of email address:

- Previous email address: [*insert email address*]
- Replacement email address: [*insert email address*]

Please use this replacement email address for any notice expressly required by the Contract to be given in accordance with clause 1.7.4.

Resource 7

Description	The confirmation of an oral communication, in the case of an emergency.
Type	Confirmation: Contractor or Employer to the other Party
Provision references	clause 1.7.5
Drafting notes	**Timing:** Clause 1.7.5 requires the Parties to send written confirmation to the other as soon after the oral communication as is 'reasonably practicable'. Ideally, this will be undertaken the same day.

Reference: Confirmation of oral communications in case of emergency (clause 1.7.5)

We write further to the emergency which occurred on [*insert date*], in respect of the following: [*insert details of the emergency.*]

In accordance with clause 1.7.5 (Notices and other communications), we confirm that, during this emergency, the following communications were made orally with respect to health and safety, risk of damage to property, and/or insurance matters:

[*Insert details of the communications, including the relevant parties, dates, content, whether the communications were made in person or by telephone, etc.*]

Resource 8

Description	A reminder that consents or approvals must not be unreasonably delayed or withheld.
Type	Reminder: Employer and Contractor
Provision references	clause 1.10
Drafting notes	When requesting consent or approval, the Parties are encouraged to specify the date by which that consent or approval is required, and an explanation of the potential consequences if it is not provided (for example, a delay to the progress of the Works).

Reference: Request for consent or approval (clause 1.10)

We write to remind you that, in accordance with clause 1.10 (Consents and approvals), where consent or approval of either Party is expressly required under the Conditions and is requested, such consent or approval must not be unreasonably delayed or withheld.

In accordance with clause [*insert clause reference*], we requested your [consent] / [approval] on [*insert date*] in respect of the following: [*insert details of the consent or approval requested.*]

[We have yet to receive your [consent] / [approval], and request that you give it as soon as possible, and by no later than [*insert date*], in order to avoid delay to the progress of the Works.] /

[We consider that your decision to withhold your [consent] / [approval] was unreasonable, on the following basis: [*insert explanation as to why the decision to withhold consent or approval was unreasonable.*] We therefore request that you reconsider, and give your consent or approval as soon as possible, and by no later than [*insert date*], in order to avoid delay to the progress of the Works.]

Resource 9

Description	The provision by the Contractor to the Employer of approvals received by the Contractor, in connection with the Statutory Requirements.
Type	Correspondence: Contractor to Employer
Provision references	clause 2.1.3
Drafting notes	**Good practice** Clause 2.1.3 does not require the Contractor to give this notice, but it is considered good practice to inform them of any such approvals as soon as possible.

Reference: Receipt of approval in connection with the Statutory Requirements (clause 2.1.3)

In accordance with clause 2.1.3 (Contractor's Obligations: General obligations), we attach the following approval we received from [*insert name of party providing approval*] on [*insert date*], in connection with the Statutory Requirements: [*insert details of the approval received.*]

This approval relates to the following Statutory Requirements: [*insert details of the relevant Statutory Requirements.*]

Resource 10

Description	A tracker template for recording approvals in connection with the Statutory Requirements.
Type	Tracker: Contractor and Employer
Provision references	clause 2.1.3

Below is a tracker template for recording approvals received in connection with the Statutory Requirements:

- Approval reference: [*insert reference*]
- Approval description: [*insert description*]
- Statutory Requirements: [*insert applicable Statutory Requirements*]
- Contract Documents reference: [*insert reference*]
- Date approval required: [*insert date*]
- Date approval received: [*insert date*]
- Contact: [*insert name and email address*]
- Document link: [*insert link to documents*]
- Status / action required: [*insert status of the approval, and any action required*]

Resource 11

Description	A reminder from the Employer to the Contractor of the Contractor's obligations to comply with Employer's instructions, and to be bound by the Employer's decisions.
Type	Reminder: Employer to Contractor
Provision references	clause 2.1.4
Drafting notes	**Good practice** Clause 2.1.4 does not require the Employer to give this reminder, but it is considered good practice if the Employer considers that the Contractor is not complying with a particular instruction or decision.

Reference: Obligation to comply with Employer's instructions (clause 2.1.4)

In accordance with clause 2.1.4 (Contractor's Obligations: General obligations), we remind you of your obligation to comply with the Employer's instructions, and to be bound by the Employer's decisions, which are issued or made under or pursuant to the Conditions.

In particular, we remind you of your obligation to comply with the following [instruction] / [decision], dated [*insert date*], with reference [*insert reference*]: [*insert details of the instruction or decision.*]

This [instruction] / [decision] will have effect, except to the extent that we vary it, or it is varied under the dispute resolution procedures of the Contract.

Resource 12

Description	A proposed change to the Works by the Contractor, to improve environmental performance and sustainability.
Type	Proposed change: Contractor to Employer
Provision references	clause 2.1.5

Reference: Sustainable development and environmental considerations (clause 2.1.5)

In accordance with clause 2.1.5 (Contractor's Obligations: General obligations), we propose the following economically viable amendment to the Works which, if instructed as a Change, may result in an improvement in environmental performance and sustainability, and a reduction in environmental impact [in the carrying out of the Works] / [of the completed Works]: [*insert details of the proposed change, and the potential improvement in environmental performance and sustainability, and reduction in environmental impact.*]

Resource 13

Description	A request by the Contractor for the Employer's consent to substitute materials or goods.
Type	Request for consent: Contractor to Employer
Provision references	clause 2.2.1

Reference: Request for consent to substitute materials or goods (clause 2.2.1)

In accordance with clause 2.2.1 (Materials, goods and workmanship), we request your consent to substitute certain materials or goods described in the Employer's Requirements, Contractor's Proposals, or other Contractor's Design Documents.

In particular, we request your consent to substitute the following:

Details of materials or goods: [*insert description of materials or goods.*]

Reference: [*insert reference within the Employer's Requirements, Contractor's Proposals, or other Contractor's Design Documents.*]

We propose to substitute those materials or goods with the following: [*insert description of proposed replacement materials or goods.*]

Further details of the proposed materials or goods are attached.

We have proposed this substitution for the following reasons: [*insert reasons or potential benefits of the substitution to the Parties.*]

Please give your consent by [*insert date*].

We understand that, if your consent is given, this will not relieve us of our other obligations under the Contract.

Resource 14

Description	Consent given by the Employer for the Contractor to substitute materials or goods.
Type	Request for consent given: Employer to Contractor
Provision references	clause 2.2.1
Drafting notes	**Good practice** Clause 2.2.1 does not require the Employer's consent to be given by way of an instruction, but it is considered good practice, to avoid any uncertainty. The instruction should specify: • precisely which materials or goods are being substituted, to what extent and in which locations, and with appropriate reference to the Employer's Requirements, Contractor's Proposals, or other Contractor's Design Documents; • corresponding details of the replacement materials or goods; and • any deviation between the Contractor's request and the Employer's instruction.

Reference: Request for consent to substitute materials or goods; Employer's instruction [*insert reference*] (clause 2.2.1)

We write in reply to your request dated [*insert date*], with reference [*insert reference*], for our consent to substitute certain materials or goods described in the Employer's Requirements, Contractor's Proposals, or other Contractor's Design Documents.

We give our consent to this substitution, in accordance with the attached instruction.

Resource 15

Description	Consent withheld by the Employer for the Contractor to substitute materials or goods.
Type	Consent withheld: Employer to Contractor
Provision references	clause 2.2.1
Drafting notes	Clause 1.10 requires that the Employer's consent must not be unreasonably delayed or withheld.

Reference: Request for consent to substitute materials or goods (clause 2.2.1)

We write in reply to your request dated [*insert date*], with reference [*insert reference*], for our consent to substitute certain materials or goods described in the Employer's Requirements, Contractor's Proposals, or other Contractor's Design Documents.

We withhold our consent to this substitution, for the following reasons: [*insert reasons, to demonstrate why the consent was not unreasonably delayed or withheld.*]

Resource 16

Description	A request by the Employer for information on the environmental impact of materials and goods selected by the Contractor.
Type	Request: Employer to Contractor
Provision references	clause 2.2.2

Reference: Sustainable development and environmental considerations (clause 2.2.2)

We write in relation to the supply and use of the following materials and/or goods which you have selected: [*insert details of the materials and/or goods.*]

In accordance with clause 2.2.2 (Materials, goods and workmanship), we request that you provide us with the following information regarding the environmental impact of these materials and/or goods: [*insert details of the information required by the Employer.*]

Resource 17

Description	The provision of samples by the Contractor to the Employer.
Type	Correspondence: Contractor to Employer
Provision references	clause 2.2.4
Drafting notes	**Good practice** This correspondence is not required by clause 2.2.4, but it is considered good practice to give the Employer as much notice as possible of any samples.

Reference: Provision of samples (clause 2.2.4)

In accordance with clause 2.2.4 (Materials, goods and workmanship), before we carry out the relevant work and/or order the relevant materials or goods, we are required to provide the Employer with such samples of the standard of workmanship, or the quality of the materials or goods, which we intend to provide, as are specifically referred to in the Employer's Requirements or the Contractor's Proposals.

The [Employer's Requirements] / [Contractor's Proposals] require us to provide the following samples: [*insert description of samples, along with references from the Employer's Requirements or Contractor's Proposals.*]

These samples will be available for inspection from [*insert date*] in the following location: [*insert location, for example, site offices, a marketing suite, offsite workshops or other premises, etc.*]

Please confirm your availability.

We intend to [commence the relevant work] / [order the relevant materials or goods] by [*insert date*], in order to avoid delay to the progress of the Works.

Resource 18

Description	A tracker template for recording the provision of samples.
Type	Tracker: Contractor and Employer
Provision references	clause 2.2.4

Below is a tracker template for recording samples:

- Reference: [*insert sample reference*]
- Description: [*insert sample description*]
- Product details: [*insert details*]
- Supplier / Manufacturer: [*insert details*]
- Location: [*insert sample location*]
- Dates available for inspection: [*insert dates*]
- Order of relevant materials or goods date: [*insert date*]
- Commencement of relevant work date: [*insert date*]

- Document link: [*insert link to documents*]
- Employer comments: [*insert any comments on behalf of the Employer*]
- Status / action required: [*insert status, and any action required*]

Resource 19

Description	A request by the Employer for reasonable proof from the Contractor that materials and goods are compliant.
Type	Request: Employer to Contractor
Provision references	clause 2.2.5

Reference: Materials, goods, and workmanship; request for proof of compliance (clause 2.2.5)

In accordance with clause 2.2.5 (Materials, goods and workmanship), we request that you provide us with reasonable proof that the following materials and goods comply with clause 2.2:

[*insert description of materials and goods.*]

In particular, we seek confirmation that these materials and goods comply with the following provisions contained within the Employer's Requirements, Contractor's Proposals, or other Contractor's Design Documents: [*insert references to specifications, drawings, product data sheets, etc.*]

Resource 20

Description	The provision by the Contractor to the Employer of reasonable proof that materials and goods are compliant.
Type	Reply: Contractor to Employer
Provision references	clause 2.2.5
Drafting notes	The Contractor is encouraged to reply as soon as possible to the Employer's request.

Reference: Materials, goods, and workmanship; request for proof of compliance (clause 2.2.5)

Further to your request dated [*insert date*], we confirm that the materials and goods referenced in your request comply with clause 2.2 (Materials, goods and workmanship), on the following basis: [*insert explanation, demonstrating compliance.*]

We attach the following supporting documents: [*insert description of documents, for example, manufacturer correspondence, product data sheets, etc.*]

Please let us know if there is anything further we can assist with on this matter.

Resource 21

Description	Confirmation by the Contractor to the Employer of steps taken to encourage Contractor's Persons to be CSCS registered cardholders.
Type	Confirmation: Contractor to Employer
Provision references	clause 2.2.6
Drafting notes	**Good practice** This correspondence is not required by clause 2.2.6, but it is considered good practice to demonstrate the efforts the Contractor has made to encourage Contractor's Persons to be registered cardholders.

Reference: Construction Skills Certification Scheme (clause 2.2.6)

In accordance with clause 2.2.6 (Materials, goods and workmanship), we encourage all Contractor's Persons to be registered cardholders under the Construction Skills Certification Scheme (CSCS), or to be qualified under an equivalent recognised qualification scheme.

We have taken the following action in this regard: [*insert details, including details of any relevant correspondence with Contractor's Persons.*]

At the time of writing, we can confirm that [*insert percentage*] of Contractor's Persons are CSCS registered cardholders, or are qualified under an equivalent recognised qualification scheme.

Resource 22

Description	Confirmation by the Employer that it has given possession of the site (or part of it) to the Contractor.
Type	Confirmation: Employer to Contractor
Provision references	clause 2.3
Drafting notes	**Good practice** This correspondence is not required by clause 2.3, but it is considered good practice for the Employer to confirm the details of site possession.

Reference: Confirmation of site possession (clause 2.3)

We write to confirm that, in accordance with clause 2.3 (Date of Possession – progress), the Employer gave you possession of [the site] / [the relevant part of the site relating to Section [*insert references*]] on the following date: [*insert date*]

This date is also the [relevant] Date of Possession.

Please begin the construction of [the Works] / [the relevant Section], and regularly and diligently proceed with and complete them on or before the [relevant] Completion Date, namely [*insert date*].

For Works insurance purposes, you will retain possession of [the site and the Works up to and including the date of issue of the Practical Completion Statement] / [the Section and the relevant part of the site up to and including the date of issue of the Section Completion Statement for that Section and, in respect of any balance of the site, up to and including the date of issue of the Practical Completion Statement.]

We appreciate that, subject to clause 2.30 (Partial Possession by Employer: Contractor's consent) and section 8 (Termination), the Employer will not be entitled to take possession of any part or parts of the Works or Section until this date.

Resource 23

Description	Confirmation by the Contractor that it has taken possession of the site (or part of it) from the Employer.
Type	Confirmation: Contractor to Employer
Provision references	clause 2.3
Drafting notes	**Good practice** This correspondence is not required by clause 2.3, but it is considered good practice for the Contractor to confirm the details of site possession.

Reference: Confirmation of site possession (clause 2.3)

We write to confirm that, in accordance with clause 2.3 (Date of Possession – progress), the Employer gave us possession of [the site] / [the relevant part of the site relating to Section [*insert references*]] on the following date: [*insert date*]

This date is also the [relevant] Date of Possession.

We will begin the construction of [the Works] / [relevant Section], and regularly and diligently proceed with and complete them on or before the relevant Completion Date, namely [*insert date*].

For Works insurance purposes, we will retain possession of [the site and the Works up to and including the date of issue of the Practical Completion Statement] / [the Section and the relevant part of the site up to and including the date of issue of the Section Completion Statement for that Section and, in respect of any balance of the site, up to and including the date of issue of the Practical Completion Statement.]

We remind you that, subject to clause 2.30 (Partial Possession by Employer: Contractor's consent) and section 8 (Termination), the Employer will not be entitled to take possession of any part or parts of the Works or Section until this date.

Resource 24

Description	A notice from the Employer that it is deferring giving possession of the site (or part of it) to the Contractor.
Type	Notice: Employer to Contractor
Provision references	clause 2.4
Drafting notes	**Good practice** This correspondence is not required by clause 2.4, but it is considered good practice for the Employer to confirm any deferment of possession of the site to the Contractor.

Reference: Deferment of possession (clause 2.4)

We write to notify you that, in accordance with clause 2.4 (Deferment of possession), the Employer will defer the giving of possession of the site (or relevant part of it) as follows:

[Section: [*insert relevant Sections*]]

Date of Possession: [*insert date*]

Date on which the Employer intends to give possession: [*insert date*]

Duration of deferment: [*insert duration*]

Resource 25

Description	A request by the Employer for the Contractor's consent for early use.
Type	Request: Employer to Contractor
Provision references	clause 2.5.1
Drafting notes	The Employer should first check which Insurance Option applies.

Reference: Request for consent for early use (clause 2.5.1)

In accordance with clause 2.5.1 (Early use by Employer), we request your consent for the Employer to use or occupy the site or the Works or part of them, before the date of issue of the [Practical Completion Statement] / [relevant Section Completion Statement] as follows:

Area: [*insert details, for example, the site, relevant Sections, etc.*]

Purpose: [*insert description of why the Employer wishes to use or occupy the area.*]

Dates: [*insert proposed dates*]

[*Insurance Option A:* Prior to giving your consent, and given that Insurance Option A applies, please first notify the insurers of the Works Insurance Policy [and in relation to the Existing Structures], and request that they confirm that such use or occupation will not prejudice the insurance.] /

[*Insurance Option B or C:* Given that Insurance Option [B] / [C] applies, we have notified the insurers of the Works Insurance Policy [and in relation to the Existing Structures], and attach their confirmation that such use or occupation will not prejudice the insurance.]

Please give your consent by [*insert date*].

Resource 26

Description	Consent given by the Contractor to the Employer for early use.
Type	Request for consent given: Contractor to Employer
Provision references	clause 2.5.1

Reference: Request for consent for early use (clause 2.5.1)

We write in reply to your request dated [*insert date*], for our consent for the Employer to use or occupy the site or the Works or part of them, before the date of issue of the [Practical Completion Statement] / [relevant Section Completion Statement].

[*Insurance Option A:* Given that Insurance Option A applies, we attach confirmation from the insurers of the Works Insurance Policy [and in relation to the Existing Structures] that such use or occupation will not prejudice the insurance.] /

[*Insurance Option B or C:* We acknowledge the confirmation you provided from the insurers of the Works Insurance Policy [and in relation to the Existing Structures] that such use or occupation will not prejudice the insurance.]

[*Unconditional consent*: We give our consent.] / [*Conditional consent*: We give our consent, subject to the following conditions: [*insert conditions.*]]

Resource 27

Description	Consent withheld by the Contractor to the Employer for early use.
Type	Request for consent withheld: Contractor to Employer
Provision references	clause 2.5.1
Drafting notes	Clause 1.10 requires that the Contractor's consent must not be unreasonably delayed or withheld.

Reference: Request for consent for early use (clause 2.5.1)

We write in reply to your request dated [*insert date*], for our consent for the Employer to use or occupy the site or the Works or part of them, before the date of issue of the [Practical Completion Statement] / [relevant Section Completion Statement].

We withhold our consent, for the following reasons: [*insert reasons, to demonstrate why the consent was not unreasonably delayed or withheld.*]

Resource 28

Description	A request by the Contractor or Employer for confirmation from the insurers under the Works Insurance Policy (and in relation to the Existing Structures) that early use will not prejudice the insurance.
Type	Request: Contractor or Employer to insurer
Provision references	clause 2.5.1
Drafting notes	The Party requesting confirmation from the insurers should provide a copy of clause 2.5, to provide the insurers with context for the request.

Reference: Works insurance policy reference [*insert policy reference number*]

We refer to the following insurance policy:

[*Insert details of the Works Insurance Policy, and if applicable, any relating to the Existing Structures.*]

In accordance with clause 2.5.1 (Early use by Employer) of the building contract (the 'Contract') dated [*insert date*] between [*insert name*] (the 'Contractor') and [*insert name*] (the 'Employer'), the particulars of which are attached, the Employer wishes to use or occupy the site or the Works or part of them, before the date of issue of the [Practical Completion Statement] / [relevant Section Completion Statement] as follows:

Area: [*insert details, for example, the site, relevant Sections, etc.*]

Purpose: [*insert description of why the Employer wishes to use or occupy the area.*]

Dates: [*insert proposed dates*]

Please confirm that such use or occupation will not prejudice the insurance.

Resource 29

Description	A notice from the Contractor to the Employer of an additional insurance premium for early use.
Type	Notice: Contractor to Employer
Provision references	clause 2.5.2
Drafting notes	This notice is only relevant where Insurance Option A applies and/or where the Contractor is to cover its own or any other risks in relation to any Existing Structures.

Reference: Early use by Employer; additional insurance premium (clause 2.5.2)

We write in reply to your request dated [*insert date*], for our consent for the Employer to use or occupy the site or the Works or part of them, before the date of issue of the [Practical Completion Statement] / [relevant Section Completion Statement].

[Given that Insurance Option A applies, we have requested confirmation from the insurers of the Works Insurance Policy that such use or occupation will not prejudice the insurance.] /

[Given that we are to cover [our own risks] / [certain risks] in relation to the Existing Structures, we have requested confirmation from the insurers in relation to the Existing Structures that such use or occupation will not prejudice the insurance.] /

[Given that Insurance Option A applies, and we are to cover [our own risks] / [certain risks] in relation to the Existing Structures, we have requested confirmation from those insurers that such use or occupation will not prejudice the insurance.]

In accordance with clause 2.5.2 (Early use by Employer), we attach correspondence from the insurers, stating that their confirmation is conditional upon an additional premium of [*insert amount*] (exclusive of VAT) being paid.

Please confirm whether the Employer still requires such use or occupation, in which case, this additional premium will be added to the Contract Sum, and we will provide a corresponding receipt.

Resource 30

Description	An acceptance by the Employer of an additional insurance premium for early use.
Type	Confirmation: Employer to Contractor
Provision references	clause 2.5.2

Reference: Early use by Employer; additional insurance premium (clause 2.5.2)

We write in reply to your notice dated [*insert date*], issued in accordance with clause 2.5.2 (Early use by Employer).

We confirm that the Employer still requires such use or occupation referred to in our correspondence dated [*insert date*].

We acknowledge that the insurers' confirmation is conditional upon an additional premium of [*insert amount*] (exclusive of VAT) being paid, and that this will be added to the Contract Sum.

We request that you make payment of this additional premium to the insurers, and provide us with the corresponding receipt.

Resource 31

Description	A reminder from the Employer to the Contractor of work not forming part of the Contract.
Type	Reminder: Employer to Contractor
Provision references	clause 2.6.1
Drafting notes	**Good practice** This correspondence is not required by clause 2.6.1, but it is considered good practice to remind the Contractor of work not forming part of the Contract.

Reference: Work not forming part of the Contract (clause 2.6.1)

We write in regard to the following work not forming part of the Contract which the Employer requires to be carried out (whether by the Employer directly, or by any Employer's Persons):

[*Insert description of work not forming part of the Contract, including relevant details relating to the programme, the scope of works, access requirements, etc.*]

We remind you that, in respect of this work, the Contract Documents provide the information necessary to enable you to carry out and complete [the Works] / [the relevant Section(s)] in accordance with the Contract. In particular, we refer you to the following Contract Documents:

[*Insert references to the Contract Documents relating to the work not forming part of the Contract.*]

In accordance with clause 2.6.1 (Work not forming part of the Contract), you are required to permit the execution of such work.

Resource 32

Description	Permission given by the Contractor to the Employer for the execution of work not forming part of the Contract.
Type	Confirmation: Contractor to Employer
Provision references	clause 2.6.1
Drafting notes	**Good practice** This correspondence is not required by clause 2.6.1, but it is considered good practice for the Contractor to confirm that it permits the execution of work not forming part of the Contract.

Reference: Work not forming part of the Contract (clause 2.6.1)

We write in reply to your correspondence dated [*insert date*], relating to certain work not forming part of the Contract, which the Employer requires to be carried out.

We agree that, in respect of this work, the Contract Documents provide the information necessary to enable us to carry out and complete [the Works] / [the relevant Section(s)] in accordance with the Contract.

Therefore, in accordance with clause 2.6.1 (Work not forming part of the Contract), we confirm that we permit the execution of such work.

Resource 33

Description	A request by the Employer for the Contractor's consent for the Employer to arrange for the execution of work not forming part of the Contract.
Type	Request: Employer to Contractor
Provision references	clause 2.6.2
Drafting notes	The Employer is encouraged to request the Contractor's consent as early as possible, to help coordinate the works not forming part of the Contract.

Reference: Work not forming part of the Contract (clause 2.6.2)

The Employer requires the following work not forming part of the Contract to be carried out (whether by the Employer directly, or by any Employer's Persons):

[*Insert description of work not forming part of the Contract, including relevant details relating to the programme, the scope of works, access requirements, etc.*]

We accept that, in respect of this work, the Contract Documents do not provide the information necessary to enable you to carry out and complete [the Works] / [the relevant Section(s)] in accordance with the Contract.

Therefore, in accordance with clause 2.6.2, we request your consent to arrange for the execution of this work.

Please give your consent by [*insert date*].

Resource 34

Description	Consent given by the Contractor for the Employer to arrange for the execution of work not forming part of the Contract.
Type	Request for consent given: Contractor to Employer
Provision references	clause 2.6.2

Reference: Work not forming part of the Contract (clause 2.6.2)

We write in reply to your request dated [*insert date*], for our consent for the Employer to arrange the execution of certain work not forming part of the Contract.

[*Unconditional consent*: We give our consent.] / [*Conditional consent*: We give our consent, subject to the following conditions: [*insert conditions.*]]

Resource 35

Description	Consent withheld by the Contractor for the Employer to arrange for the execution of work not forming part of the Contract.
Type	Request for consent withheld: Contractor to Employer
Provision references	clause 2.6.2
Drafting notes	Clause 1.10 requires that the Contractor's consent must not be unreasonably delayed or withheld.

Reference: Work not forming part of the Contract (clause 2.6.2)

We write in reply to your request dated [*insert date*], for our consent for you to arrange the execution of certain work not forming part of the Contract.

We withhold our consent, for the following reasons: [*insert reasons, to demonstrate why the consent was not unreasonably delayed or withheld.*]

Resource 36

Description	A request by the Contractor to the Employer to inspect the Contract Documents.
Type	Request: Contractor to Employer
Provision references	clause 2.7.1

Reference: Inspection of Contract Documents (clause 2.7.1)

In accordance with clause 2.7.1 (Contract Documents), we would like to inspect the Contract Documents on the following date: [*insert date*]

In particular, we would like to inspect the following Contract Documents: [*insert references to Contract Documents.*]

Please confirm that the necessary access will be provided.

Resource 37

Description	The provision by the Employer to the Contractor of a certified copy of the Contract Documents, pre-construction information under the CDM Regulations, and building information under the Building Regulations.
Type	Correspondence: Employer to Contractor
Provision references	clause 2.7.2
Drafting notes	**Timing:** Clause 2.7.2 requires the Employer to provide these documents to the Contractor 'immediately' after the execution of the Contract.

Reference: Certified copy of Contract Documents and pre-construction information (clause 2.7.2)

In accordance with clause 2.7.2 (Contract Documents), a certified copy of the Contract Documents, together with all pre-construction information required to be provided to you under regulation 4(4) of the CDM Regulations, and all building information required to be provided to you under regulation 11A(4) of the Building Regulations, is [enclosed] / [attached] / [available to download at the following link: [*insert link*]]

Resource 38

Description	A reminder by the Contractor to the Employer's Agent of its entitlement to access copies of the Contract Documents and the Contractor's Design Documents.
Type	Correspondence: Contractor to Employer's Agent
Provision references	clause 2.7.3
Drafting notes	**Good practice** This correspondence is not required by clause 2.7.3, but it is considered good practice to remind the Employer's Agent.

Reference: Copies of Contract Documents and Contractor's Design Documents (clause 2.7.3)

In accordance with clause 2.7.3 (Contract Documents), we are required to keep on site (or accessible there) copies of the Contract Documents and all Contractor's Design Documents, for your benefit.

[We confirm that they are all available on our document management platform, to which you already have access. Please contact [*insert name and email address*] if you have any difficulties in this regard.] /

[We confirm that they are all available to download at the following link: [*insert link*]] /

[We confirm that they are available at the following location: [*insert location, for example, site office.*] Please contact [*insert name and email address*] if you would like to inspect them.]

Resource 39

Description	Confirmation from the Contractor to the Employer of the provision of Contractor's Design Documents.
Type	Correspondence: Contractor to Employer
Provision references	clause 2.8
Drafting notes	**Good practice** This correspondence is not required by clause 2.8, but it is considered good practice to remind the Employer of the Parties' respective obligations under the Design Submission Procedure.

Reference: Provision of Contractor's Design Documents (clause 2.8)

In accordance with clause 2.8 (Construction information), we will provide you with copies of the Contractor's Design Documents as and when necessary from time to time, in accordance with the Design Submission Procedure.

[This will be done by way of our document management platform, to which you already have access. Please contact [*insert name and email address*] if you have any difficulties in this regard.]

We confirm that we will not commence any work to which any Contractor's Design Document relates before the Design Submission Procedure has been complied with.

Resource 40

Description	A reminder from the Employer to the Contractor of its obligation to comply with the Design Submission Procedure.
Type	Reminder: Employer to Contractor
Provision references	clause 2.8
Drafting notes	**Good practice** This correspondence is not required by clause 2.8, but it is considered good practice to remind the Contractor that it must not commence any work to which a Contractor's Design Document relates before complying with the Design Submission Procedure.

Reference: Compliance with the Design Submission Procedure (clause 2.8)

We write to remind you that, in accordance with clause 2.8 (Construction information), you must not commence any work to which any Contractor's Design Document relates before the Design Submission Procedure has been complied with.

In particular, we note that you have commenced the following work: [*insert details of the work commenced by the Contractor.*]

We consider that this work may relate to the following Contractor's Design Documents: [*insert references to relevant Contractor's Design Documents.*]

The Design Submission Procedure has yet to be complied with in respect of these Contractor's Design Documents, on the following basis: [*insert details, for example, the Contractor's Design Documents have not yet been marked 'A' or 'B'.*]

[We further remind you that, in accordance with paragraph 6 of the Design Submission Procedure, you must not carry out any work in accordance with a Contractor's Design Document marked 'C', and the Employer will not be liable to pay for any work within the Works executed otherwise than in accordance with Contractor's Design Documents marked 'A' or 'B'.]

We request that you reply as soon as possible, to confirm whether and how you have complied with your obligations under clause 2.8 in this regard.

Resource 41

Description	Confirmation by the Employer to the Contractor of the boundaries of the site.
Type	Confirmation: Employer to Contractor
Provision references	clause 2.9

Reference: Confirmation of the site boundaries (clause 2.9)

In accordance with clause 2.9 (Site boundaries), we are required to define the boundaries of the site.

Therefore, we confirm that the boundaries of the site are defined in the following document:

[*Insert reference to the document defining the boundaries of the site.*]

Resource 42

Description	A notice from the Contractor to the Employer of a divergence between the Employer's Requirements and the definition of the site boundary, and a request for an instruction.
Type	Notice: Contractor to Employer
Provision references	clause 2.10.2
Drafting notes	**Timing:** Clause 2.10.2 requires the Parties to notify the other Party 'immediately' upon becoming aware of any such divergence.

Reference: Divergence between the Employer's Requirements and the definition of the site boundary (clause 2.10.2)

In accordance with clause 2.10.2 (Divergence in Employer's Requirements and definition of site boundary), we write to notify you of the following divergence between the Employer's Requirements and the definition of the site boundary as provided under clause 2.9 (Site boundaries):

[*Insert details of the divergence, with references to the Employer's Requirements.*]

In accordance with clause 2.10.1, please issue an instruction correcting this divergence, which will be treated as a Change.

Resource 43

Description	A notice from the Employer to the Contractor of a divergence between the Employer's Requirements and the definition of the site boundary, with an accompanying instruction.
Type	Notice: Employer to Contractor
Provision references	clause 2.10.2
Drafting notes	**Timing:** Clause 2.10.2 requires the Parties to notify the other Party 'immediately' upon becoming aware of any such divergence.

Reference: Divergence between the Employer's Requirements and the definition of the site boundary; Employer's instruction [*insert reference*] (clause 2.10.2)

In accordance with clause 2.10.2 (Divergence in Employer's Requirements and definition of site boundary), we write to notify you of the following divergence between the Employer's Requirements and the definition of the site boundary as provided under clause 2.9 (Site boundaries):

[*Insert details of the divergence, with references to the Employer's Requirements.*]

In accordance with clause 2.10.1, we attach an instruction correcting this divergence, which will be treated as a Change.

Resource 44

Description	A notice from the Contractor to the Employer of an inadequacy in the Employer's Requirements, which is dealt with in the Contractor's Proposals.
Type	Notice: Contractor to Employer
Provision references	clause 2.12, clause 2.13

Reference: Inadequacy in the design in the Employer's Requirements (clause 2.12)

In accordance with clause 2.13 (Notice of discrepancies etc.), we write to notify you of the following inadequacy in the design in the Employer's Requirements:

[*Insert details, with references to the Employer's Requirements.*]

In accordance with clause 2.11 (Preparation of Employer's Requirements), we are not responsible for the contents of the Employer's Requirements, or for verifying the adequacy of any design contained within them.

However, in accordance with clause 2.12, we can confirm that this inadequacy is dealt with in the Contractor's Proposals as follows:

[*Insert details, with references to the Contractor's Proposals.*]

Resource 45

Description	A notice from the Contractor to the Employer of an inadequacy in the Employer's Requirements (which is not dealt with in the Contractor's Proposals), and a request for an instruction.
Type	Notice: Contractor to Employer
Provision references	clause 2.12, clause 2.13
Drafting notes	**Timing:** Clause 2.13 requires the Contractor to give this notice 'immediately', upon becoming aware of such an inadequacy.

Reference: Inadequacy in the design in the Employer's Requirements (clause 2.13)

In accordance with clause 2.13 (Notice of discrepancies etc.), we write to notify you of the following inadequacy in the design in the Employer's Requirements:

[*Insert details, with references to the Employer's Requirements.*]

In accordance with clause 2.11 (Preparation of Employer's Requirements), we are not responsible for the contents of the Employer's Requirements, or for verifying the adequacy of any design contained within them.

We can also confirm that this inadequacy is not dealt with in the Contractor's proposals.

Therefore, in accordance with clause 2.12 (Employer's Requirements – inadequacy), please issue an instruction to correct, alter, or modify the Employer's Requirements accordingly, which will be treated as a Change.

Resource 46

Description	An instruction from the Employer following a notice of inadequacy in the design in the Employer's Requirements, under clause 2.12.
Type	Instruction: Employer to Contractor
Provision references	clause 2.13
Drafting notes	The Employer is encouraged to issue this instruction as soon as possible, to avoid a delay to the progress of the Works.

Reference: Inadequacy in the design in the Employer's Requirements; Employer's instruction [*insert reference*] (clause 2.13)

We write in reply to your notice dated [*insert date*], with reference [*insert reference*], relating to an inadequacy in the design in the Employer's Requirements.

In accordance with clause 2.13 (Notice of discrepancies etc.), we attach an instruction to correct, alter, or modify the Employer's Requirements, which will be treated as a Change.

Resource 47

Description	A notice from the Contractor to the Employer of a discrepancy or divergence in or between the Contractor's Proposals and/or other Contractor's Design Documents.
Type	Notice: Contractor to Employer
Provision references	clause 2.13, clause 2.14.1
Drafting notes	**Timing:** Clause 2.13 requires the Contractor to give this notice 'immediately', upon becoming aware of such a discrepancy or divergence.

Reference: Discrepancy or divergence in or between the Contractor's Proposals and/or other Contractor's Design Documents (clause 2.13)

In accordance with clause 2.13 (Notice of discrepancies etc.), we write to notify you of a discrepancy or divergence in or between the following:

- the Contractor's Proposals, and in particular, [*insert references*]; and/or
- other Contractor's Design Documents, and in particular, [*insert references*].

The details of the discrepancy or divergence are as follows:

[*Insert details.*]

In accordance with clause 2.14.1 (Discrepancies in documents), our proposed amendment to remove this discrepancy or divergence is as follows:

[*Insert details of the proposed amendment.*]

We confirm that our proposed amendment complies with the Statutory Requirements.

Please issue an instruction either to decide between the discrepant items or to accept our proposed amendment.

[Our preference is that you accept our proposed amendment.]

We accept that we will comply with your decision, or acceptance or our proposed amendment, without cost to you.

Resource 48

Description	An instruction from the Employer following a notice of discrepancy or divergence in or between the Contractor's Proposals and/or other Contractor's Design Documents.
Type	Instruction: Employer to Contractor
Provision references	clause 2.13, clause 2.14.1
Drafting notes	The Employer is encouraged to issue this instruction as soon as possible, to avoid a delay to the progress of the Works.

Reference: Discrepancy or divergence in or between the Contractor's Proposals and/or other Contractor's Design Documents; Employer's instruction [*insert reference*] (clause 2.13)

We write in reply to your notice dated [*insert date*], with reference [*insert reference*], relating to a discrepancy or divergence in or between the Contractor's Proposals and/or other Contractor's Design Documents.

In accordance with clause 2.13 (Notice of discrepancies etc.) and clause 2.14.1 (Discrepancies in documents), we attach an instruction [deciding between the discrepant items] / [accepting your proposed amendment].

This instruction will be without cost to the Employer.

Resource 49

Description	A notice from the Contractor to the Employer of a discrepancy in or between the Employer's Requirements and/or instructions.
Type	Notice: Contractor to Employer
Provision references	clause 2.13, clause 2.14.2
Drafting notes	**Timing:** Clause 2.13 requires the Contractor to give this notice 'immediately', upon becoming aware of such a discrepancy. The Contractor must first ascertain whether the Contractor's Proposals deal with the discrepancy.

Reference: Discrepancy in or between the Employer's Requirements and/or Employer's instructions (clause 2.13)

In accordance with clause 2.13 (Notice of discrepancies etc.), we write to notify you of a discrepancy in or between the following:

- [the Employer's Requirements, and in particular: [*insert references*]; and/or
- the following instructions issued by the Employer under the Conditions, which constitute a Change: [*insert references*]]

The details of the discrepancy are as follows:

[*Insert details.*]

[*Where the Contractor's Proposals deal with the discrepancy*: In accordance with clause 2.14.2 (Discrepancies in documents), the Contractor's Proposals will prevail, and remove the discrepancy as follows: [*Insert explanation of how the Contractor's Proposals remove the discrepancy.]* We accept that there will be no adjustment of the Contract Sum as a result.] /

[*Where the Contractor's Proposals do not deal with the discrepancy*: The Contractor's Proposals do not deal with the discrepancy. Therefore, in accordance with clause 2.14.2 (Discrepancies in documents), our proposed amendment to remove this discrepancy is as follows: [*Insert details of the proposed amendment.*] We confirm that our proposed amendment complies with the Statutory Requirements. Please issue an instruction either in agreement with our proposed amendment, or to decide how the discrepancy is to be dealt with. [Our preference is that you agree with our proposed amendment.] This instruction will be treated as a Change.]

Resource 50

Description	An instruction from the Employer following a notice of a discrepancy in or between the Employer's Requirements and/or instructions.
Type	Instruction: Employer to Contractor
Provision references	clause 2.13, clause 2.14.2
Drafting notes	The Employer is encouraged to issue this instruction as soon as possible, to avoid a delay to the progress of the Works.

Reference: Discrepancy in or between the Employer's Requirements and/or Employer's instructions; Employer's instruction [*insert reference*] (clause 2.13)

We write in reply to your notice dated [*insert date*], with reference [*insert reference*], relating to a discrepancy within the Employer's Requirements (including any Change to them).

In accordance with clause 2.13 (Notice of discrepancies etc.) and clause 2.14.2 (Discrepancies in documents), we attach an instruction [confirming agreement to your proposed amendment] / [deciding how the discrepancy is to be dealt with].

This instruction will be treated as a Change.

Resource 51

Description	A notice from the Contractor to the Employer of a divergence between the Statutory Requirements and the Employer's Requirements (including any Change), the Contractor's Proposals, or other Contractor's Design Documents, and a request for the Employer's consent to proceed with the Contractor's proposed amendment to remove the divergence.
Type	Notice: Contractor to Employer
Provision references	clause 2.15.1
Drafting notes	**Timing:** Clause 2.15.1 requires the Contractor to give this notice 'immediately', upon becoming aware of such a divergence.

Reference: Divergence from Statutory Requirements (clause 2.15.1)

In accordance with clause 2.15.1 (Divergences from Statutory Requirements), we have become aware of the following divergence between the Statutory Requirements and [the Employer's Requirements (including any Change)] / [the Contractor's Proposals or other Contractor's Design Documents]:

[*Insert details of the divergence, with references to the Statutory Requirements and relevant Contract Documents.*]

We propose the following amendment to remove the divergence:

[*Insert details of the proposed amendment.*]

We request your consent to complete the design and construction of the Works in accordance with this amendment, which we accept would be entirely at our own cost.

Please give your consent by [*insert date*], in order to avoid a delay to the progress of the Works.

Resource 52

Description	A notice from the Employer to the Contractor of a divergence between the Statutory Requirements and the Employer's Requirements (including any Change), the Contractor's Proposals, or other Contractor's Design Documents, and a request for a proposed amendment to remove the divergence.
Type	Notice: Employer to Contractor
Provision references	clause 2.15.1
Drafting notes	**Timing:** Clause 2.15.1 requires the Employer to give this notice 'immediately', upon becoming aware of such a divergence.

Reference: Divergence from Statutory Requirements (clause 2.15.1)

In accordance with clause 2.15.1 (Divergences from Statutory Requirements), we have become aware of the following divergence between the Statutory Requirements and [the Employer's Requirements (including any Change)] / [the Contractor's Proposals or other Contractor's Design Documents]:

[*Insert details of the divergence, with references to the Statutory Requirements and relevant Contract Documents.*]

Please notify us of your proposed amendment to remove the divergence.

Resource 53

Description	Consent given by the Employer to the Contractor, for the Contractor's proposed amendment, following a notice of a divergence between the Statutory Requirements and the Employer's Requirements (including any Change), the Contractor's Proposals, or other Contractor's Design Documents.
Type	Consent given: Employer to Contractor
Provision references	clause 2.15.1

Reference: Divergence from Statutory Requirements; Employer's instruction [*insert reference*] (clause 2.15.1)

We write in reply to your notice dated [*insert date*], with reference [*insert reference*], given in accordance with clause 2.15.1 (Divergences from Statutory Requirements), and which proposed an amendment to remove the stated divergence from the Statutory Requirements.

Further to your request, we attach an instruction giving our consent for you to complete the design and construction of the Works in accordance with your proposed amendment (entirely at your own cost), and we will note this amendment on the Contract Documents.

Resource 54

Description	Consent withheld by the Employer to the Contractor, for the Contractor's proposed amendment, following a notice of a divergence between the Statutory Requirements and the Employer's Requirements (including any Change), the Contractor's Proposals, or other Contractor's Design Documents.
Type	Consent withheld: Employer to Contractor
Provision references	clause 2.15.1
Drafting notes	Clause 1.10 requires that the Employer's consent must not be unreasonably delayed or withheld.

Reference: Divergence from Statutory Requirements (clause 2.15.1)

We write in reply to your notice dated [*insert date*], with reference [*insert reference*], given in accordance with clause 2.15.1 (Divergences from Statutory Requirements), and which proposed an amendment to remove the stated divergence from the Statutory Requirements.

Further to your request, we withhold our consent for you to complete the design and construction of the Works in accordance with your proposed amendment, for the following reasons: [*insert reasons, to demonstrate why the consent was not unreasonably delayed or withheld.*]

Please propose an alternative amendment to remove the divergence, taking the reasons above into consideration.

Resource 55

Description	Confirmation by the Contractor to the Employer of a change in the Statutory Requirements after the Base Date which necessitates an alteration or modification to the Works, and a request for an instruction.
Type	Confirmation and request: Contractor to Employer
Provision references	clause 2.15.2.1
Drafting notes	Although clause 2.15.2.1 does not require the Contractor to issue this correspondence, it is encouraged to do so as soon as it becomes aware of the change in the Statutory Requirements, to avoid a delay to the progress of the Works.

Reference: Change in the Statutory Requirements after the Base Date (clause 2.15.2.1)

We write to notify you of the following change in the Statutory Requirements, made after the Base Date (namely [*insert Base Date*]):

[*Insert details of the change in the Statutory Requirements, including when it came into effect.*]

This change will necessitate the following alteration or modification to the Works:

[*Insert details of the necessary alteration or modification of the Works.*]

Please issue an instruction confirming this alteration or modification to the Works, to avoid a delay to the progress of the Works.

In accordance with clause 2.15.2.1 (Divergences from Statutory Requirements), this alteration or modification will be treated as a Change.

Resource 56

Description	Confirmation by the Employer to the Contractor of a change in the Statutory Requirements after the Base Date which necessitates an alteration or modification to the Works, and an instruction.
Type	Confirmation and instruction: Employer to Contractor
Provision references	clause 2.15.2.1
Drafting notes	Although clause 2.15.2.1 does not require the Employer to issue this correspondence, it is encouraged to do so as soon as it becomes aware of the change in the Statutory Requirements, to avoid a delay to the progress of the Works.

Reference: Change in the Statutory Requirements after the Base Date (clause 2.15.2.1)

We write to notify you of the following change in the Statutory Requirements, made after the Base Date (namely [*insert Base Date*]):

[*Insert details of the change in the Statutory Requirements, including when it came into effect.*]

This change will necessitate the following alteration or modification to the Works:

[*Insert details of the necessary alteration or modification of the Works.*]

We attach an instruction confirming this alteration or modification to the Works.

In accordance with clause 2.15.2.1 (Divergences from Statutory Requirements), this alteration or modification will be treated as a Change.

Resource 57

Description	Confirmation from the Contractor to the Employer of an amendment to the Contractor's Proposals for conformity with the Development Control Requirements, and a request for an instruction.
Type	Confirmation: Contractor to Employer
Provision references	clause 2.15.2.2
Drafting notes	Although clause 2.15.2.2 does not require the Contractor to issue this correspondence, it is encouraged to do so as soon as it becomes aware of the need to amend the Contractor's Proposals, to avoid a delay to the progress of the Works. Whilst the Employer cannot issue an instruction to amend the Contractor's Proposals, for the purposes of record keeping, it is encouraged to issue an instruction acknowledging the amendment.

Reference: Amendment to the Contractor's Proposals for conformity to the Development Control Requirements (clause 2.15.2.2)

We write to notify you of the following permission or approval made by a decision of [*insert name of relevant authority*], the relevant authority, after the Base Date (namely [*insert Base Date*]) for the purposes of Development Control Requirements:

[*Insert details of the permission or approval, including when it was received.*]

The following amendment to the Contractor's Proposals has become necessary, to conform with the terms of this permission or approval:

[*Insert details of the necessary amendment to the Contractor's Proposals.*]

Please issue an instruction acknowledging this amendment to the Contractor's Proposals.

In accordance with clause 2.15.2.2 (Divergences from Statutory Requirements), this amendment will be treated as a Change (a treatment not precluded in the Employer's Requirements).

Resource 58

Description	Confirmation from the Employer to the Contractor of an amendment to the Contractor's Proposals for conformity with the Development Control Requirements, and an instruction.
Type	Confirmation and instruction: Employer to Contractor
Provision references	clause 2.15.2.2
Drafting notes	Although clause 2.15.2.2 does not require the Employer to issue this correspondence, it is encouraged to do so as soon as it becomes aware of the need to amend the Contractor's Proposals, to avoid a delay to the progress of the Works. Whilst the Employer cannot issue an instruction to amend the Contractor's Proposals, for the purposes of record keeping, it is encouraged to issue an instruction acknowledging the amendment.

Reference: Amendment to the Contractor's Proposals for conformity to the Development Control Requirements; Employer's instruction [*insert reference*] (clause 2.15.2.2)

We write to notify you of the following permission or approval made by a decision of [*insert name of relevant authority*], the relevant authority, after the Base Date (namely [*insert Base Date*]) for the purposes of Development Control Requirements:

[*Insert details of the permission or approval, including when it was received.*]

The following amendment to the Contractor's Proposals has become necessary, to conform with the terms of this permission or approval:

[*Insert details of the necessary amendment to the Contractor's Proposals.*]

We attach an instruction acknowledging this amendment to the Contractor's Proposals.

In accordance with clause 2.15.2.2 (Divergences from Statutory Requirements), this amendment will be treated as a Change (a treatment not precluded in the Employer's Requirements).

Resource 59

Description	Confirmation from the Contractor to the Employer of an amendment to the Employer's Requirements to conform with the Statutory Requirements, and a request for an instruction.
Type	Confirmation and request for instruction: Contractor to Employer
Provision references	clause 2.15.2.3
Drafting notes	Although clause 2.15.2.3 does not require the Contractor to issue this correspondence, it is encouraged to do so as soon as it becomes aware of the need to amend the Employer's Requirements, to avoid a delay to the progress of the Works.

Reference: Divergence from Statutory Requirements (clause 2.15.2.3)

We refer to the following part of the Employer's Requirements: [*insert reference*]

This part of the Employer's Requirements states specifically that they comply with the Statutory Requirements, as follows: [*insert the relevant extract from the Employer's Requirements.*]

However, an amendment to this part of the Employer's Requirements has become necessary, in order to conform with the Statutory Requirements. In particular:

[*Insert details of the divergence between the Employer's Requirements and the Statutory Requirements.*]

In accordance with clause 2.15.2.3 (Divergences from Statutory Requirements), please issue an instruction requiring a Change, to amend the Employer's Requirements to conform with the Statutory Requirements.

Resource 60

Description	Confirmation from the Employer to the Contractor of an amendment to the Employer's Requirements to conform with the Statutory Requirements, and an instruction.
Type	Confirmation and instruction: Employer to Contractor
Provision references	clause 2.15.2.3
Drafting notes	Although clause 2.15.2.3 does not require the Employer to issue this correspondence, it is encouraged to do so as soon as it becomes aware of the need to amend the Employer's Requirements, to avoid a delay to the progress of the Works.

Reference: Divergence from Statutory Requirements; Employer's instruction [*insert reference*] (clause 2.15.2.3)

We refer to the following part of the Employer's Requirements: [*insert reference*]

This part of the Employer's Requirements states specifically that they comply with the Statutory Requirements, as follows: [*insert the relevant extract from the Employer's Requirements.*]

However, an amendment to this part of the Employer's Requirements has become necessary, in order to conform with the Statutory Requirements. In particular:

[*Insert details of the divergence between the Employer's Requirements and the Statutory Requirements.*]

In accordance with clause 2.15.2.3 (Divergences from Statutory Requirements), we attach an instruction requiring a Change, which amends the Employer's Requirements to conform with the Statutory Requirements.

Resource 61

Description	A tracker template for recording inadequacies, divergences, and discrepancies under clause 2.11 to clause 2.15.
Type	Tracker: Contractor and Employer
Provision references	clause 2.11 to clause 2.15

Below is a tracker template for recording inadequacies, divergences, and discrepancies under clause 2.11 to clause 2.15:

- Reference: [*insert reference*]
- Date notified: [*insert date*]
- Applicable clause: [*insert clause 2.11, clause 2.12, clause 2.13, clause 2.14, and/ or clause 2.15*]
- Description: [*insert description of the inadequacy, divergence, or discrepancy*]
- Contract Documents reference: [*insert reference*]
- Employer's instruction date: [*insert date*]
- Employer's instruction reference: [*insert reference*]
- Document link: [*insert link to documents*]
- Status / action required: [*insert status, and any action required*]

Resource 62

Description	A notice from the Contractor to the Employer of an emergency compliance with the Statutory Requirements.
Type	Notice: Contractor to Employer
Provision references	clause 2.16.2
Drafting notes	**Timing:** Clause 2.16.2 requires the Contractor to give this notice 'forthwith' after the emergency.

Reference: Emergency compliance with Statutory Requirements (clause 2.16.2)

In accordance with clause 2.16.2 (Emergency compliance with Statutory Requirements), we write to notify you of the following emergency: [*insert details of the emergency, including when it occurred.*]

In order to comply with the Statutory Requirements during this emergency, we were required to supply materials and/or execute work before receiving your consent under clause 2.15.1 (Divergences from Statutory Requirements), as follows:

[*Insert details of the materials supplied and/or work executed, in order to comply with the Statutory Requirements.*]

We consider that these steps were reasonably necessary to secure immediate compliance with the Statutory Requirements during the emergency.

Resource 63

Description	A notice from the Contractor to the Employer of a potential infringement of patent rights.
Type	Notice: Contractor to Employer
Provision references	clause 2.20.2
Drafting notes	**Timing:** Clause 2.20.2 requires the Contractor to give this notice 'promptly'.

Reference: Potential infringement of patent rights (clause 2.20.2); Employer's instruction [*insert reference*]

We write in relation to the Employer's instruction dated [*insert date*], with reference [*insert reference*].

In accordance with clause 2.20.2 (Patent rights – Instructions), we have become aware that this instruction may infringe the following patent rights: [*insert details of the patent rights which may be infringed.*]

We therefore confirm that we have yet to comply with this instruction.

We request that the Parties meet as soon as possible to discuss this matter, to enable you to confirm whether this instruction will take effect.

Resource 64

Description	A notice from the Employer to the Contractor of a potential infringement of patent rights.
Type	Notice: Employer to Contractor
Provision references	clause 2.20.2
Drafting notes	**Timing:** Clause 2.20.2 requires the Contractor to give this notice 'promptly'.

Reference: Potential infringement of patent rights (clause 2.20.2)

We write in relation to the Employer's instruction dated [*insert date*], with reference [*insert reference*].

In accordance with clause 2.20.2 (Patent rights – Instructions), we have become aware that this instruction may infringe the following patent rights: [*insert details of the patent rights which may be infringed.*]

Therefore, please do not yet comply with this instruction.

We request that the Parties meet as soon as possible to discuss this matter, following which we will confirm whether this instruction will take effect.

Resource 65

Description	Confirmation or withdrawal of an instruction by the Employer, relating to a potential infringement of patent rights.
Type	Confirmation: Employer to Contractor
Provision references	clause 2.20.2
Drafting notes	The Employer is encouraged to provide any attendance note or meeting minutes from the consultation.

Reference: Potential infringement of patent rights; Employer's instruction [*insert reference*] (clause 2.20.2)

We write further to the Employer's instruction dated [*insert date*], with reference [*insert reference*], and our consultation to discuss the potential infringement of patent rights, held on [*insert date*].

[We attach [the meeting minutes] / [our attendance note] from that consultation.]

[*Instruction will take effect:* In accordance with clause 2.20.2 (Patent rights – Instructions), we confirm that this instruction will take effect.] /

[*Instruction will not take effect:* In accordance with clause 2.20.2 (Patent rights – Instructions), we confirm that the Employer no longer wishes for this instruction to take effect. [An alternative instruction is attached, which addresses the concerns relating to the potential infringement of patent rights.]]

Resource 66

Description	A request from the Contractor to the Employer for consent to remove Site Materials from storage on or adjacent to the Works.
Type	Request: Contractor to Employer
Provision references	clause 2.21

Reference: Removal of Site Materials from storage on or adjacent to the Works (clause 2.21)

In accordance with clause 2.21 (Materials and goods – on site), we request your consent to remove the following Site Materials from storage on or adjacent to the Works:

[*Insert description of Site Materials.*]

We wish to temporarily remove the Site Materials until [*insert date*] for the following reasons: [*insert reasons for removing the Site Materials.*]

The Site Materials would be securely stored and available for inspection by the Employer at the following location: [*insert address.*]

Please give your consent by [*insert date*].

Resource 67

Description	Consent given by the Employer for the Contractor to remove Site Materials from storage on or adjacent to the Works.
Type	Consent given: Employer to Contractor
Provision references	clause 2.21

Reference: Removal of Site Materials from storage on or adjacent to the Works (clause 2.21)

We write in reply to your request dated [*insert date*] for our consent to remove certain Site Materials from storage on or adjacent to the Works, in accordance with clause 2.21 (Materials and goods – on site).

[*Unconditional consent*: We give our consent.] / [*Conditional consent*: We give our consent, subject to the following conditions: [*insert conditions.*]]

Resource 68

Description	Consent withheld by the Employer for the Contractor to remove Site Materials from storage on or adjacent to the Works.
Type	Consent withheld: Employer to Contractor
Provision references	clause 2.21
Drafting notes	Clause 1.10 requires that the Employer's consent must not be unreasonably delayed or withheld.

Reference: Removal of Site Materials from storage on or adjacent to the Works (clause 2.21)

We write in reply to your request dated [*insert date*] for our consent to remove certain Site Materials from storage on or adjacent to the Works, in accordance with clause 2.21 (Materials and goods – on site).

We withhold our consent, for the following reasons: [*insert reasons, to demonstrate why the consent was not unreasonably delayed or withheld.*]

Resource 69

Description	A tracker template for recording Listed Items.
Type	Tracker: Contractor and Employer
Provision references	clause 2.22

Below is a tracker template for recording Listed Items:

- Description of Listed Item: [*insert description*]
- Employer's Requirements reference: [*insert reference*]
- Location: [*insert address*]
- Insurance in place: [*Yes / No*]
- Value included in Interim Payment (exclusive of VAT): [*insert amount*]
- Date delivered to storage: [*insert date*]
- Date to be delivered to the Works: [*insert date*]
- Vesting certificate provided: [*yes / no*]
- Document link: [*insert link to documents*]
- Status / action required: [*insert status, and any action required*]

Resource 70

Description	A notice from the Contractor to the Employer of a delay to the progress of the Works, and particulars of its expected effects.
Type	Notice: Contractor to Employer
Provision references	clause 2.24
Drafting notes	**Timing:** Clause 2.24.1 requires the Contractor to give this notice 'forthwith', if any whenever it becomes reasonably apparent that the progress of the Works or any Section is being or is likely to be delayed.

Reference: Delay notice [*insert reference*] (clause 2.24.1 and clause 2.24.2)

In accordance with clause 2.24.1 (Notice by Contractor of delay to progress), we write to notify you of a delay to the progress of the Works.

[*Sections:* This delay affects the following Sections: [*insert Section references*].]

The material circumstances, including the cause of the delay, are as follows:

[*Insert a description of the material circumstances, and identify the cause or causes of the delay.*]

[*If there is a Relevant Event:* We consider that the following event constitutes a Relevant Event: [*Insert description of event.*]

In particular, we consider that it constitutes a Relevant Event under the following clause: [*Insert applicable provision from clause 2.26 (Relevant Events).*]]

In accordance with clause 2.24.2, the expected effects of this delay are as follows:

[*Insert description of the expected effects of the delay.*]

We estimate the expected delay in the completion of [*Works:* the Works] / [*Sections:* the Sections affected] beyond the relevant Completion Date as follows:

[*Insert estimate of the delay. For example:*

- *Section 1: 2 weeks.*
- *Section 2: 4 weeks.*
- *Section 3: 6 weeks.*]

[We attach a programme (reference [*insert reference*]) demonstrating the expected effects of this delay.]

In accordance with clause 2.24.3, we will notify you of any material change in the estimated delay, or in any other particulars.

Please let us know if there is any further information you require in relation to this notice.

In the meantime, we can confirm the following:

- In accordance with clause 2.25.6.1 (Fixing Completion Date), we continue to constantly use our best endeavours to prevent delay in the progress of the Works [(or any Section)], however caused, and to prevent the completion of the Works [(or Section)] being delayed, or further delayed, beyond the relevant Completion Date. [In particular, we have undertaken the following: *[insert examples.*]]
- In accordance with clause 2.25.6.2, we will continue to do all that may reasonably be required, to the satisfaction of the Employer, to proceed with the Works [(or Section)].

Please let us know if there is anything further we can do in respect of these obligations.

Resource 71

Description	A notice from the Contractor to the Employer of a material change in the estimated delay, or in other particulars.
Type	Notice: Contractor to Employer
Provision references	clause 2.24.3
Drafting notes	**Timing:** Clause 2.24.3 requires the Contractor to give this notice 'forthwith', following any material change in the estimated delay or in any other particulars.

Reference: Delay notice [*insert reference*]; Material change in the estimated delay / other particulars (clause 2.24.3)

We write further to our notice dated [*insert date*], given in accordance with clause 2.24 (Notice by Contractor of delay to progress).

[*Material change in the estimated delay:* In accordance with clause 2.24.3, we write to notify you of the following material change in the estimated delay: [*Insert details of the material change in the estimated delay. For example:*

- *Section 1: our initial estimate of the expected delay in the completion of Section 1 was 2 weeks. Our revised estimate is 4 weeks.*
- *Section 2: our initial estimate of the expected delay in the completion of Section 2 was 4 weeks. Our revised estimate is 8 weeks.*
- *Section 3: our initial estimate of the expected delay in the completion of Section 3 was 6 weeks. This estimate remains unchanged.*]]

[We attach a programme (reference [*insert reference*]) demonstrating the expected effects of this delay.]] /

[*Material change in other particulars:* In accordance with clause 2.24.3, we write to notify you of the following material change in the following particulars relating to the delay:

[*Insert details of the material change in the particulars.*]]

We will notify you of any further material change in the estimated delay, or in any other particulars.

Resource 72

Description	A notice from the Employer to the Contractor requesting further information to enable the Employer to reach a decision under clause 2.25.1.
Type	Notice: Employer to Contractor
Provision references	clause 2.24.4
Drafting notes	**Timing:** Clause 2.24.4 requires the Employer to give this notice 'not later than 14 days' from receipt of the particulars under clause 2.24.2, or any notification under clause 2.24.3. The Employer is therefore encouraged to undertake an initial review of the information provided by the Contractor immediately, such that this notice can be given as soon as possible.

Reference: Delay notice [*insert reference*]; Notice to request further information (clause 2.24.4)

We write further to the following correspondence:

- your notice dated [*insert date*], with reference [*insert reference*], given in accordance with clause 2.24.1 (Notice by Contractor of delay to progress);
- the associated particulars, dated [*insert date*], given in accordance with clause 2.24.2;
- [your notification of a material change in [the estimated delay] / [certain particulars] dated [*insert date*], given in accordance with clause 2.24.3.]

In accordance with clause 2.24.4, we write to notify you that we require the following further information, which is reasonably necessary to enable us to reach a decision under clause 2.25.1 (Fixing Completion Date):

[*Insert details of further information required.*]

We require this further information for the following reasons:

[*Insert reasons why the information is reasonably necessary to enable the Employer to reach a decision under clause 2.25.1.*]

Resource 73

Description	The provision by the Contractor to the Employer of further information, further to the Employer's notice under clause 2.24.4.
Type	Correspondence: Contractor to Employer
Provision references	clause 2.24.4
Drafting notes	**Timing:** The Contractor is encouraged to provide the further information as soon as possible.

Reference: Delay notice [*insert reference*]; Provision of further information (clause 2.24.4)

We write in reply to your notice dated [*insert date*], given in accordance with clause 2.24.4 (Notice by Contractor of delay to progress).

The further information you requested is [attached] / [available to download at the following link: [*insert link*]].

[Please let us know if there is any further information you reasonably require to enable you to reach a decision under clause 2.25.1 (Fixing Completion Date).]

Resource 74

Description	A reminder from the Contractor to the Employer of its obligation to notify the Contractor of its decision under clause 2.25.1 as soon as reasonably practicable.
Type	Reminder: Contractor to Employer
Provision references	clause 2.25.2, clause 2.24
Drafting notes	**Good practice** Clause 2.25.2 does not require this correspondence to be issued, but it is considered good practice to remind the Employer of its obligations, and to check whether it requires any further information.

Reference: Delay notice [*insert reference*]; Employer's obligations under clause 2.25.2

We write further to the following correspondence:

- our notice dated [*insert date*], with reference [*insert reference*], given in accordance with clause 2.24.1 (Notice by Contractor of delay to progress);
- the associated particulars, dated [*insert date*], given in accordance with clause 2.24.2;
- [our notification of a material change in [the estimated delay] / [certain particulars] dated [*insert date*], given in accordance with clause 2.24.3;]
- [the further information we provided on [*insert date*], further to your notice dated [*insert date*], given in accordance with clause 2.24.4.]

We remind you that, in accordance with clause 2.25.2 (Fixing Completion Date), you are required to notify us of your decision as soon as reasonably practicable, and in any event not later than 8 weeks from [the date of receipt of the particulars or notification] / [the date of receipt of the further information required under clause 2.24.4].

[Given that the period from receipt of our notice to the Completion Date (namely [*insert date*]) is less than 8 weeks, you are further required to endeavour to notify us of your decision prior to the Completion Date.]

[Please let us know if there is any further information you reasonably require to enable you to reach a decision under clause 2.25.1.]

Resource 75

Description	A notice from the Employer to the Contractor, giving an extension of time.
Type	Notice: Employer to Contractor
Provision references	clause 2.25.1
Drafting notes	**Timing:** Clause 2.25.2 requires the Employer to notify the Contractor of its decision in respect of any notice under clause 2.24 as soon as is reasonably practicable, and in any event not later than 8 weeks from the later of (i) the date of receipt of the particulars under clause 2.24.2, or notification under clause 2.24.3, or (ii) the date of receipt of the further information required under clause 2.24.4. Furthermore, where the period from receipt to the Completion Date is less than 8 weeks, it must endeavour to do so prior to the Completion Date. The Employer is encouraged to provide as much detail as possible on the reasoning for its decision, and any analysis undertaken, to help avoid disputes between the Parties.

Reference: Delay notice [*insert reference*]; Extension of time given (clause 2.25.1)

We write further to the following correspondence:

- your notice dated [*insert date*], with reference [*insert reference*], given in accordance with clause 2.24.1 (Notice by Contractor of delay to progress);
- the associated particulars, dated [*insert date*], given in accordance with clause 2.24.2;
- [your notification of a material change in [the estimated delay] / [certain particulars] dated [*insert date*], given in accordance with clause 2.24.3;]
- [the further information you provided on [*insert date*], further to our notice dated [*insert date*], given in accordance with clause 2.24.4.]

We consider that the following event constitutes a Relevant Event:

[*Insert details.*]

We further consider that completion of [the Works] / [Section [*insert Section references*]] is likely to be delayed thereby beyond the relevant Completion Date.

In accordance with clause 2.25.1 (Fixing Completion Date), we therefore write to give an extension of time as we estimate to be fair and reasonable, by fixing the Completion Date(s) as follows:

[*Insert details of the extension of time given. For example*:

Section	Date for Completion	Revised Completion Date	Extension of time	Relevant Event
Section 1	*03 February 2028*	*17 February 2028*	*2 weeks*	*Employer's instruction 10, dated 05 July 2027 (clause 2.26.1)*
Section 2	*03 March 2028*	*31 March 2028*	*4 weeks*	*Employer's instruction 10, dated 05 July 2027 (clause 2.26.1)*
Section 3	*07 April 2028*	*19 May 2028*	*6 weeks*	*Employer's instruction 10, dated 05 July 2027 (clause 2.26.1)]*

[We also attach the reasons for this decision.]

Please let us know if you would like to discuss this further.

Resource 76

Description	A notice from the Employer to the Contractor, not giving an extension of time.
Type	Notice: Employer to Contractor
Provision references	clause 2.25.1
Drafting notes	**Timing:** Clause 2.25.2 requires the Employer to notify the Contractor of its decision in respect of any notice under clause 2.24 as soon as is reasonably practicable, and in any event not later than 8 weeks from the later of (i) the date of receipt of the particulars under clause 2.24.2, or notification under clause 2.24.3, or (ii) the date of receipt of the further information required under clause 2.24.4. Furthermore, where the period from receipt to the Completion Date is less than 8 weeks, it must endeavour to do so prior to the Completion Date. The Employer is encouraged to provide as much detail as possible on the reasoning for its decision, and any analysis undertaken, to help avoid disputes between the Parties.

Reference: Delay notice [*insert reference*]; Extension of time not given (clause 2.25.1)

We write further to the following correspondence:

- your notice dated [*insert date*], with reference [*insert reference*], given in accordance with clause 2.24.1 (Notice by Contractor of delay to progress);
- the associated particulars, dated [*insert date*], given in accordance with clause 2.24.2;
- [your notification of a material change in [the estimated delay] / [certain particulars] dated [*insert date*], given in accordance with clause 2.24.3;]
- [the further information you provided on [*insert date*], further to our notice dated [*insert date*], given in accordance with clause 2.24.4.]

[*Where the Employer considers that there was a Relevant Event, but that it will not delay the completion of the Works or any Section beyond the Completion Date*: We consider that the following event constitutes a Relevant Event: [*Insert details.*]

However, we do not consider that this Relevant Event will delay the completion of [the Works] / [Section [*insert Section references*]] beyond the relevant Completion Date, for the following reasons: [*Insert reasons.*]

In accordance with clause 2.25.1 (Fixing Completion Date), we therefore do not give any extension of time in respect of this Relevant Event.] /

[*Where the Employer considers that there was no Relevant Event*: We do not consider that any of the events referred to constitute a Relevant Event, for the following reasons: [*Insert reasons.*]]

In accordance with clause 2.25.1 (Fixing Completion Date), we therefore do not give any extension of time in respect of this event.]

[We also attach the reasons for this decision.]

Please let us know if you would like to discuss this further.

Resource 77

Description	A notice from the Employer to the Contractor of the final review of the Completion Dates.
Type	Notice: Employer to Contractor
Provision references	clause 2.25.5
Drafting notes	**Timing:** Clause 2.25.5 requires the Employer to give this notice no later than the expiry of 12 weeks after the date of practical completion of the Works or relevant Section. The Employer is encouraged to provide as much detail as possible on the reasoning for its decision, and any analysis undertaken, to help avoid disputes between the Parties.

Reference: Final review of Completion Dates (clause 2.25.5)

In accordance with clause 2.25.5 (Fixing Completion Date), we write to notify you of the following:

[We have fixed Completion Date(s) for [the Works] / [each Section] later than that previously fixed as follows:

[*Insert details of later Completion Dates.*]

We have had regard to the following Relevant Events: [*Insert details of Relevant Events considered.*]] /

[Having regard to any instructions for Relevant Omissions issued after the last occasion on which a new Completion Date was fixed for [the Works] / [each Section], we have fixed Completion Date(s) earlier than that previously fixed as follows:

[*Insert details of earlier Completion Dates.*]] /

[We confirm the Completion Date(s) previously fixed.]

[We attach the reasons for our decision.]

Please let us know if you would like to discuss this further.

Resource 78

Description	A reminder from the Employer to the Contractor of its obligation to use best endeavours to prevent delay.
Type	Reminder: Employer to Contractor
Provision references	clause 2.25.6.1
Drafting notes	**Good practice** This correspondence is not required under clause 2.25.6.1, but it is considered good practice, if the Employer has concerns that the Contractor is not complying with its obligations in this regard.

Reference: Delay notice [*insert reference*]; Contractor's obligations under clause 2.25.6.1

We write further to the following correspondence:

- your notice dated [*insert date*], with reference [*insert reference*], given in accordance with clause 2.24.1 (Notice by Contractor of delay to progress);
- the associated particulars, dated [*insert date*], given in accordance with clause 2.24.2;
- [your notification of a material change in [the estimated delay] / [certain particulars] dated [*insert date*], given in accordance with clause 2.24.3;]
- [the further information you provided on [*insert date*], further to our notice dated [*insert date*], given in accordance with clause 2.24.4.]

We remind you that, in accordance with clause 2.25.6.1 (Fixing Completion Date), you are required to constantly use your best endeavours to prevent delay in the progress of the Works [(or any Section)], however caused, and to prevent the completion of the Works [(or Section)] being delayed or further delayed beyond the relevant Completion Date.

We are concerned that you may not have complied with your obligations under this clause 2.25.6.1, in the following respects:

[*Insert details of how the Contractor may have failed to comply with its obligations under clause 2.25.6.1.*]

Therefore, please confirm the steps you are taking to comply with your obligations under this clause 2.25.6.1, as soon as possible.

Resource 79

Description	A reminder from the Employer to the Contractor to do all that may reasonably be required to the satisfaction of the Employer to proceed with the Works.
Type	Reminder: Employer to Contractor
Provision references	clause 2.25.6.2

Reference: Delay notice [*insert reference*]; Contractor's obligations under clause 2.25.6.2

We write further to the following correspondence:

- your notice dated [*insert date*], with reference [*insert reference*], given in accordance with clause 2.24.1 (Notice by Contractor of delay to progress);
- the associated particulars, dated [*insert date*], given in accordance with clause 2.24.2;
- [your notification of a material change in [the estimated delay] / [certain particulars] dated [*insert date*], given in accordance with clause 2.24.3;]
- [the further information you provided on [*insert date*], further to our notice dated [*insert date*], given in accordance with clause 2.24.4.]

In accordance with clause 2.25.6.2 (Fixing Completion Date), you are required to do all that may reasonably be required to our satisfaction to proceed with the Works [(or Section)].

In that regard, we require you to do the following:

[*Insert details of what further the Employer requires to be done.*]

Please let us know if you would like to discuss this further.

Resource 80

Description	A checklist for the Contractor for drafting delay notices and providing further particulars.
Type	Checklist: Contractor
Provision references	clause 2.24, clause 2.25

Below is a checklist to assist the Contractor to comply with its obligations under clause 2.24 (Notice by Contractor of delay to progress) and clause 2.25 (Fixing Completion Date).

If it has become reasonably apparent that the progress of the Works (or any Section) is either being delayed, or is likely to be delayed, the Contractor should undertake the following as soon as possible:

Item	Description	Provision	Status / action required
1	Notify the Employer of the material circumstances of the delay, including the cause or causes.	clause 2.24.1	Provided on [insert date]. / Not yet provided.
2	Identify in the notice any event which the Contractor considers is a Relevant Event, as set out in clause 2.26.	clause 2.24.1	Provided on [insert date]. / Not yet provided.

(*Continued*)

Item	Description	Provision	Status / action required
3	Give particulars of the expected effects of the delay.	clause 2.24.2	Provided on [insert date]. / Not yet provided.
4	Provide an estimate of any expected delay in the completion of the Works (or any Section) beyond the relevant Completion Date.	clause 2.24.2	Provided on [insert date]. / Not yet provided.
5	If there is a material change in the estimated delay, or in any other particulars, give a further notice to the Employer.	clause 2.24.3	No material change. / Provided on [insert date]. / Not yet provided.
6	Provide any further information required by the Employer as is reasonably necessary to enable it to reach a decision under clause 2.25.1.	clause 2.24.4	No information requested. / Provided on [insert date]. / Not yet provided.
7	Record how the Contractor has used its best endeavours to prevent delay in the progress of the Works (or any Section), however caused, and to prevent the completion of the Works (or any Section) being delayed or further delayed beyond the relevant Completion Date.	clause 2.25.6.1	Refer to the following documents or correspondence: [insert reference]
8	Record how the Contractor has done all that may reasonably be required to proceed with the Works (or any Section), to the Employer's satisfaction.	clause 2.25.6.2	Refer to the following documents or correspondence: [insert reference]

Resource 81

Description	A checklist for the Employer for giving extensions of time.
Type	Checklist: Employer
Provision references	clause 2.25

Below is a checklist to assist the Employer to comply with its obligations under clause 2.25 (Fixing Completion Date).

Item	Description	Provision	Status / action required
1	Has the Contractor provided the notice and particulars required under clause 2.24.1 and clause 2.24.2?	clause 2.25.1	Provided on [insert date]. / Not yet provided.
2	Has the Contractor notified the Employer of any material change in the estimated delay or in any other particulars?	clause 2.25.1	Not applicable. / Provided on [insert date]. / Not yet provided.
3	Has the Contractor supplied the Employer with any further information required by the Employer as is reasonably necessary to enable it to reach a decision under clause 2.25.1?	clause 2.24.4	No further information required. / Provided on [insert date]. / Not yet provided.
4	If the Contractor has complied with the requirements in questions 1 to 3, are any of the events stated to be a cause of the delay a Relevant Event, as set out in clause 2.26?	clause 2.25.1.1	Yes / No
5	Is completion of the Works (or any Section) likely to be delayed by this Relevant Event?	clause 2.25.1.2	Yes / No
6	If the answer to both question 4 and 5 is 'yes', then give an extension of time by fixing a later Completion Date as the Employer estimates to be fair and reasonable. If the answer to either question 4 or 5 is 'no', then do not give an extension of time.	clause 2.25.1	Extension of time given / not given.

Resource 82

Description	A tracker template for delay notices, for both the Contractor and the Employer.
Type	Tracker: Contractor and Employer
Provision references	clause 2.24, clause 2.25

Below is a delay notice tracker template, for use by both the Contractor and the Employer.

- Reference: [*insert reference*]
- Description: [*insert description of delay*]
- Date of notice (clause 2.24.1): [*insert date*]

- Relevant Event, if applicable (clause 2.24.1): [*insert reference*]
- Date particulars and estimate of delay provided (clause 2.24.2): [*insert date*]
- Any material change in the estimated delay, or in other particulars (clause 2.24.3): [*insert details*]
- Any further information requested by the Employer (clause 2.24.4): [*insert details*]
- Latest date for Employer's decision (clause 2.25.2): [*insert date*]
- Document link: [*insert link to documents*]
- Status / action required: [*insert status, and any action required*]

Ref.	**Description**	**Date of notice**	**Relevant Event (if applicable)**	**Date particulars and estimate of delay provided**	**Any material change, or further information requested?**	**Latest date for Employer's decision**	**Status / action required**
	clause reference	**clause 2.24.1**	**clause 2.24.1**	**clause 2.24.2**	**clause 2.24.3, clause 2.24.4**	**clause 2.25.2**	
1	*Employer's instruction 10 (Change of kitchen tile specification)*	*03 February 2028*	*clause 2.26.1*	*03 February 2028* *(6-week delay)*	*Yes, further details of supplier lead times requested on 10 February 2025, provided on 12 February 2025.*	*31 March 2028*	*[Insert status or action required.]*
2	*Delay in procurement of bathroom vanity units*	*07 April 2028*	*Not applicable.*	*07 April 2028* *(1-week delay)*	*No.*	*Not applicable.*	*[Insert status or action required.]*
3	*Statutory Provider delay (Gas provider)*	*12 May 2028*	*clause 2.26.7*	*12 May 2028* *(6-week delay)*	*Yes, copy of contract with Statutory Provider requested, not yet provided.*	*To be confirmed.*	*[Insert status or action required.]*

Resource 83

Description	A reminder from the Contractor to the Employer of the Contractor's anticipated date of practical completion.
Type	Reminder: Contractor to Employer
Provision references	clause 2.27
Drafting notes	**Good practice** This correspondence is not required by clause 2.27, but it is considered good practice for the Contractor to regularly update the Employer on the anticipated dates of practical completion.

Reference: Anticipated date of practical completion (clause 2.27)

We write to advise you on the date(s) we anticipate practical completion of [the Works] / [each Section] will be achieved, as follows:

[*Insert anticipated dates of practical completion.*]

By then, we consider that practical completion will be achieved, and we will have complied sufficiently with our obligations under clause 2.37 (As-built Drawings) and clause 3.16 (CDM Regulations and Part 2A of the Building Regulations), and all other requirements for practical completion under the Contract, such that the [*the Works:* Practical Completion Statement] / [*Sections:* Section Completion Statements] can be issued in accordance with clause 2.27 (Practical completion).

We wish to provide you with as much notice as possible, to enable you to undertake any inspections of the Works and all associated documents and information.

We will advise you of any change to the anticipated date(s) of practical completion.

Resource 84

Description	A Practical Completion Statement template, from the Employer to the Contractor.
Type	Employer to Contractor: Practical Completion Statement
Provision references	clause 2.27
Drafting notes	**Timing:** Clause 2.27 requires the Employer to issue the Practical Completion Statement 'forthwith', when practical completion of the Works is achieved.

Reference: Practical Completion Statement (clause 2.27)

In accordance with clause 2.27 (Practical completion), we confirm that:

- practical completion of the Works was achieved; and
- you have complied sufficiently with clause 2.37 (As-built Drawings) and clause 3.16 (CDM Regulations and Part 2A of the Building Regulations), in respect of the supply of documents and information,
  on the following date: [*insert date.*]

The Rectification Period will commence from [*insert date*], and will expire on [*insert date*].

Resource 85

Description	A Section Completion Statement template, from the Employer to the Contractor.
Type	Employer to Contractor: Section Completion Statement
Provision references	clause 2.27
Drafting notes	**Timing:** Clause 2.27 requires the Employer to issue the Section Completion Statement 'forthwith', when practical completion of the relevant Section is achieved.

Reference: Section Completion Statement (clause 2.27)

In accordance with clause 2.27 (Practical completion), we confirm that:

- practical completion of Section [*insert reference*] was achieved; and
- you have complied sufficiently with clause 2.37 (As-built Drawings) and clause 3.16 (CDM Regulations and Part 2A of the Building Regulations), in respect of the supply of documents and information,
  on the following date: [*insert date*]

The Rectification Period for this Section will commence from [*insert date*], and will expire on [*insert date*].

Resource 86

Description	A checklist for the requirements of practical completion.
Type	Checklist: Contractor and Employer
Provision references	clause 2.27

Below is a checklist for determining whether practical completion of the Works or a Section has been achieved:

Item	Description	Provision	Status / action required
1	Practical completion: are the Works or Section practically complete?	2.27	Yes / No
2	Clause 2.37 (As-built Drawings): has the Contractor supplied for retention and use by the Employer such Contractor's Design Documents and related information as is specified in the Contract Documents, or as the Employer may reasonably require, that show or describe the Works as built, or relate to the maintenance and operation of them, or their installations?	2.27 / 2.37	Provided on [insert date]. / Not yet provided.
3	Clause 3.16 (CDM Regulations and Part 2A of the Building Regulations): where the Contractor is and while it remains the Principal Designer for the purposes of the CDM Regulations, has it prepared and delivered the health and safety file to the Employer?	2.27 / 3.16	Not applicable. / Provided on [insert date]. / Not yet provided.

(*Continued*)

Item	Description	Provision	Status / action required
4	Clause 3.16 (CDM Regulations and Part 2A of the Building Regulations): has the Contractor prepared and delivered the relevant information in accordance with: • regulation 11F (Competence: general requirement) • regulation 11J (Duties of dutyholders: General duty) • regulation 11K (Duties of dutyholders: Additional duties of designers), including 11K(3) and, if applicable, 11K(5) • regulation 11L (Duties of dutyholders: Additional duties of contractors), including 11L(3) and, if applicable, 11K(5)	2.27 / 3.16	Provided on [insert date]. / Not yet provided.
5	Clause 3.16 (CDM Regulations and Part 2A of the Building Regulations): where the Contractor is the Principal Contractor for the purposes of the Building Regulations, has the Contractor prepared and delivered the relevant information in accordance with regulation 11N (Duties of dutyholders: Additional duties of a principal contractor), including 11N(3), 11N(5).	2.27 / 3.16	Not applicable. / Provided on [insert date]. / Not yet provided.
6	Other requirements for practical completion: [*insert any other requirements of practical completion included in the Contract Documents.*]	n/a	Not applicable. / Provided on [insert date]. / Not yet provided.

Resource 87

Description	A Non-Completion Notice template, from the Employer to the Contractor.
Type	Notice: Employer to Contractor
Provision references	clause 2.28
Drafting notes	Clause 2.28 does not expressly state a time by which this notice must be given, but the Employer is encouraged to give it as soon as possible after the relevant Completion Date. However, if the Employer has yet to make a decision under clause 2.25 in respect of any notice and particulars from the Contractor under clause 2.24.1 and clause 2.24.2 or (where applicable) a notification under clause 2.24.3, it is recommended to make this decision first, before issuing the Non-Completion Notice.

Reference: Non-Completion Notice (clause 2.28)

In accordance with clause 2.28 (Non-Completion Notice), we write to notify you that you have failed to complete [the Works] / [Section [*insert reference*]] by the relevant Completion Date, namely [*insert date*].

We confirm that, if a new Completion Date is fixed after the issue of this notice, such fixing will cancel this notice, and where necessary, we will issue a further notice.

Resource 88

Description	A further Non-Completion Notice template from the Employer to the Contractor, if a new Completion Date was fixed after the issue of the initial Non-Completion Notice.
Type	Notice: Employer to Contractor
Provision references	clause 2.28
Drafting notes	Clause 2.28 does not expressly state a time by which this notice must be issued, but the Employer is encouraged to issue it as soon as possible after the relevant revised Completion Date.

Reference: (Further) Non-Completion Notice (clause 2.28)

In accordance with clause 2.28 (Non-Completion Notice), we write to notify you that you have failed to complete [the Works] / [Section [*insert reference*]] by the relevant Completion Date, namely [*insert date*].

Further to our correspondence dated [*insert date*], the Completion Date was revised from [*insert date*] to [*insert date*], thereby cancelling our initial Non-Completion Notice dated [*insert date*].

(We confirm that, if a new Completion Date is fixed again after the issue of this notice, such fixing will cancel this notice, and where necessary, we will issue a further notice.)

Resource 89

Description	A notice from the Employer to the Contractor that the Employer may require payment of, or may withhold or deduct, liquidated damages.
Type	Notice: Employer to Contractor
Provision references	clause 2.29.1.2
Drafting notes	**Timing:** If the Employer wishes to give this notice, clause 2.29.1.2 requires them to do so before the due date for the final payment under clause 4.24.5 (Final Date and Final Payment), or (if applicable) the Termination Payment under clause 8.7.5 (Consequences of termination under clauses 8.4 to 8.6), clause 8.8.2 (Employer's decision not to complete the Works), or clause 8.12.4 (Consequences of Termination under clauses 8.9 to 8.11, etc.). However, the Employer is encouraged to give this notice much earlier, as soon as the decision whether or not to deduct liquidated damages has been taken.

Reference: Liquidated damages (clause 2.28.1.2)

We write further to our Non-Completion Notice for [the Works] / [Section [*insert reference*]], dated [*insert date*].

In accordance with clause 2.29.1.2 (Payment or allowance of liquidated damages), we write to notify you that the Employer may require payment of, or may withhold or deduct, liquidated damages.

Resource 90

Description	A notice from the Employer to the Contractor that the Employer requires payment of, or will withhold or deduct, liquidated damages.
Type	Notice: Employer to Contractor
Provision references	clause 2.29.1
Drafting notes	**Timing:** If the Employer wishes to give this notice, clause 2.29.1 requires them to do so not later than 5 days before the final date for payment of the amount payable under clause 4.24 (Final Date and Final Payment), or (if applicable) the Termination Payment under clause 8.7.6 (Consequences of termination under clauses 8.4 to 8.6), clause 8.8.3 (Employer's decision not to complete the Works), or clause 8.12.5 (Consequences of Termination under clauses 8.9 to 8.11, etc.). However, the Employer is encouraged to give this notice much earlier, as soon as the decision whether or not to deduct liquidated damages has been taken.

Reference: Liquidated damages (clause 2.29.1)

We write further to the following correspondence, in respect of [the Works] / [Section [*insert reference*]]:

- our Non-Completion Notice dated [*insert date*], issued in accordance with clause 2.28 (Non-Completion Notice); and
- our notice dated [*insert date*], given in accordance with clause 2.29.1.2 (Payment or allowance of liquidated damages).

In accordance with clause 2.29.1 and clause 2.29.2, we write to notify you that, for the period between the Completion Date and the date of practical completion:

- the Employer requires you to pay liquidated damages [at the rate stated in the Contract Particulars] / [at a lesser rate than that stated in the Contract Particulars]; and/or
- the Employer will withhold or deduct liquidated damages [at the rate stated in the Contract Particulars] / [at a lesser rate than that stated in the Contract Particulars], from sums due to you.

The relevant particulars are as follows:

- Completion Date: [*insert date*]
- Date of practical completion: [*insert date*]
- Duration: [*insert period in days or weeks, as appropriate*]
- Rate of liquidated damages stated in the Contract Particulars: [*insert rate*]
- Rate the Employer requires you to pay and/or will withhold or deduct from sums due to you: [*insert rate, which must be equal to or less than the rate stated in the Contract Particulars*]

We confirm that we will give a Pay Less Notice in respect of the above, in accordance with clause 4.9.5 (Interim and final payments – final date and amount).

Resource 91

Description	A reminder from the Contractor to the Employer to repay liquidated damages following an extension of time.
Type	Reminder: Contractor to Employer
Provision references	clause 2.29.3
Drafting notes	**Good practice** This correspondence is not required by clause 2.29.3, but it is considered good practice for the Contractor to remind the Employer of any amounts which need to be repaid.

Reference: Repayment of liquidated damages (clause 2.29.3)

We write further to your correspondence dated [*insert date*], which fixed a later Completion Date for [the Works] / [Section [*insert reference*]], from [*insert previous date*] to [*insert date*].

We write to remind you that, in accordance with clause 2.29.3 (Payment or allowance of liquidated damages), the Employer is required to pay or repay to us any amounts recovered, allowed or paid under clause 2.29 for the period up to the later Completion Date.

We calculate this amount to be as follows:

- Revised Completion Date: [*insert date*]
- Date of practical completion: [*insert date*]
- Duration: [*insert period in days or weeks, as appropriate*]
- Rate of liquidated damages: [*insert rate*]
- Amount to be paid or repaid: [*insert amount*]

Resource 92

Description	Correspondence from the Employer to the Contractor that the Employer no longer may require payment of, or may withhold or deduct, liquidated damages.
Type	Correspondence: Employer to Contractor
Provision references	clause 2.29.4

Reference: Liquidated damages (clause 2.29.4)

We write further to our notice dated [*insert date*], given in accordance with clause 2.29.1.2 (Payment or allowance of liquidated damages), in respect of [the Works] / [Section [*insert reference*]].

In accordance with clause 2.29.4, and contrary to that notice, we write to notify you that we no longer may require payment of, or may withhold or deduct, liquidated damages.

If our position changes, we will notify you again in accordance with clause 2.29.1.2.

Resource 93

Description	A checklist for the deduction of liquidated damages by the Employer.
Type	Checklist: Employer
Provision references	clause 2.29

Below is a checklist for the deduction of liquidated damages by the Employer:

Item	Description	Provision	Status / action required
1	Has the Employer issued a Non-Completion Notice, in accordance with clause 2.28 (Non-Completion Notice)?	2.29.1.1 / 2.28	Yes, issued on [insert date] / No
2	Has the Employer notified the Contractor that it may require payment of, or may withhold or deduct, liquidated damages, in accordance with clause 2.29.1.2 (Payment or allowance of liquidated damages)?	2.29.1.2	Yes, given on [insert date] / No
3	Has the Employer notified the Contractor that it requires payment of, or will withhold or deduct, liquidated damages, at the rate stated in the Contract Particulars (or a lesser rate), in accordance with clause 2.29.1 and clause 2.29.2?	2.29.1 / 2.29.2	Yes, given on [insert date] / No
4	If the Employer intends to withhold or deduct all or any of the liquidated damages payable, has it given a Pay Less Notice, in accordance with clause 4.9.5 (Interim and final payments – final date and amount), clause 8.13.3 (Termination Payment – final date, notices and amount) or clause 8.13.5?	2.29.2.2 / 4.9.5	Yes, given on [insert date] / No

Resource 94

Description	A request from the Employer for the Contractor's consent for the Employer to take partial possession.
Type	Request: Employer to Contractor
Provision references	clause 2.30

Reference: Request to take partial possession (clause 2.30)

In accordance with clause 2.30 (Partial Possession by Employer: Contractor's consent), we request your consent for the Employer to take possession of the following part(s) of [the Works] / [Section [*insert reference*]]:

[*Insert description of parts for which the Employer wishes to take partial possession.*]

The Employer wishes to take partial possession for the following reasons: [*insert reasons.*]

Please give your consent by [*insert date*].

[In relation to clause 2.34 (Partial Possession by Employer: Liquidated damages – Relevant Part), we consider that the value of the Relevant Part is [*insert amount*] (exclusive of VAT). We attach the basis on which this sum was calculated.]

Resource 95

Description	Consent given by the Contractor for the Employer to take partial possession.
Type	Consent given: Contractor to Employer
Provision references	clause 2.30

Reference: Request to take partial possession (clause 2.30)

We write in reply to your request dated [*insert date*] for our consent for the Employer to take partial possession of [the Works] / [Section [*insert reference*]].

[*Unconditional consent*: We give our consent.] / [*Conditional consent*: We give our consent, subject to the following conditions: [*insert conditions*.]]

[In relation to clause 2.34 (Partial Possession by Employer: Liquidated damages – Relevant Part), [we agree with your valuation of the Relevant Part.] / [we disagree with your valuation of the Relevant Part of [*insert amount*] (exclusive of VAT), and consider that the value of the Relevant Part is [*insert amount*] (exclusive of VAT). We attach the basis on which this sum was calculated.]]

Resource 96

Description	Consent withheld by the Contractor for the Employer to take partial possession.
Type	Consent withheld: Contractor to Employer
Provision references	clause 2.30
Drafting notes	Clause 1.10 requires that the Contractor's consent must not be unreasonably delayed or withheld.

Reference: Request to take partial possession (clause 2.30)

We write in reply to your request dated [*insert date*] for our consent for the Employer to take partial possession of [the Works] / [Section [*insert reference*]].

We withhold our consent, for the following reasons: [*insert reasons, to demonstrate why the consent was not unreasonably delayed or withheld.*]

Resource 97

Description	A notice from the Contractor to the Employer of partial possession.
Type	Notice: Contractor to Employer
Provision references	clause 2.30
Drafting notes	**Timing:** Clause 2.30 requires the Contractor to give this notice upon the Employer taking partial possession. The Contractor should also check which Insurance Option applies, in respect of the final paragraph.

Reference: Notice of partial possession (clause 2.30)

We write further to your request dated [*insert date*], for our consent for the Employer to take partial possession of [the Works] / [Section [*insert reference*]], and our reply dated [*insert date*], giving our consent.

In accordance with clause 2.30 (Partial Possession by Employer: Contractor's consent), we write to notify you of the parts taken into possession, and the date when the Employer took possession, as follows:

- Relevant Part: [*insert description of the parts taken into possession, and attaching any relevant drawings or documents.*]
- Relevant Date: [*insert date when the Employer took possession.*]

In accordance with clause 2.31 (Partial Possession by Employer: Practical completion date), for the purposes of clause 2.35 (Schedules of defects and instructions) and clause 4.18.2 (Retention – amounts and periods):

- practical completion of the Relevant Part has occurred; and
- the Rectification Period in respect of the Relevant Part has commenced, on the Relevant Date.

In accordance with clause 2.33 (Partial Possession by Employer: Insurance – Relevant Part), as from the Relevant Date, the Works insurance obligation under Insurance Option [A] / [B] / [C.2] will cease in respect of the Relevant Part (but not otherwise) [and the Existing Structures will from the Relevant Date be deemed to include the Relevant Part].

Resource 98

Description	A Notice of Completion of Making Good template for a Relevant Part, from the Employer to the Contractor.
Type	Notice: Employer to Contractor
Provision references	clause 2.32
Drafting notes	The Employer is encouraged to issue this notice immediately once the relevant defects, shrinkages, or other faults in the Relevant Part have been made good.

Reference: Notice of Completion of Making Good for Relevant Part (clause 2.32)

We write in relation to the following Relevant Part: [*insert description of the Relevant Part.*]

In accordance with clause 2.32 (Partial Possession by Employer: Defects etc. – Relevant Part), we write to notify you that any defects, shrinkages, or other faults in the Relevant Part which we required to be made good under clause 2.35 (Schedules of defects and instructions) have been made good on [*insert date*].

Resource 99

Description	A tracker template for recording partial possessions by the Employer.
Type	Tracker: Contractor and Employer
Provision references	clause 2.30 to clause 2.34

Below is a tracker template for recording partial possessions by the Employer of the Works or a Section:

- Relevant Part: [*insert description of the parts taken into possession, with references to any relevant drawings or documents*]
- Relevant Date: [*insert date when the Employer took possession*]
- Expiry of Rectification Period for the Relevant Part: [*insert date*]
- Date of Notice of Completion of Making Good for the Relevant Part: [*insert date*]
- Value of the Relevant Part (exclusive of VAT): [*insert amount*]
- Document link: [*insert link to documents*]
- Status / action required: [*insert status, and any action required*]

Resource 100

Description	An instruction by the Employer of a schedule of defects, shrinkages, and other faults in the Works, to be made good by the Contractor.
Type	Instruction: Employer to Contractor
Provision references	clause 2.35.1
Drafting notes	**Timing:** Clause 2.35.1 requires the Employer to issue this instruction not later than 14 days after the expiry of the relevant Rectification Period.

Reference: Defects, shrinkages, and/or other faults; Employer's instruction [*insert reference*] (clause 2.35.1)

In accordance with clause 2.35.1 (Schedules of defects and instructions), we attach an instruction appending a schedule of defects, shrinkages, and other faults in [the Works] / [Section [*insert reference*]], which have appeared within the Rectification Period, due to your failure to comply with your obligations under the Contract.

We instruct you to make good these defects, shrinkages, and other faults within a reasonable time, unless we subsequently instruct you otherwise. We remind you that, in accordance with clause 2.35, this will be at no cost to the Employer.

[Please contact [*insert name and email address*] to arrange the necessary access.]

[Schedule of defects:

- Reference: [*insert reference*]
- Description of defect, shrinkage, or other fault: [*insert description*]
- Package / element of works (for example, mechanical, electrical, etc.): [*insert description*]
- Location: [*insert location*]
- Relevant provision from the Contract Documents: [*insert reference*]
- Making good required: [*insert description of the making good required*]
- Date making good required: [*insert date by which the making good is required*]
- Status / action required: [*insert status of the defect, and any action required*]]

Resource 101

Description	An instruction from the Employer requiring a particular defect, shrinkage, or other fault in the Works to be made good by the Contractor.
Type	Instruction: Employer to Contractor
Provision references	clause 2.35.2
Drafting notes	**Timing:** If the Employer wishes to issue these instructions, clause 2.35.2 permits them to do so (i) prior to the issue of the schedule of defects under clause 2.35.1, and (ii) no later than 14 days after the expiry of the relevant Rectification Period. The Employer should therefore consider this restriction in mind before issuing the schedule of defects under clause 2.35.1.

Reference: Defects, shrinkages, and/or other faults; Employer's instruction [*insert reference*] (clause 2.35.2)

In accordance with clause 2.35.2 (Schedules of defects and instructions), we instruct you to make good the following defect, shrinkage, or other fault, which has appeared within the Rectification Period for [the Works] / [Section [*insert reference*]], due to your failure to comply with your obligations under the Contract:

[*Insert details of the defect, including a description, its location, the relevant provisions from the Contract Documents, the making good required, and, if applicable, the date by which the defect should be made good.*]

We remind you that, in accordance with clause 2.35, this will be at no cost to the Employer.

In accordance with clause 2.35.1, we may also issue a schedule of defects as an instruction, not later than 14 days after the expiry of the Rectification Period.

Resource 102

Description	An instruction from the Employer requiring a particular defect, shrinkage, or other fault in the Works not to be made good by the Contractor.
Type	Instruction: Employer to Contractor
Provision references	clause 2.35.2

Reference: Defects, shrinkages, and/or other faults; Employer's instruction [*insert reference*] (clause 2.35.2)

In accordance with clause 2.35 (Schedules of defects and instructions), we instruct you not to make good the following defect, shrinkage, or other fault, which has appeared within the Rectification Period for [the Works] / [Section [*insert reference*]], due to your failure to comply with your obligations under the Contract:

[*Insert details of the defect, including a description, its location, and the relevant provisions from the Contract Documents.*]

We will therefore make an appropriate deduction from the Contract Sum, in the amount of [*insert amount*] (exclusive of VAT), and we attach the basis on which this sum was calculated.

Resource 103

Description	A reminder from the Contractor to the Employer of its obligation to issue a Notice of Completion of Making Good.
Type	Reminder: Contractor to Employer
Provision references	clause 2.36
Drafting notes	**Good practice** This correspondence is not required by clause 2.36, but it is considered good practice to advise the Employer that all defects, shrinkages, and other faults have been made good, to enable the Notice of Completion of Making Good to be issued. This in turn will help with the release of the remaining Retention (see clause 4.18.2).

Reference: Request for Notice of Completion of Making Good (clause 2.36)

We write to advise you that we have now made good all the defects, shrinkages, and other faults in [the Works] / [Section [*insert reference*]] which appeared within the Rectification Period, and which you required to be made good in accordance with clause 2.35 (Schedules of defects and instructions).

We therefore request that, in accordance with clause 2.36 (Notice of Completion of Making Good), you issue a Notice of Completion of Making Good without delay.

Resource 104

Description	A Notice of Completion of Making Good template, from the Employer to the Contractor.
Type	Notice: Employer to Contractor
Provision references	clause 2.36
Drafting notes	**Timing:** Clause 2.36 requires the Employer to issue the Notice of Completion of Making Good when the defects, shrinkages, or other faults which the Employer required to be made good have been made good, and states that it must not be unreasonably delayed or withheld. The Employer is therefore encouraged to issue the Notice of Completion of Making Good as soon as possible thereafter.

Reference: Notice of Completion of Making Good (clause 2.36)

In accordance with clause 2.36 (Notice of Completion of Making Good), we write to notify you that on [*insert date*], you have made good all of the defects, shrinkages, or other faults in [the Works] / [Section [*insert reference*]] which appeared within the Rectification Period, and which under clause 2.35 (Schedules of defects and instructions) we required to be made good.

Resource 105

Description	A tracker template for the making good of defects, shrinkages, or other faults in the Works.
Type	Tracker: Contractor and Employer
Provision references	clause 2.35, clause 2.36

Below is a tracker template for recording the making good of defects, shrinkages, or other faults in the Works or a Section:

- Works / Section: [*insert Section reference, if applicable*]
- Reference: [*insert defect reference*]
- Schedule of defects or Employer's instruction reference: [*insert schedule of defects, or the Employer's instruction reference*]
- Description of defect, shrinkage, or other fault: [*insert description*]
- Package / element of works (for example, mechanical, electrical, etc.): [*insert description*]
- Location: [*insert location*]
- Relevant provision from the Contract Documents: [*insert reference*]
- Making good required: [*insert description of the making good required*]

- Date making good required: [*insert date by which the making good is required*]
- Document link: [*insert link to documents*]
- Status / action required: [*insert status of the defect, and any action required*]

Resource 106

Description	A request from the Employer to the Contractor for further documents to show or describe the Works as built, or relate to the maintenance and operation of them, or their installations.
Type	Request for further documents: Employer to Contractor
Provision references	clause 2.37

Reference: As-built Drawings (clause 2.37)

In accordance with clause 2.37 (As-built Drawings), and in addition to what is already specified in the Contract Documents, we require the following Contractor's Design Documents and related information that show or describe the Works as built, or relate to the maintenance and operation of them, or their installations:

[*Insert details of the further information required from the Contractor.*]

We require this information for the following reasons: [*insert reasons.*]

Please provide this information by [*insert date*].

Resource 107

Description	A tracker template for recording the collation of Contractor's Design Documents, as-built drawings, and documents for operation and maintenance.
Type	Tracker: Contractor and Employer
Provision references	clause 2.37

Below is a tracker template for recording the collation of Contractor's Design Documents and related information as is specified in the Contract Documents or as the Employer may reasonably require that show or describe the Works as built, or relate to the maintenance and operation of them, or their installations.

- Document description: [*insert description*]
- Contract Documents reference: [*insert reference*]
- Document revision: [*insert revision*]
- Document author: [*insert author*]
- Date provided to the Employer: [*insert date*]
- Document link: [*insert link to documents*]
- Status / action required: [*insert status, and any action required*]

Resource 108

Description	A request from the Employer to the Contractor for access for the Employer's Agent and the Employer to the Contractor's workshops and other premises.
Type	Request: Employer to Contractor
Provision references	clause 3.1

Reference: Request for access (clause 3.1)

In accordance with clause 3.1 (Access for Employer's Agent), we require access to the Works and to your workshops or other premises where work is being prepared for the Contract, as follows:

[*Insert details of the access required, including the relevant parties, locations, proposed dates, and what the parties wish to inspect.*]

Please confirm that this access will be provided.

Resource 109

Description	A request from the Contractor to the Employer for approval to appoint a Site Manager.
Type	Request: Contractor to Employer
Provision references	clause 3.2

Reference: Request to appoint Site Manager (clause 3.2)

In accordance with clause 3.2 (Site Manager), we request your approval to appoint a full-time Site Manager.

We propose to appoint [*insert name*], and attach details of their background and experience. Given the importance of the post, we will arrange an introductory meeting as soon as possible, to answer any questions you may have.

They will act as our representative on site, in charge of the Works. We will ensure that they, or a competent deputy, are on site at all material times.

Please give your approval by [*insert date*].

Resource 110

Description	A request from the Contractor to the Employer for approval to replace the Site Manager.
Type	Request: Contractor to Employer
Provision references	clause 3.2
Drafting notes	**Timing:** Clause 3.2 requires the Contractor appoint a replacement Site Manager 'promptly' upon the previous Site Manager ceasing to hold their post.

Reference: Request to appoint replacement Site Manager (clause 3.2)

We write to notify you that our previous appointee to the post of Site Manager, [*insert name*], has ceased to hold that post, for the following reasons: [*insert reasons.*]

In accordance with clause 3.2 (Site Manager), we request your approval to appoint a replacement Site Manager.

We propose to appoint [*insert name*], and attach details of their experience. Given the importance of the post, we will arrange an introductory meeting as soon as possible, to answer any questions you may have.

They will act as our representative on site, in charge of the Works. We will ensure that they, or a competent deputy, are on site at all material times.

Please give your approval by [*insert date*].

Resource 111

Description	Approval given by the Employer for the Contractor to appoint a Site Manager (or a replacement Site Manager).
Type	Approval given: Employer to Contractor
Provision references	clause 3.2

Reference: Request to appoint replacement Site Manager (clause 3.2)

We write in reply to your request dated [*insert date*], for our approval for you to appoint [*insert name*] as [replacement] Site Manager, in accordance with clause 3.2 (Site Manager).

We give our approval for this appointment.

We remind you that any instructions issued to the Site Manager, or their deputy, will be treated as issued to the Contractor.

Resource 112

Description	Approval withheld by the Employer for the Contractor to appoint a Site Manager (or a replacement Site Manager).
Type	Approval withheld: Employer to Contractor
Provision references	clause 3.2
Drafting notes	Clause 1.10 requires that the Employer's approval must not be unreasonably delayed or withheld.

Reference: Request to appoint replacement Site Manager (clause 3.2)

We write in reply to your request dated [*insert date*], for our approval for you to appoint [*insert name*] as [replacement] Site Manager, in accordance with clause 3.2 (Site Manager).

We withhold our approval for this appointment, for the following reasons: [*insert reasons, to demonstrate why the approval was not unreasonably delayed or withheld. For example, the proposed appointee had insufficient experience.*]

Please propose an alternative replacement Site Manager for our approval, taking the reasons above into consideration.

Resource 113

Description	A request from the Contractor to the Employer for consent to sub-contract the whole or any part of the Works or the design for the Works.
Type	Request: Contractor to Employer
Provision references	clause 3.3.1

Reference: Request to sub-contract (clause 3.3.1)

In accordance with clause 3.3.1, we request the Employer's consent to:

- sub-contract [the whole] / [part] of the Works; and/or
- sub-contract the design for [part of] the Works.

In particular, we request the Employer's consent to sub-contract the following:

[*Insert details of the elements of the Works or design which the Contractor wishes to sub-contract.*]

We propose to appoint [*insert name of sub-contractor*] to carry out [the Works] / [this part of the Works] / [the design for the Works] / [the design for this part of the Works], and attach details of their experience.

Please give your consent by [*insert date*], to avoid a delay to the progress of the Works.

We appreciate that, in no case, will any such consent or any sub-contracting in any way affect our obligations under any other provision of the Contract.

Resource 114

Description	Consent given by the Employer for the Contractor to sub-contract the whole or any part of the Works or the design for the Works.
Type	Consent given: Employer to Contractor
Provision references	clause 3.3.1

Reference: Request to sub-contract (clause 3.3.1)

We refer to your request dated [*insert date*], with reference [*insert reference*], made in accordance with clause 3.3.1 (Consent to sub-contracting).

We give our consent for you to sub-contract [the Works] / [this part of the Works] / [the design for the Works] / [the design for this part of the Works] to [*insert name of sub-contractor*].

We remind you that this consent, nor any sub-contracting, will not in any way affect your obligations under any other provision of the Contract.

Resource 115

Description	Consent withheld by the Employer for the Contractor to sub-contract the whole or any part of the Works or the design for the Works.
Type	Consent withheld: Employer to Contractor
Provision references	clause 3.3.1
Drafting notes	Clause 1.10 requires that the Employer's consent must not be unreasonably delayed or withheld. The final sentence has been included in the event that the Employer's objection relates to the identity of the proposed sub-contractor, rather than the principle of sub-contracting.

Reference: Request to sub-contract (clause 3.3.1)

We refer to your request dated [*insert date*], with reference [*insert reference*], made in accordance with clause 3.3.1 (Consent to sub-contracting).

We withhold our consent for you to sub-contract [the Works] / [this part of the Works] / [the design for the Works] / [the design for this part of the Works] to [*insert name of sub-contractor*], for the following reasons: [*insert reasons, to demonstrate why the consent was not unreasonably delayed or withheld. For example, the proposed sub-contractor had insufficient experience.*]

[Please propose an alternative [sub-contractor] / [design consultant], taking the reasons above into consideration.]

Resource 116

Description	A tracker template for the appointment of sub-contractors in accordance with clause 3.3 and clause 3.4.
Type	Tracker: Contractor and Employer
Provision references	clause 3.3, clause 3.4

Below is a tracker template for recording the appointment of sub-contractors (including consultants):

- Name: [*insert name of sub-contractor*]
- Company number: [*insert sub-contractor's company number*]
- Registered address: [*insert address*]
- Contact: [*insert sub-contractor contact name and email address*]
- Description of the sub-contract works: [*insert description of the sub-contract works*]
- Date consent requested: [*insert date*]
- Date consent provided: [*insert date*]
- Date of sub-contract: [*insert date*]
- Professional indemnity insurance: [*insert level of cover*]
- Third party rights or collateral warranty required: [*insert whether third party rights or collateral warranties are required*]
- [Sub-contract sum (exclusive of VAT): [*insert amount*]]
- Document link: [*insert link to documents*]
- Status / action required: [*insert status, and any action required*]

Resource 117

Description	A notice from the Contractor to the Employer of a reasonable objection to an Employer's instruction.
Type	Notice: Contractor to Employer
Provision references	clause 3.5.1
Drafting notes	Clause 3.5.1 requires that the Contractor's objection must be reasonable. The Contractor is therefore encouraged to apply this to the relevant part of clause 5.1.2. For example, the instruction would impose such severe limitations of working space as to materially impede the carrying out and completion of the Works. The Contractor is encouraged to give this notice as soon as possible after receiving the instruction.

Reference: Notice of objection to Employer's instruction (clause 3.5.1)

In accordance with clause 3.5.1 (Compliance with instructions), we write to notify you of an objection to the Employer's instruction dated [*insert date*], with reference [*insert reference*].

This instruction requires a Change of the type referred to in clause 5.1.2 (Definition of Changes), as follows:

[*Insert details of how the instruction constitutes a Change under clause 5.1.2, namely:*

'the imposition by the Employer of any obligations or restrictions in regard to the following matters or any addition to or alteration or omission of any such

obligations or restrictions that are so imposed or are imposed in the Employer's Requirements in regard to:

1. *access to the site or use of any specific parts of the site;*
2. *limitations of working space;*
3. *limitations of working hours; or*
4. *the execution or completion of the work in any specific order.'*]

We object to this instruction for the following reasons: [*insert reasons.*]

We therefore need not comply with this instruction.

We propose that the Parties meet as soon as possible to discuss proposals for an alternative instruction, taking the concerns above into consideration.

Resource 118

Description	A notice from the Employer to the Contractor requiring compliance with an instruction.
Type	Notice: Employer to Contractor
Provision references	clause 3.6

Reference: Notice requiring compliance with Employer's instruction (clause 3.6)

We write to notify you that we require compliance with the Employer's instruction dated [*insert date*], with reference [*insert reference*].

In accordance with clause 3.6 (Non-compliance with instructions), if within 7 days after receipt of this notice you do not comply, the Employer may employ and pay other persons to execute work of any kind that may be necessary to give effect to that instruction. You will be liable for all additional costs incurred by the Employer in connection with such employment, and an appropriate deduction may be made from the Contract Sum.

Resource 119

Description	A reply from the Contractor to the Employer following a notice from the Employer requiring compliance with an instruction.
Type	Reply: Contractor to Employer
Provision references	clause 3.6
Drafting notes	The Contractor is encouraged to give this reply as soon as possible after receipt of the Employer's notice under clause 3.6, to demonstrate its compliance with the Employer's instruction.

Reference: Notice requiring compliance with Employer's instruction (clause 3.6)

We write in reply to your notice dated [*insert date*], requiring compliance with the Employer's instruction dated [insert date], with reference [*insert reference*].

We confirm that we are complying with this instruction, and have taken the following steps to date: [*insert details of the extent to which the Contractor has complied with the instruction.*]

Please let us know if there is any further information we can provide in this regard.

Resource 120

Description	A further notice from the Employer to the Contractor, following the Employer's initial notice under clause 3.6, that the Employer will employ and pay other persons.
Type	Notice: Employer to Contractor
Provision references	clause 3.6
Drafting notes	**Timing:** If the Employer wishes to give this notice, it must wait at least 7 days from receipt of the first notice under clause 3.6 before doing so.

Reference: Notice requiring compliance with Employer's instruction (clause 3.6)

We write further to our notice dated [*insert date*], requiring compliance with the Employer's instruction dated [*insert date*], with reference [*insert reference*].

You have not complied with this instruction within 7 days after receipt of that notice.

Therefore, in accordance with clause 3.6, the Employer will employ and pay other persons to execute work of any kind that may be necessary to give effect to this instruction. You will be liable for all additional costs incurred by the Employer in connection with such employment, and an appropriate deduction may be made from the Contract Sum.

Resource 121

Description	A further notice from the Employer to the Contractor, following the Employer's initial notice under clause 3.6, that the Employer has employed and paid other persons.
Type	Notice: Employer to Contractor
Provision references	clause 3.6

Reference: Notice requiring compliance with Employer's instruction (clause 3.6)

We write further to the following correspondence, given in accordance with clause 3.6 (Non-compliance with instructions):

- our notice dated [*insert date*], requiring compliance with the Employer's instruction dated [*insert date*], with reference [*insert reference*]; and
- our correspondence dated [*insert date*], where we advised you that, given that you did not comply with this instruction within 7 days after receipt of the notice above, the Employer would employ and pay other persons to execute work of any kind that may be necessary to give effect to this instruction.

The Employer has now employed and paid other persons to execute this work, and has incurred a total cost of [*insert amount*] (exclusive of VAT), particulars of which are attached.

A corresponding deduction will therefore be made from the Contract Sum.

Resource 122

Description	Confirmation by the Contractor to the Employer of an instruction given otherwise than in writing.
Type	Confirmation: Contractor to Employer
Provision references	clause 3.7.1
Drafting notes	**Timing:** Clause 3.7.1 requires the Contractor to confirm the terms of the Employer's instruction otherwise than in writing within 7 days.

Reference: Confirmation of instruction otherwise than in writing (clause 3.7.1)

In accordance with clause 3.7.1 (Instructions other than in writing), we write to confirm the terms of an Employer's instruction given otherwise than in writing:

- Date: [*insert date*]
- Method of service: [*insert details of how the instruction was given, for example, orally, by way of a telephone call*]
- Terms of the instruction: [*insert details of the instruction*]

This instruction will be of no immediate effect, but if you do not dissent by notice within 7 days from receipt of this confirmation (namely [*insert date*]), this instruction will take effect from the expiry of this 7-day period.

Resource 123

Description	A notice from the Employer to the Contractor dissenting from the Contractor's confirmation of an instruction given otherwise than in writing.
Type	Notice: Employer to Contractor
Provision references	clause 3.7.1
Drafting notes	**Timing:** If the Employer wishes to give this notice, clause 3.7.1 requires them to do so within 7 days from receipt of the Contractor's confirmation. Although not required by clause 3.7, the Employer is encouraged to give a reason for its dissent, to avoid uncertainty.

Reference: Instruction otherwise than in writing (clause 3.7.1)

In accordance with clause 3.7.1 (Instructions other than in writing), we write to notify you of our dissent to your confirmation dated [*insert date*], relating to an instruction otherwise than in writing.

We dissent for the following reasons: [*insert reasons for the dissent, for example, the terms do not accurately reflect the Employer's intention in respect of the instruction.*]

This instruction will therefore not take effect.

[In accordance with clause 1.7.1 (Notices and other communications), we will issue a separate instruction in writing, relating to this matter.]

Resource 124

Description	A notice from the Employer to the Contractor confirming the terms of an instruction given otherwise than in writing.
Type	Confirmation: Employer to Contractor
Provision references	clause 3.7.2
Drafting notes	**Timing:** Clause 3.7.2 provides that if the Employer confirms the terms of the instruction in writing prior to or within 7 days from receipt of the Contractor's confirmation, the instruction will take effect from the date and in the terms of the Employer's confirmation. The Employer should carefully review the Contractor's confirmation of the terms of the instruction, to ensure that it aligns precisely with its intention. If it does not, the Employer should clarify this in its confirmation.

Reference: Instruction otherwise than in writing; Employer's instruction (clause 3.7.2)

We write in reply to your confirmation dated [*insert date*], relating to an instruction otherwise than in writing.

In accordance with clause 3.7.2 (Instructions other than in writing), [we confirm the terms of this instruction, as stated in your confirmation] / [we confirm the terms of this instruction as follows: [*insert the terms of this instruction, and clarify how they differ from that set out in the Contractor's confirmation.*]]

This instruction will take effect from the date and in the terms of this confirmation.

Resource 125

Description	A notice from the Employer to the Contractor retrospectively confirming the terms of an instruction given otherwise than in writing.
Type	Confirmation: Employer to Contractor
Provision references	clause 3.7.3

Reference: Instruction otherwise than in writing; Employer's instruction (clause 3.7.3)

We write further to the following Employer's instruction, which was given otherwise than in writing:

- Date: [*insert date*]
- Method of service: [*insert details of how the instruction was given, for example, orally, by way of a telephone call*]
- Terms of the instruction: [*insert details of the instruction*]

In accordance with clause 3.7.3 (Instructions other than in writing), given that neither you nor we confirmed this instruction in the manner and time stated under clause 3.7.1 and clause 3.7.2, and that you have nevertheless complied with it, we confirm this instruction with retrospective effect.

Resource 126

Description	A request from the Contractor to the Employer to notify which of the Conditions empowers the Employer to issue a particular instruction.
Type	Request: Contractor to Employer
Provision references	clause 3.8

Reference: Provisions empowering Employer's instruction (clause 3.8)

We write in relation to the Employer's instruction dated [*insert date*], with reference [*insert reference*].

In accordance with clause 3.8 (Provisions empowering instructions), we request that you notify us which provision of the Conditions empowers the issue of this instruction.

Resource 127

Description	A notice from the Employer to the Contractor of the provision of the Conditions which empowers the Employer to issue a particular instruction.
Type	Notice: Employer to Contractor
Provision references	clause 3.8
Drafting notes	**Timing:** Clause 3.8 requires the Employer to comply with this request from the Contractor 'forthwith'.

Reference: Provisions empowering Employer's instruction (clause 3.8)

We write in reply to your request dated [*insert date*], made in accordance with clause 3.8 (Provisions empowering instructions), and in respect of the Employer's instruction dated [*insert date*], with reference [*insert reference*].

We write to notify you that the following provision of the Conditions empowers the issue of this instruction: [*insert provision reference, for example, clause 3.10 (Postponement of work).*]

Please let us know if there is any further information we can provide in this regard.

Resource 128

Description	A request from the Employer for the Contractor's consent for the Employer to issue an instruction requiring a Change which is, or makes necessary, an alteration or modification in the design of the Works.
Type	Request: Employer to Contractor
Provision references	clause 3.9.1

Reference: Request to issue an instruction requiring a Change (clause 3.9.1)

We wish to issue an instruction requiring a Change, a draft of which is attached.

This Change would be, or would make necessary, an alteration or modification in the design of the Works, as follows: [*insert details of how the Change would alter or modify the design of the Works.*]

In accordance with clause 3.9.1 (Instructions requiring Changes), we request your consent for us to issue this instruction, effecting this Change.

Please give your consent by [*insert date*].

Resource 129

Description	Consent given by the Contractor for the Employer to issue an instruction requiring a Change which is, or makes necessary, an alteration or modification in the design of the Works.
Type	Consent given: Contractor to Employer
Provision references	clause 3.9.1

Reference: Request to issue an instruction requiring a Change (clause 3.9.1)

We write in reply to your request for our consent dated [*insert date*], in respect of an instruction requiring a Change, which would be, or which would make necessary, an alteration or modification in the design of the Works.

[*Unconditional consent*: We give our consent for you to issue this instruction requiring a Change, on the terms set out in your request.] /

[*Conditional consent*: We give our consent for you to issue this instruction requiring a Change, on the terms set out in your request, subject to the following conditions: [*insert conditions.*]]

Resource 130

Description	Consent withheld by the Contractor for the Employer to issue an instruction requiring a Change which is, or makes necessary, an alteration or modification in the design of the Works.
Type	Consent withheld: Contractor to Employer
Provision references	clause 3.9.1
Drafting notes	Clause 1.10 requires that the Contractor's consent must not be unreasonably delayed or withheld.

Reference: Request to issue an instruction requiring a Change (clause 3.9.1)

We write in reply to your request for our consent dated [*insert date*], in respect of an instruction requiring a Change, which would be, or which would make necessary, an alteration or modification in the design of the Works.

We withhold our consent, for the following reasons: [*insert reasons, to demonstrate why the consent was not unreasonably delayed or withheld.*]

We propose that the Parties meet as soon as possible to discuss proposals for an alternative instruction, taking the reasons above into consideration.

Resource 131

Description	A notice from the Contractor to the Employer of an objection to an instruction, in its capacity as Principal Designer or Principal Contractor.
Type	Notice: Contractor to Employer
Provision references	clause 3.9.4
Drafting notes	This applies to the Contractor in its capacity as either Principal Designer or Principal Contractor, both under the CDM Regulations and the Building Regulations.

Reference: Notice of objection to Employer's instruction (clause 3.9.4)

In accordance with clause 3.9.4 (Instructions requiring Changes), we write to notify you, in our capacity as [Principal Designer and/or Principal Contractor], of the following objection to the Employer's instruction dated [*insert date*], with reference [insert reference]:

[*Insert details of the objection to the instruction effecting a Change, or an instruction regarding the expenditure of a Provisional Sum included in the Employer's Requirements.*]

Please revise the terms of this instruction so as to remove this objection.

We propose the following revision: [*insert proposal for removing the objection.*]

In the meantime, we will not be required pursuant to clause 2.1.1 to clause 2.1.4 (Contractor's Obligations: General obligations) to comply with this instruction.

Resource 132

Description	A varied instruction from the Employer to the Contractor, following the Contractor's notice under clause 3.9.4.
Type	Instruction: Employer to Contractor
Provision references	clause 3.9.4
Drafting notes	This applies to the Contractor in its capacity as either Principal Designer or Principal Contractor, both under the CDM Regulations and the Building Regulations.

Reference: Notice of objection to Employer's instruction; revised Employer's instruction [*insert reference*] (clause 3.9.4)

We write in reply to your notice dated [*insert date*], given in accordance with clause 3.9.4 (Instructions requiring Changes), and in relation to the Employer's instruction dated [*insert date*], with reference [insert reference].

We have revised the terms of this instruction as follows, so as to remove your objection as stated in that notice: [*insert description of how the instruction has been varied, to remove the Contractor's objection.*]

A copy of the revised instruction is attached.

Please confirm whether, to your reasonable satisfaction, we have revised the terms of the instruction so as to remove your objection.

Resource 133

Description	Confirmation by the Contractor to the Employer of the varied instruction, following the Contractor's notice under clause 3.9.4.
Type	Confirmation: Contractor to Employer
Provision references	clause 3.9.4
Drafting notes	This applies to the Contractor in its capacity as either Principal Designer or Principal Contractor, both under the CDM Regulations and the Building Regulations.

Reference: Notice of objection to Employer's instruction (clause 3.9.4)

We write in reply to your correspondence dated [*insert date*], given in accordance with clause 3.9.4 (Instructions requiring Changes), and which attached a revised instruction, dated [*insert date*], with reference [*insert reference*].

We are satisfied that the terms of this instruction have been revised so as to remove our objection, as stated in our notice dated [*insert date*].

We therefore confirm that we will comply with this instruction, in accordance with clause 2.1.1 to clause 2.1.4 (Contractor's Obligations: General obligations).

Resource 134

Description	An instruction from the Employer to postpone any work to be executed under the Contract.
Type	Instruction: Employer to Contractor
Provision references	clause 3.10
Drafting notes	If necessary, the Employer should also clarify any related work which is to continue, and is not subject to the instruction to postpone.

Reference: Postponement of work; Employer's instruction [*insert reference*] (clause 3.10)

In accordance with clause 3.10 (Postponement of work), [with immediate effect] / [from [*insert date*]], we instruct you to postpone the following elements of the work to be executed under the Contract:

[*Insert description of the work to be postponed.*]

We will issue a further instruction when we require you to resume this work.

Resource 135

Description	Confirmation by the Contractor of an Employer's instruction regarding the postponement of any work to be executed under the Contract.
Type	Confirmation: Contractor to Employer
Provision references	clause 3.10
Drafting notes	**Good practice** This confirmation is not required by clause 3.10, but it is considered good practice for the Contractor to seek clarification on any correspondence from the Employer which, although it may not expressly refer to 'postponement', may have the same effect.

Reference: Postponement of work (clause 3.10)

We write in reply to your correspondence dated [*insert date*], and in particular, the following statement:

[*Insert statement from the Employer, which the Contractor considers constitutes an instruction under clause 3.10 to postpone an element of the work. For example, a request to 'stand down', or 'hold off' from carrying out certain work.*]

We consider that this constitutes an instruction under clause 3.10 (Confirmation of postponement) for us to postpone the following elements of the work to be executed under the Contract:

[*Insert description of the work to be postponed.*]

Please confirm as soon as possible the Employer's intentions in this regard.

Resource 136

Description	An instruction from the Employer regarding the expenditure of Provisional Sums.
Type	Instruction: Employer to Contractor
Provision references	clause 3.11
Drafting notes	Under clause 4.2.5 (Items included in Adjustments), the Contract Sum will be adjusted by the deduction of all Provisional Sums included in the Employer's Requirements.

Reference: Employer's instruction on Provisional Sums [*insert reference*] (clause 3.11)

In accordance with clause 3.11 (Instructions on Provisional Sums), and in respect of Provisional Sum reference [*insert reference*] contained within the Employer's Requirements (reference [*insert reference*]), we instruct you as follows:

[Insert description of the instruction regarding the expenditure of the Provisional Sum.]

Resource 137

Description	A Provisional Sums tracker template.
Type	Tracker: Contractor and Employer
Provision references	clause 3.11

In accordance with clause 3.11 (Instructions on Provisional Sums), the Employer is required to issue instructions in regard to the expenditure of Provisional Sums included in the Employer's Requirements.

Below is a tracker template to help record such instructions:

- Provisional Sum reference: [*insert reference*]
- Employer's Requirements reference: [*insert reference to the Employer's Requirements*]
- Description of Provisional Sum: [*insert description*]
- Value of Provisional Sum (exclusive of VAT): [*include amount*]
- Latest date for issue of instruction: [*insert date*]
- Instruction reference: [*insert reference*]
- Instruction date: [*insert date*]
- Document link: [*insert link to documents*]
- Status / action required: [*insert status, and any action required*]

Resource 138

Description	An instruction from the Employer to open up work for inspection or testing.
Type	Instruction: Employer to Contractor
Provision references	clause 3.12

Reference: Employer's instruction to open up for inspection or testing [*insert reference*] (clause 3.12)

In accordance with clause 3.12 (Inspection – tests), we instruct you to [open up for inspection the following work covered up] / [arrange for, or carry out, the following test of the following materials or goods (whether or not already incorporated in the Works), or executed work]:

[*Insert description of work to be opened up for inspection, or test to be carried out.*]

The cost of this [opening up] / [testing] (including the cost of making good) will be added to the Contract Sum (given that it is not provided for in the Employer's Requirements nor in the Contractor's Proposals), unless the inspection or test shows that the materials, goods, or work are not in accordance with the Contract.

Resource 139

Description	Confirmation from the Employer to the Contractor that the costs of opening up or inspection will be added to the Contract Sum, where the materials, goods, or work were in accordance with the Contract.
Type	Confirmation: Employer to Contractor
Provision references	clause 3.12
Drafting notes	**Good practice** This confirmation is not required under clause 3.12, but it is considered good practice, particularly in the context of materials, goods, or work not in accordance with the Contract.

Reference: Employer's instruction to open up for inspection or testing [*insert reference*] (clause 3.12)

Further to the Employer's instruction dated [*insert date*], with reference [*insert reference*], and issued in accordance with clause 3.12 (Inspection – tests), requiring you to [open up for inspection] / [test] certain materials, goods, or work, we consider that the [inspection] / [test] shows that the materials, goods, or work were in accordance with the Contract.

Please therefore provide details on the cost of the [opening up] / [inspection], including the cost of making good, such that these can be added to the Contract Sum.

Resource 140

Description	Confirmation from the Employer to the Contractor that the costs of opening up or inspection will not be added to the Contract Sum, where the materials, goods, or work were not in accordance with the Contract.
Type	Confirmation: Employer to Contractor
Provision references	clause 3.12
Drafting notes	**Good practice** This confirmation is not required under clause 3.12, but it is considered good practice, particularly in the context of materials, goods, or work not in accordance with the Contract.

Reference: Employer's instruction to open up for inspection or testing (clause 3.12)

Further to the Employer's instruction dated [*insert date*], with reference [*insert reference*], and issued in accordance with clause 3.12 (Inspection – tests), requiring you to [open up for inspection] / [test] certain materials, goods, or work, we consider that the [inspection] / [test] shows that the materials, goods, or work were not in accordance with the Contract, for the following reasons: [*insert reasons why the materials, goods, or work were not in accordance with the Contract.*]

The cost of the [opening up] / [inspection] will therefore not be added to the Contract Sum.

Resource 141

Description	An instruction from the Employer to remove from the site all or any work, materials, or goods not in accordance with the Contract.
Type	Instruction: Employer to Contractor
Provision references	clause 3.13.1

Reference: Work not in accordance with the Contract (clause 3.13.1)

In accordance with clause 3.13.1 (Work not in accordance with the Contract), we instruct you to remove from the site the following work, materials, or goods which are not in accordance with the Contract:

[*Insert description of work, materials, or goods which are not in accordance with the Contract.*]

Resource 142

Description	Confirmation from the Contractor of compliance with the Employer's instruction to remove from the site all or any work, materials, or goods not in accordance with the Contract.
Type	Confirmation: Contractor to Employer
Provision references	clause 3.13.1
Drafting notes	**Good practice** This confirmation is not required under clause 3.13.1, but it is considered good practice, particularly in the context of materials, goods, or work not in accordance with the Contract.

Reference: Work not in accordance with the Contract (clause 3.13.1)

We write to confirm that we have complied with the Employer's instruction dated [*insert date*], with reference [*insert reference*], and issued in accordance with clause 3.13.1 (Work not in accordance with the Contract).

We have removed the relevant work, materials, or goods which were not in accordance with the Contract, and will promptly [undertake this work] / [procure these materials or goods] again, in accordance with the Contract.

Resource 143

Description	An instruction from the Employer, as a reasonably necessary consequence of any instruction issued under clause 3.13.1.
Type	Instruction: Employer to Contractor
Provision references	clause 3.13.2
Drafting notes	The Employer is encouraged to provide any attendance note or meeting minutes from the consultation, as part of this instruction.

Reference: Work not in accordance with the Contract; Employer's instruction [*insert reference*] (clause 3.13.2)

Further to our consultation dated [*insert date*] [([the meeting minutes] / [our attendance note] from which are attached)], and in accordance with clause 3.13.2 (Work not in accordance with the Contract), we instruct you as follows:

[*Insert description of instruction requiring a Change.*]

This Change instruction is a reasonably necessary consequence of the Employer's instruction dated [*insert date*], with reference [*insert reference*], issued in accordance with clause 3.13.1.

Therefore, no addition will be made to the Contract Sum, and no extension of time will be given, in respect of this instruction.

Resource 144

Description	Confirmation from the Contractor of compliance with the Employer's instructions issued as a reasonably necessary consequence of any instruction under clause 3.13.1.
Type	Confirmation: Contractor to Employer
Provision references	clause 3.13.2
Drafting notes	**Good practice** This confirmation is not required under clause 3.13.2, but it is considered good practice, particularly in the context of work not in accordance with the Contract.

Reference: Work not in accordance with the Contract (clause 3.13.2)

We write to confirm that we have complied with the Employer's instruction dated [*insert date*], with reference [*insert reference*], and issued in accordance with clause 3.13.2 (Work not in accordance with the Contract).

We accept that this Change instruction is a reasonably necessary consequence of the Employer's instruction dated [*insert date*], with reference [*insert reference*], issued in accordance with clause 3.13.1.

We therefore further accept that no addition will be made to the Contract Sum, and no extension of time will be given.

Resource 145

Description	An instruction from the Employer to open up for inspection or to test the likelihood or extent of any further similar non-compliance.
Type	Instruction: Employer to Contractor
Provision references	clause 3.13.3
Drafting notes	Before issuing this instruction, clause 3.13.3 requires the Employer to have due regard to the Code of Practice, set out in Schedule 4.

Reference: Work not in accordance with the Contract; Employer's instruction [*insert reference*] (clause 3.13.3)

Having had due regard to the Code of Practice set out in Schedule 4, in accordance with clause 3.13.3 (Work not in accordance with the Contract), and clause 3.12 (Inspection – tests), we instruct you to [open up for inspection the following work covered up] / [arrange for, or carry out, the following test of the following materials or goods (whether or not already incorporated in the Works), or executed work]:

[*Insert description of work to be opened up for inspection, or test to be carried out.*]

We consider that this is reasonable, in all the circumstances, to establish to our reasonable satisfaction the likelihood or extent, as appropriate to the circumstances, of any further similar non-compliance.

Therefore, whatever the results of the opening up, no addition will be made to the Contract Sum in respect of this instruction, but clause 2.25 (Fixing Completion Date) and clause 2.26.2.3 (Relevant Events) will apply, unless the inspection or test shows that the work, materials, or goods are not in accordance with the Contract.

Resource 146

Description	Confirmation by the Contractor of compliance with the Employer's instruction made under clause 3.13.3 and clause 3.12.
Type	Confirmation: Contractor to Employer
Provision references	clause 3.13.3
Drafting notes	**Good practice** This confirmation is not required under clause 3.13.3, but it is considered good practice, particularly in the context of work not in accordance with the Contract.

Reference: Work not in accordance with the Contract (clause 3.13.3)

We write to confirm that we have complied with the Employer's instruction dated [*insert date*], with reference [*insert reference*], and issued in accordance with clause 3.13.3 (Work not in accordance with the Contract), and clause 3.12 (Inspection – tests).

We accept that this instruction was reasonable, in all the circumstances, to establish to the Employer's reasonable satisfaction the likelihood or extent, as appropriate to the circumstances, of any further similar non-compliance.

We therefore further accept that no addition will be made to the Contract Sum in respect of this instruction.

[*Work, materials, or goods are in accordance with the Contract:* However, given that the inspection or test shows that the work, materials, or goods are in accordance with the Contract, clause 2.25 (Fixing Completion Date) and clause 2.26.2.3 (Relevant Events) apply, and we may therefore give the Employer a notice in accordance with clause 2.24 (Notice by Contractor of delay to progress).] /

[*Work, materials, or goods are not in accordance with the Contract:* Given that the inspection or test shows that the work, materials, or goods are not in accordance with the Contract, we further accept that no extension of time will be given in respect of this instruction.]

Resource 147

Description	An instruction from the Employer as a result of the Contractor's failure to comply with clause 2.1.1 to clause 2.1.4.
Type	Instruction: Employer to Contractor
Provision references	clause 3.14
Drafting notes	The Employer is encouraged to provide any attendance note or meeting minutes from the consultation.

Reference: Workmanship not in accordance with the Contract; Employer's instruction [*insert reference*] (clause 3.14)

We write further to our consultation dated [*insert date*] [([the meeting minutes] / [our attendance note] from which are attached)], where we discussed the following failure to comply with clause 2.1.1 to clause 2.1.4 (Contractor's Obligations: General obligations) in regard to the carrying out of work in a proper and workmanlike manner, or in accordance with the Construction Phase Plan:

[*Insert description of how the Contractor has failed to comply with clause 2.1.1 to clause 2.1.4.*]

In accordance with clause 3.14 (Workmanship not in accordance with the Contract), we instruct you as follows:

[*Insert details of the instruction, whether requiring a Change or otherwise, as are reasonably necessary as a consequence of the Contractor's failure to comply with clause 2.1.1 to clause 2.1.4.*]

We consider that this instruction is reasonably necessary, as a consequence of this failure to comply with clause 2.1.1 to clause 2.1.4. Therefore, no addition will be made to the Contract Sum, and no extension of time will be given, in respect of this instruction.

Resource 148

Description	Confirmation from the Contractor of compliance with an Employer's instruction issued as a result of the Contractor's failure to comply with clause 2.1.1 to clause 2.1.4.
Type	Confirmation: Contractor to Employer
Provision references	clause 3.14
Drafting notes	**Good practice** This confirmation is not required under clause 3.14, but it is considered good practice, particularly in the context of workmanship not in accordance with the Contract.

Reference: Workmanship not in accordance with the Contract; Employer's instruction [*insert reference*] (clause 3.14)

We write to confirm that we have complied with the Employer's instruction dated [*insert date*], with reference [*insert reference*], and issued in accordance with clause 3.14 (Workmanship not in accordance with the Contract).

We accept that this instruction was reasonably necessary, as a consequence of our failure to comply with clause 2.1.1 to clause 2.1.4 in regard to the carrying out of certain work in a proper and workmanlike manner, or in accordance with the Construction Phase Plan.

We therefore further accept that no addition will be made to the Contract Sum, and no extension of time will be given, in respect of this instruction.

Resource 149

Description	An attendance note template for consultations between the Contractor and the Employer, relating to materials, goods, work, or workmanship not in accordance with the Contract.
Type	Template: Attendance Note
Provision references	clause 3.12 to clause 3.14

Below is an attendance note template for consultations between the Contractor and the Employer, relating to work or workmanship not in accordance with the Contract under clause 2.12 (Inspection – tests), clause 2.13 (Work not in accordance with the Contract), and clause 2.14 (Workmanship not in accordance with the Contract):

- Date: [*insert date*]
- Attended: [*insert names of those who attended the consultation*]
- Relevant provision: [*insert whether clause 2.12, clause 2.13, and/or clause 2.14 applies*]
- Relevant Contract Documents: [*insert references to the relevant Contract Documents*]
- Description of relevant work, materials, or goods: [*insert description and location of relevant work, materials, or goods*]
- Supporting evidence: [*details of supporting evidence, for example, photographs, video, documents, etc.*]
- Employer's representations: [*insert summary of Employer's opinion*]
- Contractor's representations: [*insert summary of Contractor's opinion*]
- Conclusions: [*insert summary of any agreed conclusions*]
- Actions required by the Parties: [*insert descriptions of any further actions required, including the issuing of instructions, remedial works, etc.*]

Resource 150

Description	A notice from the Contractor to the Employer of the discovery and precise location of any fossils, antiquities, or other objects of interest or value.
Type	Notice: Contractor to Employer
Provision references	clause 3.15.1
Drafting notes	**Timing:** Clause 3.15.1 requires the Contractor to give this notice 'forthwith'.

Reference: Fossils, antiquities, and other objects of interest or value (clause 3.15.1)

In accordance with clause 3.15.1.3 (Antiquities, asbestos, contaminated material and unexploded ordnance), we have discovered the following fossil, antiquity, or other object of interest of value, which will become the Employer's property: [*insert description.*]

This was discovered at the site [in excavating it during the progress of the Works], at the following precise location: [*insert description of location.*]

In accordance with clause 3.15.1.1, we will use our best endeavours not to disturb the object, and we will cease work if and insofar as its continuance would endanger the object, or prevent or impede its excavation or removal.

In accordance with clause 3.15.1.2, we will take all steps necessary to preserve the object in the exact position and condition in which it was found.

Please issue an instruction as to any action to be taken concerning this object, in accordance with clause 3.15.2.

Resource 151

Description	An instruction from the Employer concerning the object reported by the Contractor under clause 3.15.1.
Type	Instruction: Employer to Contractor
Provision references	clause 3.15.2

Reference: Fossils, antiquities, and other objects of interest or value; Employer's instruction [*insert reference*] (clause 3.15.2)

In accordance with clause 3.15.2 (Antiquities, asbestos, contaminated material and unexploded ordnance), and further to your notice dated [*insert date*] reporting the discovery of the object, we instruct you as follows:

[*Insert details of the instruction, which may require the Contractor to permit the examination, excavation, or removal of the object by a third party.*]

Resource 152

Description	A notice from the Contractor to the Employer of the discovery of any asbestos, contaminated material, or unexploded ordnance.
Type	Notice: Contractor to Employer
Provision references	clause 3.15.3
Drafting notes	**Timing:** Clause 3.15.3 requires the Contractor to give this notice 'forthwith'.

Reference: Asbestos, contaminated material, or unexploded ordnance (clause 3.15.3)

In accordance with clause 3.15.3.2 (Antiquities, asbestos, contaminated material and unexploded ordnance), we have discovered [asbestos] / [contaminated material] / [an unexploded ordnance] as follows: [*insert description.*]

This was discovered at the site [in excavating it during the progress of the Works], at the following location: [*insert description of location.*]

In accordance with clause 3.15.3.1, we confirm the following:

- we are using our best endeavours not to disturb the material or item; and
- we have ceased work insofar as continuing work would endanger health and safety and/or life and property, or prevent or impede the disposal or removal of the material or item.

We have taken the following steps in that regard: [*insert details of how the Contractor has used its best endeavours, and the extent to which the Contractor has ceased work.*]

Please issue an instruction in accordance with clause 3.15.4, with respect to this material or item.

Resource 153

Description	An instruction from the Employer concerning the material or item reported by the Contractor under clause 3.15.3.
Type	Instruction: Employer to Contractor
Provision references	clause 3.15.4

Reference: Asbestos, contaminated material, or unexploded ordnance (clause 3.15.4)

In accordance with clause 3.15.4 (Antiquities, asbestos, contaminated material and unexploded ordnance), and further to your notice dated [*insert date*] reporting the discovery of [asbestos] / [contaminated material] / [an unexploded ordnance], we instruct you as follows:

[*Insert details of the instruction, which may require the Contractor to permit the investigation, disposal, or removal of such material or item by a third party.*]

Resource 154

Description	A tracker template for recording Employer's instructions.
Type	Tracker: Contractor and Employer
Provision references	clause 3.5 to clause 3.15

Below is a tracker template for recording Employer's instructions:

- Instruction reference: [*insert reference*]
- Instruction date: [*insert date*]
- Description: [*insert a brief description of the instruction*]
- Applicable provision: [*insert which provision of the Conditions empowers its issue, for example, clause 3.9 (Instructions requiring Changes)*]
- Adjustment to the Contract Sum (exclusive of VAT): [*insert details of any adjustment*]

- Adjustment to the Completion Date: [*insert details of any adjustment*]
- Document link: [*insert link to documents*]
- Status / action required: [*insert status, and any action required*]

Resource 155

Description	A checklist for the Contractor upon receiving an Employer's instruction.
Type	Checklist: Contractor
Provision references	clause 3.5 to clause 3.15

Below is a checklist for the Contractor upon receiving an Employer's instruction:

Item	Description	Particulars	Status / action required
1	Date received:	[Insert date]	
2	Date acknowledged:	[Insert date]	
3	Which provision empowers its issue / under which provision was it issued:	[Insert reference]	
4	Does it constitute a Change under clause 5.1, or any other provision:	Yes, under clause [insert reference] / No	
5	Does it constitute a Relevant Event under clause 2.26:	Yes, under clause [insert reference] / No	
6	Does it constitute a Relevant Matter under clause 4.21.1:	Yes, under clause [insert reference] / No	
7	Estimated value:	[Insert amount]	
8	Agreed value / Valuation:	[Insert amount]	

Resource 156

Description	The provision from the Contractor to the Employer of the health and safety file.
Type	Correspondence: Contractor to Employer
Provision references	clause 3.16.2

Reference: Health and safety file (clause 3.16.2)

In accordance with clause 3.16.2 (CDM Regulations and Part 2A of the Building Regulations), and in our capacity as Principal Designer for the purposes of the CDM Regulations, the health and safety file is [attached] / [available to download at the following link: [*insert link*]].

Resource 157

Description	A notice from the Employer to the Contractor of the appointment of a replacement Principal Designer or Principal Contractor, for the purposes of either the CDM Regulations or the Building Regulations.
Type	Notice: Employer to Contractor
Provision references	clause 3.16.4
Drafting notes	**Timing:** Clause 3.16.4 requires the Employer to give this notice 'immediately'.

Reference: Appointment of replacement [Principal Designer] / [Principal Contractor] (clause 3.16.4)

In accordance with clause 3.16.4 (CDM Regulations and Part 2A of the Building Regulations), we write to inform you that we have appointed a replacement [Principal Designer] / [Principal Contractor], for the purposes of the [CDM Regulations] / [Building Regulations].

[Effective immediately] / [From [*insert date*]], the Employer has appointed [*insert name of replacement Principal Designer or Principal Contractor*] as [Principal Designer] / [Principal Contractor], in place of [*insert name of previous Principal Designer or Principal Contractor*].

Their contact details are as follows: [*insert contact details.*]

We will arrange an introductory meeting as soon as possible.

Resource 158

Description	A checklist for other amounts to be added to, or deducted from, the Contract Sum, as referred to in clause 4.2.6.
Type	Checklist: Contractor and Employer
Provision references	clause 4.2.6

Clause 4.2.6 (Items included in adjustments) provides that the Contract Sum will be adjusted by 'any other amount which under this Contract is to be added to the Contract Sum or may be deducted from it'.

Below is a checklist of the provisions which refer to an adjustment of the Contract Sum:

- clause 2.5.2 (Early use by Employer)
- clause 2.18 (Fees or charges legally demandable)
- clause 2.20.1 (Patent rights – Instructions)
- clause 2.35 (Schedules of defects and instructions)
- clause 3.6 (Non-compliance with instructions)

- clause 3.12 (Inspection – tests)
- clause 4.22 (Loss and Expense: Amounts ascertained – addition to Contract Sum)
- clause 6.10.2 (Terrorism Cover – policy extensions and premiums)
- clause 6.10.3 (Terrorism Cover – policy extensions and premiums)
- clause 6.11.3 (Terrorism Cover – non-availability – Employer's options)
- clause 6.12.2 (Evidence of insurance)
- clause 6.19.2 (Breach of Joint Fire Code – Remedial Measures)
- clause 6.20 (Joint Fire Code – amendments/revisions)
- Supplemental Provision 2 paragraph 2.6 (Valuation of Changes – Contractor's estimates)
- Supplemental Provision 3 paragraph 3.5 (Loss and expense – Contractor's estimates)
- Supplemental Provision 3 paragraph 3.6 (Loss and expense – Contractor's estimates)
- Supplemental Provision 4 paragraph 4.4 (Acceleration Quotation)
- Supplemental Provision 6 paragraph 7.3 (Cost savings and value improvements)

Resource 159

Description	A reminder from the Contractor to the Employer to take adjustments to the Contract Sum into account in the next Interim Payment.
Type	Reminder: Contractor to Employer
Provision references	clause 4.3
Drafting notes	**Good practice** This reminder is not required by clause 4.3, but it is considered good practice for the Contractor to remind the Employer of the need to take any adjustments into account in the next Interim Payment.

Reference: Taking adjustments to the Contract Sum into account (clause 4.3)

We write to remind you that, in accordance with clause 4.3 (Taking adjustments into account), where the Conditions provide that an amount is to be added to, deducted from, or dealt with by adjustment of the Contract Sum, then, as soon as the amount is ascertained in whole or in part, the ascertained amount will be taken into account in the next Interim Payment.

[*Insert Conditions reference*] provides that the following amount is to be added to the Contract Sum: [*insert details of the addition to the Contract Sum.*]

[*Ascertained in whole:* This amount has been ascertained in whole as [*insert amount*] (exclusive of VAT).] /

[*Ascertained in part:* At the time of writing, this amount has been ascertained in part as [*insert amount*] (exclusive of VAT).]

This amount should therefore be taken into account in the next Interim Payment.

Resource 160

Description	A reminder from the Contractor to the Employer in respect of a VAT exemption.
Type	Reminder: Contractor to Employer
Provision references	clause 4.4.2
Drafting notes	**Good practice** This reminder is not required by clause 4.4.2, but in such circumstances, the Contractor is encouraged to remind the Employer as soon as possible after the relevant VAT exemption is introduced.

Reference: VAT (clause 4.4.2)

In accordance with clause 4.4.2 (VAT), if after the Base Date (namely [*insert date*]) the supply of any goods or services to the Employer becomes exempt from VAT, there will be paid to us an amount equal to the input tax on the supply to us of goods and services that contribute to the Works, which as a consequence of that exemption, we cannot recover.

From [*insert date*], the supply of the following goods or services to the Employer became exempt from VAT: [*insert details of the relevant goods or services.*]

As a consequence of this exemption, we cannot recover the input tax on the supply to us of goods and services that contribute to the Works, in the amount of [*insert amount*]. We attach the basis on which this sum was calculated.

We are therefore entitled to be paid an amount equal to this input tax.

Resource 161

Description	Confirmation from the Employer to the Contractor of the Employer becoming a 'contractor' for the purposes of the Construction Industry Scheme.
Type	Confirmation: Employer to Contractor
Provision references	clause 4.5
Drafting notes	**Good practice** This correspondence is not required by clause 4.5, but it is considered good practice for the Employer to advise the Contractor of any change to its status for the purposes of the CIS.

Reference: Construction Industry Scheme (clause 4.5)

We write to advise you that, from [*insert date*], we became a 'contractor' for the purposes of the CIS.

Therefore, in accordance with clause 4.5 (Construction Industry Scheme (CIS)), our obligation to make any payment under the Contract is subject to the provisions of the CIS.

Resource 162

Description	A tracker template to record the reimbursement to the Employer of an advance payment made to the Contractor.
Type	Tracker: Contractor and Employer
Provision references	clause 4.6

In accordance with clause 4.6 (Advance payment), where the Contract Particulars state that clause 4.6 applies, and an advance payment is to be made, it will be paid to the Contractor on the date and reimbursed to the Employer on the terms stated in the Contract Particulars.

Below is a tracker template to record this reimbursement:

- Amount of advance payment: [*insert amount*]
- Date advance payment made: [*insert date*]

Payment Notices

- Reimbursement sum: [*insert amount*]
- Cumulative reimbursement to date: [*insert amount*]
- Amount remaining to be reimbursed: [*insert amount*]

Note: All sums stated are exclusive of VAT.

Resource 163

Description	An Interim Payment Application template, from the Contractor to the Employer.
Type	Template (Contractor): Interim Payment Application
Provision references	clause 4.7.3

Reference: Interim Payment Application (clause 4.7.3)

In accordance with clause 4.7.3 (Interim Payments – Contractor's Interim Payment Applications, due dates and Payment Notices), we attach an Interim Payment Application, which states the sum we consider to be due to us at the due date, and the basis on which that sum has been calculated.

Interim Payment Application

- Date of Interim Payment Application: [*insert date*]
- Reference: [*insert reference*]
- Interim Valuation Date: [*insert date*]
- Due date: [*insert due date*]

At the due date stated above, we consider the sum due to us to be [*insert sum*] (exclusive of VAT).

We attach the basis on which this sum was calculated.

In accordance with clause 4.7.4, we also attach the further information specified in the Employer's Requirements.

Resource 164

Description	A Payment Notice template, from the Employer to the Contractor.
Type	Template: Payment Notice
Provision references	clause 4.7.5
Drafting notes	**Timing:** Clause 4.7.5 requires the Employer to give the Payment Notice not later than 5 days after each due date. In accordance with clause 4.10.3, in relation to the requirements for the giving of Payment Notices, it is immaterial that the amount then considered by the Employer to be due to the Contractor may be zero.

Reference: Payment Notice (clause 4.7.5)

In accordance with clause 4.7.5 (Interim Payments – Contractor's Interim Payment Applications, due dates and Payment Notices), we attach a Payment Notice, which states the sum we consider to be due to you at the due date, and the basis on which that sum has been calculated.

Payment Notice

- Payment Notice date: [*insert date*]
- Payment Notice reference: [*insert date*]
- Interim Payment Application date: [*insert date*]
- Interim Payment Application reference: [*insert reference*]
- Interim Valuation Date: [*insert date*]
- Due date: [*insert due date*]
- Final date for payment: [*insert final date for payment*]

At the due date stated above, we consider the sum due to you to be [*insert sum*] (exclusive of VAT).

We attach the basis on which this sum was calculated, which includes the following amounts:

- Gross Valuation (calculated in accordance with [Alternative A (Stage Payments)] / [Alternative B (Periodic Payments]: [*insert amount*]
- Less Retention: [*insert amount*]
- Less cumulative reimbursement of advance payment (if applicable): [*insert amount*]
- Less total of amounts paid in previous Interim Payments: [*insert amount*]
- Amount due: [*insert amount*]

The sum due in this Payment Notice is: [*insert amount in words*]

Note: All sums stated are exclusive of VAT, and the Employer will in addition pay the amount of VAT properly chargeable.

Resource 165

Description	A Final Payment Notice template.
Type	Final Payment Notice: Contractor or Employer to the other Party
Provision references	clause 4.8
Drafting notes	**Timing:** Clause 4.8 requires the Party issuing the Final Payment Notice to do so not later than 5 days after the due date for the final payment fixed under clause 4.24.5, and notwithstanding any dispute regarding the final statement. In accordance with clause 4.10.3, in relation to the requirements for the giving of the Final Payment Notice, it is immaterial that the amount then considered by the Employer to be due may be zero.

Reference: Final Payment Notice

In accordance with clause 4.8 (Relevant statement and Final Payment Notice), we attach a Final Payment Notice, which states the sum we consider to be due to [the Contractor] / [the Employer], and the basis on which that sum has been calculated.

Final Payment Notice

- Due date: [*insert due date*]
- Final date for payment: [*insert final date for payment*]

At the due date stated above, we consider the sum due to [the Contractor] / [the Employer] to be [*insert sum*].

We attach the basis on which this sum was calculated, which includes the following amounts:

- The Contract Sum, adjusted in accordance with clause 4.2: [*insert amount*]
- Less the sum of amounts already paid by the Employer to the Contractor: [*insert amount*]
- Amount due: [*insert amount*]

At the due date stated above, the sum due in this Final Payment Notice is: [*insert amount in words*], as a balance due from [the Employer to the Contractor] / [the Contractor to the Employer].

Note: All sums stated are exclusive of VAT, and the Employer will in addition pay the amount of VAT properly chargeable.

Resource 166

Description	A Pay Less Notice template.
Type	Pay Less Notice: Employer and Contractor
Provision references	clause 4.9.5
Drafting notes	**Timing:** Clause 4.9.5 requires the Employer to give the Pay Less Notice not later than 5 days before the final date for payment. In accordance with clause 4.10.3, in relation to the requirements for the giving of the Pay Less Notice, it is immaterial that the amount then considered by the Employer to be due to the Contractor may be zero. In most instances, Pay Less Notices will be given to the Contractor, but they may also be given to the Employer in respect of the Final Payment Notice, under clause 4.9.5.2 (Interim and final payments – final date and amount).

Reference: Pay Less Notice

In accordance with clause 4.9.5 (Interim and final payments – final date and amount) and clause 4.10.1 (Pay Less Notices and other general provisions), we attach a Pay Less Notice, which states the sum we consider to be due to [the Contractor] / [the Employer], and the basis on which that sum has been calculated.

Pay Less Notice

- [Payment Notice date: [*insert date*]
- Payment Notice reference: [*insert reference*]] /
- [Interim Application date: [*insert date*]
- Interim Application reference: [*insert reference*]] /
- Final Payment Notice dated: [*insert date*] /
- Final Statement dated: [*insert date*] /
- Employer's Final Statement dated: [*insert date*]

We consider that the sum due to [the Contractor] / [the Employer] on the date of this notice is: [*insert amount*]

We attach the basis on which this sum was calculated.

The final date for payment is: [*insert final date for payment*]

Note: All sums stated are exclusive of VAT, and the Employer will in addition pay the amount of VAT properly chargeable.

Resource 167

Description	A notice from the Contractor to the Employer of the Contractor's intention to suspend the performance of its obligations under the Contract.
Type	Notice: Contractor to Employer
Provision references	clause 4.11.1
Drafting notes	The Contractor is encouraged to first check with the Employer why the relevant payment was not made (for example, an administrative error), before issuing this notice.

Reference: Notice of intention to suspend (clause 4.11.1)

In accordance with clause 4.11.1 (Contractor's right of suspension), given that you have failed to pay a sum payable to us in accordance with clause 4.9 (Interim and final payments – final date and amount), together with any VAT properly chargeable in respect of that payment, by the final date for payment, we write to notify you of our intention to suspend the performance of our obligations under the Contract.

The grounds for such suspension are as follows:

- Sum due: [*insert amount*] (exclusive of VAT)
- Final date for payment: [*insert date*]
- [Applicable relevant Payment Notice, Pay Less Notice, or Interim Payment Application reference: [*insert reference*]]

If this failure continues for 7 days after receipt of this notice, we (without affecting our other rights and remedies) may suspend performance of any or all of our obligations, until payment is made in full.

In accordance with clause 4.11.2, where we exercise this right of suspension, we will be entitled to a reasonable amount in respect of costs and expenses reasonably incurred by us, as a result of exercising the right.

Resource 168

Description	A reply from the Employer to the Contractor's notice of intention to suspend the performance of its obligations under the Contract.
Type	Reply: Employer to Contractor
Provision references	clause 4.11.1
Drafting notes	**Good practice** This notice is not required by clause 4.11.1, but it is considered good practice for the Employer to explain to the Contractor why the payment has not been made, and to inform them by when it will be made.

Reference: Notice of intention to suspend (clause 4.11.1)

We write in reply to your notice dated [*insert date*], given in accordance with clause 4.11.1 (Contractor's right of suspension).

We acknowledge that we have failed to pay a sum payable to you in accordance with clause 4.9 (Interim and final payments – final date and amount), together with any VAT properly chargeable in respect of that payment, by the final date for payment.

The reason for our delay in making payment is as follows: [*insert reasons.*]

We anticipate that we will make payment by no later than [*insert date*], and will keep you updated.

Resource 169

Description	Confirmation from the Contractor to the Employer of the Contractor's intention to suspend the performance of its obligations under the Contract.
Type	Confirmation: Contractor to Employer
Provision references	clause 4.11.1
Drafting notes	**Good practice** This notice is not required by clause 4.11.1, but it is considered good practice to notify the Employer of the date and extent of the Contractor's suspension. The Contractor must not give this notice unless the failure has continued for 7 days after it gave notice under clause 4.11.1. The notice gives the Contractor the option of suspending the performance of all its obligations under the Contract, or certain obligations.

Reference: Notice of intention to suspend (clause 4.11.1)

We write further to our notice dated [*insert date*], given in accordance with clause 4.11.1 (Contractor's right of suspension).

As we stated in that notice, you have failed to pay a sum payable to us in accordance with clause 4.9 (Interim and final payments – final date and amount), together with any VAT properly chargeable in respect of that payment, by the final date for payment.

That failure has continued for 7 days since we gave that notice.

[*Suspension of all obligations:* Therefore, we will suspend the performance of all our obligations under the Contract until payment is made in full.] /

[*Suspension of certain obligations:* Therefore, we will suspend the performance of the following obligations under the Contract, until payment is made in full: [*insert list of obligations, for example, certain works.*]]

Resource 170

Description	An application from the Contractor to the Employer for payment of costs and expenses incurred as a result of the Contractor's suspension of the performance of its obligations under the Contract.
Type	Application: Contractor to Employer
Provision references	clause 4.11.3
Drafting notes	The Contractor is encouraged to submit this application as soon as possible after the period of suspension has ended, and the costs and expenses have been incurred.

Reference: Notice of intention to suspend; application for payment of costs and expenses (clause 4.11.3)

We write in relation to the recent exercise of our right of suspension, and further to our correspondence dated [*insert dates*].

In accordance with clause 4.11.2 (Contractor's right of suspension), where we exercise this right of suspension under clause 4.11.1, we are entitled to a reasonable amount in respect of costs and expenses reasonably incurred by us, as a result of exercising this right.

In accordance with clause 4.11.3, we attach our application in respect of such costs and expenses, in the amount of [*insert amount*] (exclusive of VAT), along with details of them as are reasonably necessary for ascertaining this amount.

Please let us know if there is anything further we can provide in this regard.

Resource 171

Description	A request from the Employer to the Contractor for details of costs and expenses incurred as a result of the Contractor's suspension of the performance of its obligations under the Contract.
Type	Request: Employer to Contractor
Provision references	clause 4.11.3

Reference: Notice of intention to suspend; application for payment of costs and expenses (clause 4.11.3)

We write further to your application for costs and expenses dated [*insert date*], incurred as a result of exercising your right of suspension under clause 4.11.1 (Contractor's right of suspension).

In accordance with clause 4.11.3, please submit the following further details, which are reasonably necessary for us to ascertain the amount in question: [*insert further details and information reasonably required.*]

Resource 172

Description	A checklist for calculating the Gross Valuation in accordance with Alternative A (Stage Payments).
Type	Checklist: Contractor and Employer
Provision references	clause 4.12

Below is a checklist for calculating the Gross Valuation for each Interim Payment, in accordance with Alternative A (Stage Payments):

Calculation: (the total of the amounts in clause 4.12.1) + (the total of the amounts in clause 4.12.2) – (the amounts referred to in clause 4.12.3).

Provision	Description	Amount
clause **4.12.1**	**the total of the following, which are subject to Retention:**	
clause 4.12.1.1	the cumulative value of stages completed;	
clause 4.12.1.2	the value of any Changes or other work referred to in clause 5.2 (Valuation of Changes and provisional sum work) that are relevant to the Interim Payment (whether agreed pursuant to clause 5.2 or valued under the Valuation Rules) but excluding any amounts referred to in clause 4.12.2.4;	
clause 4.12.1.3	the value of any Listed Items for which the conditions set out in clause 4.15 (Listed Items) are satisfied; and	
clause 4.12.1.4	where JCT Fluctuations Option C is applicable and where in accordance with the Formula Rules amounts in the Value of Work are to be allocated to lift installations, structural steelwork installations or catering equipment installations, the total value of Site Materials of those descriptions, provided that their value shall only be included if they are adequately protected against weather and other casualties and they are not on the Works prematurely.	
	These values shall be adjusted, where appropriate, in accordance with any applicable Fluctuations Provision or any Acceleration Quotation for which there has been Confirmed Acceptance.	
	Total:	
clause **4.12.2**	**the total of the following, which are not subject to Retention:**	

(*Continued*)

Provision	Description	Amount
clause 4.12.2.1	any amounts to be included in accordance with clause 4.3 (Contract Sum and Adjustments: Taking adjustments into account) by the Employer as a result of payments made or costs incurred by the Contractor under: • clause 2.5.2 (Early use by Employer), • clause 2.20 (Patent rights – Instructions), • clause 3.12 (Inspection – tests), • clause 6.10.2 or clause 6.10.3 (Terrorism Cover – policy extensions and premiums),	
	• clause 6.11.3 (Terrorism Cover – non-availability – Employer's options), • clause 6.12.2 (Insurance of the Works and Existing Structures: Evidence of insurance), or • clause 6.20 (Joint Fire Code – amendments/ revisions);	
clause 4.12.2.2	any amounts payable under clause 4.11.2 (Contractor's right of suspension);	
clause 4.12.2.3	the amount of any loss and/or expense to which the Contractor is entitled under clause 4.19.1 (Loss and Expense: Matters materially affecting regular progress) or by a Confirmed Acceptance;	
clause 4.12.2.4	where Insurance Option B or C applies or to the extent that the work is under clause 6.13.5.3 (Insurance of the Works and Existing Structures: Loss or damage – insurance claims and reinstatement) to be treated as a Change, any amounts in respect of reinstatement work under clause 6.13.4; and	
clause 4.12.2.5	any amount payable to the Contractor under any applicable Fluctuations Provision, other than by means of an adjustment made under clause 4.12.1;	
	Total:	
clause **4.12.3**	**the following deductions:**	
clause 4.12.3.1	any amounts deductible under: • clause 2.35 (Schedules of defects and instructions), • clause 3.6 (Non-compliance with instructions), • clause 6.12.2 (Insurance of the Works and Existing Structures: Evidence of insurance), or • clause 6.19.2 (Breach of Joint Fire Code – Remedial Measures); and	
clause 4.12.3.2	any amount allowable by the Contractor under clause 6.10.2 (Terrorism Cover – policy extensions and premiums) or under any applicable Fluctuations Provision, other than by means of an adjustment made under clause 4.12.1.	
	Total:	

Resource 173

Description	A checklist for calculating the Gross Valuation in accordance with Alternative B (Periodic Payments).
Type	Checklist: Contractor and Employer
Provision references	clause 4.13

Below is a checklist for calculating the Gross Valuation for each Interim Payment, in accordance with Alternative B (Periodic Payments):

Calculation: (the total of the amounts in clause 4.13.1) + (the total of the amounts in clause 4.13.2) – (the amounts referred to in clause 4.13.3).

Provision	Description	Amount
clause 4.13.1	**the total values of the following, which are subject to Retention:**	
clause 4.13.1.1	work properly executed including any design work carried out by the Contractor and work so executed for which a value has been agreed pursuant to clause 5.2 or which has been valued under the Valuation Rules, but excluding any amounts referred to in clause 4.13.2.4;	
clause 4.13.1.2	Site Materials provided they are adequately protected against weather and other casualties and they are not on the Works prematurely; and	
clause 4.13.1.3	Listed Items (if any) for which the conditions set out in clause 4.15 (Listed Items) are satisfied;	
	These values shall be adjusted, where appropriate, in accordance with any applicable Fluctuations Provision or any Acceleration Quotation for which there has been Confirmed Acceptance.	
	Total:	
clause 4.13.2	**the total of the following, which are not subject to Retention:**	
clause 4.13.2.1	any amounts to be included in accordance with clause 4.3 by the Employer as a result of payments made or costs incurred by the Contractor under: • clause 2.5.2 (Early use by Employer), • clause 2.20 (Patent rights – Instructions), • clause 3.12 (Inspection – tests), • clause 6.10.2 or clause 6.10.3 (Terrorism Cover – policy extensions and premiums), • clause 6.11.3 (Terrorism Cover – non-availability – Employer's options), • clause 6.12.2 (Insurance of the Works and Existing Structures: Evidence of insurance), or • clause 6.20 (Joint Fire Code – amendments/ revisions);	

(*Continued*)

Provision	Description	Amount
clause 4.13.2.2	any amounts payable under clause 4.11.2 (Contractor's right of suspension);	
clause 4.13.2.3	the amount of any loss and/or expense to which the Contractor is entitled under clause 4.19.1 (Loss and Expense: Matters materially affecting regular progress) or by a Confirmed Acceptance;	
clause 4.13.2.4	where Insurance Option B or C applies or to the extent that the work is under clause 6.13.5.3 (Insurance of the Works and Existing Structures: Loss or damage – insurance claims and reinstatement) to be treated as a Change, any amounts in respect of reinstatement work under clause 6.13.4; and	
clause 4.13.2.5	any amount payable to the Contractor under any applicable Fluctuations Provision, other than by means of an adjustment made under clause 4.13.1;	
	Total:	
clause 4.13.3	**the following deductions:**	
clause 4.13.3.1	any amounts deductible under: • clause 2.35 (Schedules of defects and instructions), • clause 3.6 (Non-compliance with instructions), • clause 6.12.2 (Insurance of the Works and Existing Structures: Evidence of insurance), or • clause 6.19.2 (Breach of Joint Fire Code – Remedial Measures); and	
clause 4.13.3.2	any amount allowable by the Contractor under clause 6.10.2 (Terrorism Cover – policy extensions and premiums) or under any applicable Fluctuations Provision, other than by means of an adjustment made under clause 4.13.1.	
	Total:	

Resource 174

Description	A tracker template for recording Interim Payments.
Type	Tracker: Contractor and Employer
Provision references	clause 4.6 to clause 4.14

Below is a tracker template for recording Interim Payments:

Interim Payment Applications

- Interim Payment Application reference: [*insert reference*]
- Interim Payment Application date: [*insert date*]
- Interim Payment Application amount: [*insert amount*]

- Due date: [*insert date*]
- Final date for payment: [*insert date*]

Payment Notices

- Final date for giving of Payment Notice: [*insert date*]
- Payment Notice reference: [*insert reference*]
- Payment Notice date: [*insert date*]
- Payment Notice amount: [*insert amount*]

Pay Less Notices

- Final date for giving of Pay Less Notice: [*insert date*]
- Pay Less Notice reference: [*insert reference*]
- Pay Less Notice date: [*insert date*]
- Pay Less Notice amount: [*insert amount*]
- Date paid: [*insert date*]

Document link: [*insert link to documents*]

Note: All sums stated are exclusive of VAT, and the Employer will in addition pay the amount of VAT properly chargeable.

Resource 175

Description	A request from the Contractor for the Employer's approval to provide a bond from a particular surety.
Type	Request: Contractor to Employer
Provision references	clause 4.15.4

Reference: Bond for uniquely identified Listed Items (clause 4.15.4)

We write in relation to the following uniquely identified Listed Items: [*insert description of relevant uniquely identified Listed Items.*]

In accordance with clause 4.15.4 (Listed Items), we request your approval to provide a bond for these uniquely identified Listed Items in your favour, in the terms set out in Part 2 of Schedule 6, and in the amount specified in the Contract Particulars, from the following surety: [*insert details of the proposed surety.*]

Please give your approval by [*insert date*].

Resource 176

Description	Approval given from the Employer for the Contractor to provide a bond from a particular surety.
Type	Approval given: Employer to Contractor
Provision references	clause 4.15.4

Reference: Bond for uniquely identified Listed Items (clause 4.15.4)

We write further to your request dated [*insert date*], made in accordance with clause 4.15.4 (Listed Items).

We give our approval for you to provide the required bond from [*insert name of surety*] as surety.

Resource 177

Description	Approval withheld from the Employer for the Contractor to provide a bond from a particular surety.
Type	Approval withheld: Employer to Contractor
Provision references	clause 4.15.4
Drafting notes	Clause 1.10 requires that the Employer's approval must not be unreasonably delayed or withheld.

Reference: Bond for uniquely identified Listed Items (clause 4.15.4)

We write further to your request dated [*insert date*], made in accordance with clause 4.15.4 (Listed Items).

We withhold our approval for you to provide the required bond from [*insert name of surety*] as surety, for the following reasons: [*insert reasons, to demonstrate why the approval was not unreasonably delayed or withheld.*]

Please propose an alternative surety for our approval, taking the reasons above into consideration.

Resource 178

Description	A checklist for the Employer to determine whether the conditions for inclusion of the value of a Listed Item in the Gross Valuation before its delivery to or adjacent to the Works have been met.
Type	Checklist: Employer
Provision references	clause 4.15

Below is a checklist to determine whether the conditions for inclusion of the value of a Listed Item in the Gross Valuation before its delivery to or adjacent to the Works (as referred to in clause 4.12.1.3 or clause 4.13.1.3) have been met, in accordance with clause 4.15 (Listed Items):

Item	Description	Provision	Status / action required
1	Is the Listed Item in accordance with the Contract?	clause 4.15.1	Yes / No
2	Has the Contractor provided the Employer with reasonable proof that the property in the Listed Item is vested in the Contractor?	clause 4.15.2	Provided on [insert date]. / Not yet provided.
3	Has the Contractor provided the Employer with reasonable proof that the Listed Item is and will remain insured against loss or damage for its full value under a policy of insurance protecting the interests of the Employer and the Contractor in respect of the Specified Perils until delivered to, or adjacent to, the Works?	clause 4.15.2	Provided on [insert date]. / Not yet provided.
4	At the premises where the Listed Item is situated pending delivery, is there in relation to that item clear identification of: • the Employer as the person to whose order it is held; and • the Works as its destination, and is each item set apart or clearly and visibly marked, individually or as a set, by letters, figures or a pre-determined code?	clause 4.15.3	Yes / No
5	For uniquely identified Listen Items, and it is stated as required in the Contract Particulars, has the Contractor provided a bond in favour of the Employer from a surety approved by the Employer in the terms set out in Part 2 of Schedule 6 ('the required bond') in the amount specified in the Contract Particulars for clause 4.15.4?	clause 4.15.4	Not applicable. / Provided on [insert date]. / Not yet provided.
6	For Listed Items which are not uniquely identified, has the Contractor provided the required bond in the amount specified in the Contract Particulars for clause 4.15.5?	clause 4.15.5	Not applicable. / Provided on [insert date]. / Not yet provided.

Resource 179

Description	A request from the Contractor to the Employer to place the Retention in a separate bank account.
Type	Request: Contractor to Employer
Provision references	clause 4.16.2
Drafting notes	If the Contractor wishes for the Retention to be placed in a separate bank account, it is encouraged to make this request at the same time as the first Interim Payment.

Reference: Request to place Retention in separate bank account (clause 4.16.2)

In accordance with clause 4.16.2 (Rules on treatment of Retention), we request that you place the Retention in a separate bank account (so designated as to identify the amount as the Retention held by the Employer on trust, as provided in clause 4.16.1), and notify us that the amount has been so placed.

We acknowledge that you are entitled to the full beneficial interest in any interest accruing on the separate bank account, and that you are under no duty to account for any such interest to us.

Resource 180

Description	A notice from the Employer to the Contractor that the Retention has been placed in a separate bank account.
Type	Notice: Employer to Contractor
Provision references	clause 4.16.2

Reference: Request to place Retention in separate bank account (clause 4.16.2)

Further to your request dated [*insert date*], and made in accordance with clause 4.16.2 (Rules on treatment of Retention), we write to notify you that, in respect of Interim Payment reference [*insert reference*], we have placed the Retention, in the amount of [*insert amount*] (exclusive of VAT) in a separate bank account, the details of which are as follows:

[*Insert details of the separate bank account, including bank name, address, sort code, account number, etc.*]

[We attach a bank statement to that effect.]

Resource 181

Description	A request from the Contractor for the Employer's approval to provide a Retention Bond from a particular surety.
Type	Request: Contractor to Employer
Provision references	clause 4.17.2

Reference: Retention Bond (clause 4.17.2)

In accordance with clause 4.17.2 (Retention Bond), we request your approval to provide a Retention Bond in your favour, in the terms set out in Part 3 of Schedule 6 (incorporating in clause 2 (maximum aggregate sum) and clause 6.3 (expiry date) of the bond the sum and date stated in the Contract Particulars), from the following surety: [*insert details of the proposed surety.*]

Please give your approval by [*insert date*].

Resource 182

Description	Approval given from the Employer for the Contractor to provide a Retention Bond from a particular surety.
Type	Approval given: Employer to Contractor
Provision references	clause 4.17.2

Reference: Retention Bond (clause 4.17.2)

We write further to your request dated [*insert date*], made in accordance with clause 4.17.2 (Retention Bond).

We give our approval for you to provide the Retention Bond from [*insert name of surety*] as surety.

Resource 183

Description	Approval withheld from the Employer for the Contractor to provide a Retention Bond from a particular surety.
Type	Approval withheld: Employer to Contractor
Provision references	clause 4.17.2
Drafting notes	Clause 1.10 requires that the Employer's approval must not be unreasonably delayed or withheld.

Reference: Retention Bond (clause 4.17.2)

We write further to your request dated [*insert date*], made in accordance with clause 4.17.2 (Retention Bond).

We withhold our approval for you to provide the Retention Bond from [*insert name of surety*] as surety, for the following reasons: [*insert reasons, to demonstrate why the approval was not unreasonably delayed or withheld.*]

Please propose an alternative surety for our approval, taking the reasons above into consideration.

Resource 184

Description	Confirmation from the Employer of the deduction of Retention by the Employer, if the Contractor fails to provide or maintain the Retention Bond.
Type	Confirmation: Employer to Contractor
Provision references	clause 4.17.3
Drafting notes	**Good practice** Clause 4.17.3 does not require this notice to be given, but it is considered good practice to inform the Employer.

Reference: Retention Bond (clause 4.17.3)

In accordance with clause 4.17.3 (Retention Bond), given your failure to provide or maintain the Retention Bond in accordance with clause 4.17.2, we write to confirm that the provisions of clause 4.14 (Sums due as Interim Payments) and clause 4.18 (Retention – amounts and periods) permitting the deduction of the Retention will apply in respect of Interim Payments given after the date of your failure to provide or maintain the bond, namely [*insert date*].

However, if you subsequently provide the required bond, we confirm that any Retention deducted during the period of failure will become due for release to you on the next due date thereafter.

Resource 185

Description	A request from the Employer for the Contractor to confirm how the additional Retention will be accounted for.
Type	Request: Employer to Contractor
Provision references	clause 4.17.4
Drafting notes	**Good practice** Clause 4.17.4 does not require this request to be made, but it is considered good practice for the Employer to give the Contractor the opportunity to adjust the aggregate sum of the Retention Bond, before such sums are deducted as Retention. Footnote [44] of the Contract states that the Employer should keep up-to-date records of the amount of Retention that would have been retained had clause 4.14 and clause 4.18 applied.

Reference: Retention Bond (clause 4.17.4)

As at [*insert date*], the amount of the Retention that would have been deducted had the provisions of clause 4.14 (Sums due as Interim Payments) and clause 4.18 (Retention – amounts and periods) applied exceeds the aggregate sum stated in the Retention Bond as follows:

- Amount of the Retention that would have been deducted had the provisions of clause 4.14 and clause 4.18 applied: [*insert amount*]
- Aggregate sum stated in the Retention Bond: [*insert amount*]
- Difference: [*insert amount*]

(Note: All amounts stated are exclusive of VAT.)

In accordance with clause 4.17.4 (Retention Bond), please arrange with the Surety for the aggregate sum to be increased to equate to the amount of the Retention that would have been deducted had the provisions of clause 4.14 and clause 4.18 applied.

If this is not possible, the amount not covered by the Retention Bond may be deducted as Retention.

Resource 186

Description	A checklist for the calculation of the Retention which may be deducted.
Type	Checklist: Contractor and Employer
Provision references	clause 4.18

Checklist (Works not divided into Sections):

Below is a checklist for calculating the Retention which may be deducted, where the Works are not divided into Sections:

- Retention Percentage: [*insert percentage*]
- Has practical completion been achieved? (If not, then the full Retention Percentage may be deducted. If so, then half the Retention Percentage may be deducted, and the Contractor will be entitled to payment of the other half.)
- Has the Notice of Completion of Making Good been issued? (If not, then half the Retention Percentage may be deducted, and the Contractor will be entitled to payment of the other half. If so, then none of the Retention Percentage may be deducted, and the Contractor will be entitled to payment of the second half.)
- Are there any Relevant Parts (for which the Employer has taken partial possession)? (If so, then only half the Retention Percentage may be deducted where it relates to each Relevant Part, and the Contractor will be entitled to payment of the other half.)
- If there are any Relevant Parts, has the corresponding Notices of Completion of Making Good been issued? (If not, then only half the Retention Percentage may be deducted where it relates to each Relevant Part, and the Contractor will be entitled to payment of the other half. If so, then none of the Retention Percentage may be deducted, and the Contractor will be entitled to payment of the second half.)

Checklist (Works divided into Sections):

Below is a checklist for calculating the Retention which may be deducted, where the Works are divided into Sections:

- Retention Percentage: [*insert percentage*]
- In respect of each Section, has practical completion of been achieved? (If not, then the full Retention Percentage may be deducted. If so, then half the Retention Percentage may be deducted, and the Contractor will be entitled to payment of the other half).
- In respect of each Section, has the corresponding Notice of Completion of Making Good been issued? (If not, then half the Retention Percentage may be deducted, and the Contractor will be entitled to payment of the other half. If so, then none of the Retention Percentage may be deducted, and the Contractor will be entitled to payment of the second half.)
- Are there any Relevant Parts (for which the Employer has taken partial possession)? (If so, then only half the Retention Percentage may be deducted, and the Contractor will be entitled to payment of the other half.)
- If there are any Relevant Parts, has the corresponding Notices of Completion of Making Good been issued? (If not, then only half the Retention Percentage may be deducted, and the Contractor will be entitled to payment of the other half. If so, then none of the Retention Percentage may be deducted, and the Contractor will be entitled to payment of the second half.)

Resource 187

Description	A notice from the Contractor to the Employer of loss and expense incurred by the Contractor, arising from a deferment of giving possession of the site.
Type	Notice: Contractor to Employer
Provision references	clause 4.20.1, clause 4.20.2
Drafting notes	**Timing:** Clause 4.20.1 requires the Contractor to give this notice as soon as the likely nature and extent of any loss and/or expense arising from a deferment of possession becomes (or should have become) reasonably apparent to him. **Timing:** Clause 4.20.2 requires the Contractor to provide its initial assessment of the loss and/or expense incurred either (i) with the notice under clause 4.20.1, or (ii) as soon as reasonably practicable thereafter. (This notice template assumes the former.)

Reference: Loss and expense; deferment of possession (clause 4.20.1 and clause 4.20.2)

In accordance with clause 4.20.1 (Loss and Expense: Notification and ascertainment), we write to notify you that we have incurred, or are likely to incur, direct loss and/or expense arising from the deferment of giving possession of the site (or part of it) under clause 2.4 (Deferment of Possession).

The likely nature and extent of this loss and/or expense is as follows: [*insert details.*]

In accordance with clause 4.20.2, we attach our initial assessment of the loss and/or expense incurred, and any further amounts likely to be incurred, together with the information reasonably necessary to enable you to ascertain the loss and/or expense incurred.

In accordance with clause 4.20.3, and in such form and manner as you may reasonably require, we will update this assessment and information at monthly intervals, until all information reasonably necessary to allow ascertainment of the total amount of such loss and expense has been supplied.

Resource 188

Description	A notice from the Contractor to the Employer of loss and expense incurred by the Contractor, arising from a Relevant Matter.
Type	Notice: Contractor to Employer
Provision references	clause 4.20.1, clause 4.20.2
Drafting notes	**Timing:** Clause 4.20.1 requires the Contractor to give this notice as soon as the likely effect of a Relevant Matter on regular progress becomes (or should have become) reasonably apparent to him. **Timing:** Clause 4.20.2 requires the Contractor to provide its initial assessment of the loss and/or expense incurred either (i) with the notice under clause 4.20.1, or (ii) as soon as reasonably practicable thereafter. (This notice template assumes the former.)

Reference: Loss and expense; Relevant Matter (clause 4.20.1 and clause 4.20.2)

In accordance with clause 4.20.1 (Loss and Expense: Notification and ascertainment), we write to notify you that the regular progress of the Works has been or is likely to be materially affected by the following matter: [*insert details.*]

This constitutes a Relevant Matter under the following clause: [*insert reference to the provision within clause 4.21 (Relevant Matters) containing the applicable Relevant Matter.*]

The likely effect of this Relevant Matter is as follows: [*insert details.*]

In accordance with clause 4.20.2, we attach our initial assessment of the loss and/or expense incurred, and any further amounts likely to be incurred, together with the information reasonably necessary to enable you to ascertain the loss and/or expense incurred.

In accordance with clause 4.20.3, and in such form and manner as you may reasonably require, we will update this assessment and information at monthly intervals, until all information reasonably necessary to allow ascertainment of the total amount of such loss and expense has been supplied.

Resource 189

Description	A requirement by the Employer of further information from the Contractor to ascertain loss and expense incurred.
Type	Requirement: Employer to Contractor
Provision references	clause 4.20.3

Reference: Loss and expense; requirement of further information (clause 4.20.3)

We write further to your notice dated [*insert date*], with reference [*insert reference*], given in accordance with clause 4.20.1 (Loss and Expense: Notification and ascertainment), and your initial assessment of the loss and/or expense incurred, given in accordance with clause 4.20.2.

In accordance with clause 4.20.3, we require you to update this assessment and information in the following form and manner: [*insert details of how the Employer wishes for the assessment and information to be updated.*]

Please provide the updated assessment and information at monthly intervals, until all information reasonably necessary to allow ascertainment of the total amount of such loss and expense has been supplied.

Resource 190

Description	A notice from the Employer to the Contractor of the ascertained amount of loss and expense incurred.
Type	Notice: Employer to Contractor
Provision references	clause 4.20.4
Drafting notes	**Timing:** Clause 4.20.4 requires the Employer to give this notice within 28 days of receipt of the initial assessment and information, and 14 days of each subsequent update of them.

Reference: Ascertainment of loss and expense (clause 4.20.4)

We write further to the following correspondence:

- your notice dated [*insert date*], with reference [*insert reference*], given in accordance with clause 4.20.1 (Loss and Expense: Notification and ascertainment);
- your initial assessment of the loss and/or expense incurred dated [*insert date*], given in accordance with clause 4.20.2; and
- the following updates of this assessment, given in accordance with clause 4.20.3: [*insert details of the monthly updates of the Contractor's initial assessment, and the dates they were provided*]

In accordance with clause 4.20.4, we write to notify you of the ascertained amount of this loss and and/or expense, the details of which are attached.

Each ascertainment has been made by reference to the information you supplied, and in sufficient detail to enable you to identify any differences between the Employer's ascertainment and your assessment.

Please let us know if you would like to discuss this further.

Resource 191

Description	A tracker template for recording applications for reimbursement of loss and expense.
Type	Tracker: Contractor and Employer
Provision references	clause 4.19 to clause 4.22

Below is a tracker template for recording the reimbursement of direct loss and/or expense:

- Date of Contractor's notice (clause 4.20.1): [*insert date*]
- Reference: [*insert reference*]
- Relevant Matter / deferment of possession: [*insert applicable Relevant Matter, or deferment of possession*]
- Date of Contractor's initial assessment (clause 4.20.2): [*insert date*]
- Amount of Contractor's initial assessment: [*insert amount*]
- Date of any updates of the Contractor's initial assessment (clause 4.20.3): [*insert date*]
- Amount of Contractor's revised assessment: [*insert amount*]
- Latest date for Employer's ascertainment (clause 4.20.4): [*insert date*]
- Amount ascertained by the Employer (clause 4.20.4): [*insert amount*]
- Difference between the Parties' assessments: [*insert amount*]
- Document link: [*insert link to documents*]
- Status / action required: *[insert status or action required]*

(Note: All amounts stated are exclusive of VAT.)

Resource 192

Description	The submission by the Contractor to the Employer of the Final Statement.
Type	Correspondence: Contractor to Employer
Provision references	clause 4.24.1
Drafting notes	The Contractor is encouraged to submit the Final Statement as soon as possible after practical completion of the Works has been achieved.

Reference: Final Statement (clause 4.24.1)

Following practical completion of the Works, which was achieved on [*insert date*], and in accordance with clause 4.24.1 (Final Statement and Final Payment), we attach the Final Statement, along with the following supporting documents: [*insert details of supporting documents.*]

Please let us know if you require any further supporting documents.

Resource 193

Description	A request by the Employer for further supporting documents required from the Contractor, in respect of the Final Statement.
Type	Request: Employer to Contractor
Provision references	clause 4.24.1
Drafting notes	The Employer is encouraged to review the Final Statement promptly upon receipt, and make this request as soon as possible thereafter.

Reference: Final Statement; requirement of further supporting documents (clause 4.24.1)

We write further to your Final Statement, submitted on [*insert date*].

In accordance with clause 4.24.1, we require the following supporting documents: [*insert details of supporting documents reasonably required.*]

Resource 194

Description	A Final Statement template, from the Contractor to the Employer.
Type	Template: Final Statement
Provision references	clause 4.24.2
Drafting notes	See also clause 1.8 (Effect of Final Statement).

Reference: Final Statement (clause 4.24.2)

- Contract Sum: [*insert Contract Sum*]
- Adjusted Contract Sum, following the adjustments made in accordance with clause 4.2 (Items included in adjustments): [*insert adjusted Contract Sum*]
- Sum of amounts already paid by the Employer to the Contractor: [*insert amount*]
- Due date: [*insert due date*]
- Final date for payment: [*insert final date for payment*]

The sum due is therefore: [*insert amount*], as the balance due [to the Contractor from the Employer] / [to the Employer from the Contractor].

We attach the basis on which this amount has been calculated, including details of all such adjustments.

(Note: All amounts stated are exclusive of VAT.)

Resource 195

Description	A notice from the Employer to the Contractor of the Employer's intention to issue the Employer's Final Statement.
Type	Notice: Employer to Contractor
Provision references	clause 4.24.3
Drafting notes	If the Contractor has not submitted the Final Statement, the Employer is encouraged to give this notice immediately following the expiry of the period 3 months after practical completion.

Reference: Employer's intention to issue a final statement (clause 4.24.3)

We write to remind you that you have not submitted the Final Statement within 3 months of practical completion of the Works, which was achieved on [*insert date*].

In accordance with clause 4.24.3 (Final Statement and Final Payment), we write to notify you that unless you issue the Final Statement within 2 months from the date of this notice, we may issue a final statement (the 'Employer's Final Statement', referred to in clause 4.24.4) in lieu of the Final Statement from you.

Resource 196

Description	The submission by the Employer to the Contractor of the Employer's Final Statement.
Type	Correspondence: Employer to Contractor
Provision references	clause 4.24.4
Drafting notes	See also clause 1.8 (Effect of Final Statement). **Timing:** Clause 4.24.3 states that the Employer may issue the Employer's Final Statement at any time after 2 months from the date of the Employer's notice under clause 4.24.3.

Reference: Employer's Final Statement (clause 4.24.4)

We write further to our notice dated [*insert date*], given in accordance with clause 4.24.3 (Final Statement and Final Payment).

You have still not submitted the Final Statement.

Therefore, in accordance with clause 4.24.4, we attach the Employer's Final Statement.

This is in the form and with the details required by clause 4.24.2, so far as we, on the information in our possession, have reasonably been able to do so.

Resource 197

Description	An Employer's Final Statement template, from the Employer to the Contractor.
Type	Template: Employer's Final Statement
Provision references	clause 4.24.4

Reference: Employer's Final Statement (clause 4.24.4)

- Contract Sum: [*insert Contract Sum*]
- Adjusted Contract Sum, following the adjustments made in accordance with clause 4.2 (Items included in adjustments): [*insert adjusted Contract Sum*]
- Sum of amounts already paid by the Employer to the Contractor: [*insert amount*]
- Due date: [*insert due date*]
- Final date for payment: [*insert final date for payment*]

The sum due is therefore: [*insert amount*], as the balance due [to the Contractor from the Employer] / [to the Employer from the Contractor].

We attach the basis on which this amount has been calculated, including details of all such adjustments.

(Note: All amounts stated are exclusive of VAT.)

Resource 198

Description	A notice disputing the Final Statement / Employer's Final Statement.
Type	Notice: Contractor or Employer to the other Party
Provision references	clause 4.24.6
Drafting notes	See also clause 1.8 (Effect of Final Statement).

Reference: Notice disputing [*Employer:* Final Statement] / [*Contractor:* Employer's Final Statement] (clause 4.24.6)

In accordance with clause 4.24.6 (Final Statement and Final Payment), we write to notify you that we dispute the contents of the [*Employer:* Final Statement] / [*Contractor:* Employer's Final Statement].

In particular, we dispute the following: [*insert details of what the Employer disputes in the Final Statement, or the Contractor disputes in the Employer's Final Statement.*]

Resource 199

Description	A checklist for establishing whether an instruction is a Change.
Type	Checklist: Contractor and Employer
Provision references	clause 5.1

The definition of the term 'Change' is included in clause 5.1 (Definition of Changes).

Below is a checklist to help establish whether an Employer's instruction constitutes a Change.

- clause 5.1.1: does the instruction require a change in the Employer's Requirements, which makes necessary the alteration or modification in the design, quality, or quantity of the Works (otherwise than such as may be reasonably necessary for the purposes of rectification pursuant to clause 3.13 (Work not in accordance with the Contract)), including the following:
  - clause 5.1.1.1: does it require the addition, omission, or substitution of any work?
  - clause 5.1.1.2: does it require the alteration of the kind or standard of any of the materials or goods to be used in the Works?
  - clause 5.1.1.3: does it require the removal from the site of any work executed or Site Materials (other than work, materials, or goods which are not in accordance with the Contract)?
- clause 5.1.2: does the instruction require the imposition by the Employer of any obligations or restrictions in regard to the following matters, or any addition to or omission of any such obligations or restrictions that are so imposed, or are imposed in the Employer's Requirements, in regard to the following:
  - clause 5.1.2.1: access to the site, or use of any specific parts of the site?
  - clause 5.1.2.2: limitations of working space?
  - clause 5.1.2.3: limitations of working hours?
  - clause 5.1.2.4: the execution or completion of the work in any specific order?

If the answer to any of the questions above is 'yes', then the instruction constitutes a Change.

Note the list in clause 5.1.1 is not exhaustive.

Resource 200

Description	A tracker template for recording Changes, work to be treated as a Change, and the expenditure of Provisional Sums.
Type	Tracker: Contractor and Employer
Provision references	clause 5.2

In accordance with clause 5.2 (Valuation of Changes and provisional sum work), the value of:

- all Changes required by Employer's instructions;
- all work which under the Conditions is to be treated as a Change; and

- all work executed by the Contractor in accordance with Employer's instructions as to the expenditure of Provisional Sums included in the Employer's Requirements,

will be either:

- such amount as is agreed by the Employer and the Contractor; or, where not agreed,
- the amount valued (a Valuation) in accordance with clause 5.4 to clause 5.7 (the Valuation Rules).

Below is a tracker template to record the matters referred to in clause 5.2:

- Reference: [*insert reference*]
- Date: [*insert date*]
- Employer's instruction, work to be treated as a Change, or expenditure of Provisional Sum (clause 5.2.1 to clause 5.2.3): [*insert applicable clause*]
- Amount agreed between Employer and Contractor (if applicable): [*insert amount*]
- Valuation: [*insert amount*]
- Status / action required: [*insert status, and any action required*]

(Note: All amounts stated are exclusive of VAT.)

<u>Matters treated as a Change</u>

Further to clause 5.2.2, the following provisions refer to matters being treated as a Change:

- clause 2.10.1 (Divergence in Employer's Requirements and definition of site boundary);
- clause 2.12.2 (Employer's Requirements – inadequacy);
- clause 2.14.2 (Discrepancies in documents);
- clause 2.15.2.1 (Divergences from Statutory Requirements);
- clause 2.15.2.2 (Divergences from Statutory Requirements);
- clause 5.6 (Change of conditions for other work);
- clause 6.13.5.3 (Loss or damage – insurance claims and reinstatement);
- clause 6.13.6 (Loss or damage – insurance claims and reinstatement);
- paragraph 1.4.2, Supplemental Provision 1 (Named Sub-Contractors); and
- paragraph 2.5, Supplemental Provision 2 (Valuation of Changes – Contractor's estimates).

Resource 201

Description	A checklist for carrying out a Valuation in accordance with the Valuation Rules.
Type	Checklist: Contractor and Employer
Provision references	clause 5.4 to clause 5.7

Below is a checklist for carrying out a Valuation in accordance with the Valuation Rules (clause 5.4 to clause 5.7).

Valuations are to be undertaken in accordance with clause 5.4 (The Valuation Rules: Measurable Work) and, so far as is relevant, the other three clauses forming part of the Valuation Rules:

- clause 5.5 (Daywork);
- clause 5.6 (Change of conditions for other work); and
- clause 5.7 (Additional provisions).

Clause 5.4: Measurable Work

- clause 5.4.1: allowance shall be made for the addition or omission of the relevant design work.
- clause 5.4.2: the valuation of additional or substituted work shall be consistent with the values of work of a similar character set out in the Contract Sum Analysis, making due allowance for any change in the conditions under which work is carried out, and/or any significant change in the quantity of the work so set out. Where there is no work of a similar character set out in the Contract Sum Analysis, a fair valuation shall be made.
- clause 5.4.3: the valuation of the omission of work set out in the Contract Sum Analysis shall be in accordance with the values therein for such work.
- clause 5.4.4: any valuation of work under clause 5.4.2 and clause 5.4.3 shall include allowance for any necessary addition to, or reduction of, the provision of:
    - site administration;
    - site facilities; and
    - temporary works.
- clause 5.4.5: allowance shall be made for any percentage or lump sum adjustments in the Contract Sum Analysis.

Clause 5.5: Daywork

Where the execution of additional or substituted work cannot be valued in accordance with clause 5.4, the Valuation shall comprise either:

- the prime cost of such work (calculated in accordance with the 'Definition of Prime Cost of Daywork carried out under a Building Contract' issued by the RICS and Construction Industry Publications Ltd as current at the Base Date), together with Percentage Additions to each section of the prime cost at the rates stated in the document identified in the Contract Particulars or, if they apply in respect of labour, at the All-Inclusive Rates stated in such document; or
- where the work is within the province of any specialist trade, and the RICS and the appropriate body representing the employers in that trade have agreed and issued a definition of prime cost of daywork, the prime cost of such work calculated in accordance with that definition current at the Base Date, together with Percentage Additions on the prime cost at the rates stated in the document identified in the Contract Particulars or, if they apply in respect of labour, at the All-Inclusive Rates stated in such document.

Provided that in any case vouchers specifying the time daily spent upon the work, the operatives' names, the plant, and the materials employed shall be delivered for verification to the Employer not later than 7 Business Days after the work has been executed.

Clause 5.6: Change of conditions for other work

If, as a result of:

- compliance with any instruction requiring a Change; or
- compliance with any instruction as to the expenditure of a Provisional Sum,

there is a substantial change in the conditions under which any other work is executed, that other work shall be treated as a Change, and shall be valued in accordance with section 5 (Changes).

Clause 5.7: Additional Provisions

To the extent that:

- a Valuation does not relate to the execution of additional or substituted work;
- a Valuation does not relate to the omission of work; or
- the valuation of any work or liabilities directly associated with a Change cannot reasonably be effected in the Valuation by the application of clause 5.4 (Measurable Work), clause 5.5 (Daywork), and clause 5.6 (Change of conditions for other work),

a 'fair valuation' shall be made.

No allowance shall be made under the Valuation Rules for any effect upon the regular progress of the Works (or any part of them), or for any other direct loss and/or expense for which the Contractor would be reimbursed by payment under any other provision in the Conditions (including clause 4.19 to clause 4.23).

Resource 202

Description	A request from the Contractor for the Employer's approval to place the insurance required under clause 6.5.1 with a particular insurer.
Type	Request: Contractor to Employer
Provision references	clause 6.5.2

Reference: Contractor's insurance of liability of Employer (clause 6.5.2)

We write in relation to the insurance required under clause 6.5.1 (Contractor's insurance of liability of Employer).

In accordance with clause 6.5.2, we request the Employer's approval to place this insurance with the following insurer: [*insert details of the proposed insurer.*]

[We attach a draft copy of the policy.]

Please give your approval by [*insert date*].

Resource 203

Description	Approval given by the Employer for the Contractor to place the insurance required under clause 6.5.1 with a particular insurer.
Type	Approval given: Employer to Contractor
Provision references	clause 6.5.2

Reference: Contractor's insurance of liability of Employer (clause 6.5.2)

We write further to your request dated [*insert date*], made in accordance with clause 6.5.2 (Contractor's insurance of liability of Employer).

We give our approval for [*insert name of insurer*] as insurer to provide this policy.

Please place this insurance and provide us with the policy.

Resource 204

Description	Approval withheld by the Employer for the Contractor to place the insurance required under clause 6.5.1 with a particular insurer.
Type	Approval withheld: Employer to Contractor
Provision references	clause 6.5.2
Drafting notes	Clause 1.10 requires that the Employer's approval must not be unreasonably delayed or withheld.

Reference: Contractor's insurance of liability of Employer (clause 6.5.2)

We write further to your request dated [*insert date*], made in accordance with clause 6.5.2 (Contractor's insurance of liability of Employer).

We withhold our approval for [*insert name of insurer*] as insurer to provide this policy, for the following reasons: [*insert reasons, to demonstrate why the approval was not unreasonably delayed or withheld.*]

Please propose an alternative insurer for our approval, taking the reasons above into consideration.

Resource 205

Description	The provision by the Contractor to the Employer of the policy for the insurance required under clause 6.5.1.
Type	Correspondence: Contractor to Employer
Provision references	clause 6.5.2

Reference: Contractor's insurance of liability of Employer (clause 6.5.2)

In accordance with clause 6.5.2 (Contractor's insurance of liability of Employer), we attach the policy for the insurance required under clause 6.5.1.

Resource 206

Description	An instruction from the Employer not to renew Terrorism Cover.
Type	Instruction: Employer to Contractor
Provision references	clause 6.10.4
Drafting notes	Clause 6.10.4 only applies in the following limited circumstances: • where Insurance Option A (New Buildings – All Risk Insurance of the Works by the Contractor) applies; • where the Employer is a Local or Public Authority; and • if at any renewal of the Terrorism Cover (of any type) there is an increase in the rate on which the premium is based.

Reference: Employer's instruction [*insert reference*]; renewal of Terrorism Cover (clause 6.10.4)

In accordance with clause 6.10.4 (Terrorism Cover – policy extensions and premiums), we instruct you not to renew the Terrorism Cover.

Clause 6.13.5.3 (Loss or damage – insurance claims and reinstatement) will therefore apply with effect from the renewal date.

Resource 207

Description	A notice of the non-availability of Terrorism Cover.
Type	Notice: Contractor or Employer to the other Party
Provision references	clause 6.11.1
Drafting notes	**Timing:** Clause 6.11.1 requires the Parties to give this notice to the other Party 'immediately' upon notice from the insurers.

Reference: Notice of non-availability of Terrorism Cover (clause 6.11.1)

In accordance with clause 6.11.1 (Terrorism Cover – non-availability – Employer's options), we write to notify you that the insurers named in the Joint Names Policy reference [*insert reference*], namely [*insert name of insurer*], have notified us that, with effect from [*insert date*], Terrorism Cover will [cease and will no longer be available] / [only continue to be available with the following reduction in the scope or level of such cover: [*insert details of the reduction in the scope or level of cover.*]]

Resource 208

Description	A notice from the Employer to the Contactor, following the non-availability of Terrorism Cover.
Type	Notice: Employer to Contractor
Provision references	clause 6.11.2
Drafting notes	**Timing:** If the Employer elects to terminate the Contractor's employment under the Contract, it must select a date between the following events: (i) after the date of the insurers' notification that Terrorism Cover will cease and will no longer be available, or will only continue to be available with a reduction in the scope or level of such cover; and (ii) on or before the date when the Terrorism Cover ceases, or continues to be available with a reduction in the scope or level of such cover.

Reference: Notice of non-availability of Terrorism Cover (clause 6.11.2)

We write further to [*Contractor*: your notice] / [*Employer*: our notice] dated [*insert date*], given in accordance with clause 6.11.1 (Terrorism Cover – non-availability – Employer's options).

[In accordance with clause 6.11.2.1, we write to notify you that, notwithstanding the cessation or reduction in scope or level of Terrorism Cover, we require that the Works continue to be carried out.] /

[In accordance with clause 6.11.2.2, we write to notify you that on [*insert date, to be after the insurers' notification, but no later than the cessation date of the Terrorism Cover*], your employment under this Contract will terminate.]

Resource 209

Description	An instruction from the Employer for the Contractor to effect and maintain an alternative or additional form of Terrorism Cover.
Type	Instruction: Employer to Contractor
Provision references	clause 6.11.3
Drafting notes	This clause 6.11.3 only applies where Insurance Option A (New Buildings – All Risk Insurance of the Works by the Contractor) applies, and the Employer has given a notice under clause 6.11.2.1 requiring continuation of the Works.

Reference: Employer's instruction [*insert reference*]; Terrorism Cover (clause 6.11.3)

We write further to our notice dated [*insert date*], given in accordance with clause 6.11.2.1 (Terrorism Cover – non-availability – Employer's options).

In accordance with clause 6.11.3 (Terrorism Cover – non-availability – Employer's options), we instruct you to effect and maintain the following alternative or additional form of Terrorism Cover, which we understand you are able to obtain: [*insert details of the alternative or additional form of Terrorism Cover.*]

We confirm that the net additional cost to you of any such cover, and its renewal, will be added to the Contract Sum.

Resource 210

Description	A request for evidence of insurance.
Type	Request: Contractor or Employer to the other Party
Provision references	clause 6.12.1

Reference: Request for documentary evidence of insurance (clause 6.12.1)

In accordance with clause 6.12 (Evidence of insurance), we request that, within 7 days, you supply documentary evidence to demonstrate that the policy or cover has been effected and remains in force in respect of the following insurance policy:

[*Insert details of the insurance policy under:*

- *clause 6.4 (Contractor's insurance of his liability), to be provided by the Contractor;*
- *clause 6.5 (Contractor's insurance of liability of Employer), to be provided by the Contractor;*
- *clause 6.7 (Insurance of the Works and Existing Structures: Insurance Options and period), to be provided by either the Contractor (where Insurance Option A applies) or the Employer (where Insurance Option B or Insurance Option C applies); and/or*
- *clause 6.10 (Terrorism Cover – policy extensions and premiums), to be provided by either the Contractor (where Insurance Option A applies) or the Employer (where Insurance Option B or Insurance Option C applies)]*

Resource 211

Description	The provision of evidence of insurance.
Type	Reply: Contractor or Employer to the other Party
Provision references	clause 6.12.1

Reference: Request for documentary evidence of insurance (clause 6.12.1)

We write further to your request dated [*insert date*], made in accordance with clause 6.12.1 (Evidence of insurance).

We attach the documentary evidence to demonstrate that the policy or cover has been effected and remains in force for the following insurance policy:

[*Insert details of the insurance policy under:*

- *clause 6.4 (Contractor's insurance of his liability), to be provided by the Contractor;*
- *clause 6.5 (Contractor's insurance of liability of Employer), to be provided by the Contractor;*
- *clause 6.7 (Insurance of the Works and Existing Structures: Insurance Options and period), to be provided by either the Contractor (where Insurance Option A applies) or the Employer (where Insurance Option B or Insurance Option C applies); and/or*
- *clause 6.10 (Terrorism Cover – policy extensions and premiums), to be provided by either the Contractor (where Insurance Option A applies) or the Employer (where Insurance Option B or Insurance Option C applies)]*

Resource 212

Description	Confirmation of a failure to provide evidence of insurance.
Type	Correspondence: Contractor or Employer to the other Party
Provision references	clause 6.12.2
Drafting notes	**Good practice** This correspondence is not required by clause 6.12.2, but it is considered good practice for the Parties to notify each other of the potential consequences from a failure to insure.

Reference: Evidence of insurance (clause 6.12.2)

We write further to our request dated [*insert date*], made in accordance with clause 6.12.1 (Evidence of insurance).

You have failed to provide the requested documentary evidence within 7 days of our request, and in accordance with clause 6.12.2, we are therefore entitled to assume that there has been a failure to insure.

We may therefore insure against any risk, liability, or expense to which we may be exposed as a consequence.

If we do, you will be liable for the costs we incur in taking out and maintaining that insurance.

[*Contractor*: Any such costs payable to us will be added to the Contract Sum.] / [*Employer*: Any such costs payable to us may be deducted from any sums due, or to become due, to you, or will be recoverable from you as a debt.]

Resource 213

Description	A notice from the Contractor to the Employer of loss or damage affecting executed work, Site Materials, or the Existing Structures.
Type	Notice: Contractor to Employer
Provision references	clause 6.13.1
Drafting notes	**Timing:** Clause 6.13.1 requires the Contractor to give this notice 'forthwith'.

Reference: Notice of loss or damage affecting work or Site Materials (clause 6.13.3)

In accordance with clause 6.13.1 (Loss or damage – insurance claims and reinstatement), we write to notify you of the following [loss or damage affecting executed work or Site Materials, occasioned by the risks covered by the Works Insurance Policy or an Excepted Risk] / [loss or damage to the Existing Structures and their contents]:

[*Insert details of the nature, location, and extent of the loss or damage.*]

This loss or damage occurred on [*insert date*].

Resource 214

Description	An authorisation from the Contractor to the insurers to pay insurance monies to the Employer.
Type	Authorisation: Contractor to insurers
Provision references	clause 6.13.3

Reference: Authorisation to pay insurance monies to Employer (clause 6.13.3)

We write in relation to the building contract (the 'Contract') dated [*insert date*] made between us (the 'Contractor') and [*insert name of Employer*] (the 'Employer').

In accordance with clause 6.13.3 (Loss or damage – insurance claims and reinstatement) of the Contract, we authorise you to pay to the Employer all monies under the [Works Insurance Policy, with policy reference [*insert reference*]] / [policy covering Existing Structures or their contents, with policy reference [*insert reference*]], as those terms are defined in the Contract.

Resource 215

Description	The provision by the Employer to the Contractor of a reinstatement work statement.
Type	Reinstatement work statements: Employer to Contractor
Provision references	clause 6.13.5.1
Drafting notes	This correspondence is only relevant where Insurance Option A (New Buildings – All Risks Insurance of the Works by the Contractor) applies. In accordance with clause 6.13.5.1, the Employer is required to pay all monies from such insurance to the Contractor by instalments under separate reinstatement work statements issued by the Employer at the same dates as those for Payment Notices under clause 4.7.5, but without deduction of Retention, and less only the amounts referred to in clause 6.13.5.2 (relating to professional fees). In accordance with clause 6.13.5.4, for the purposes of clause 2.26.1 (Relevant Events), but not otherwise except as referred to under clause 6.13.5.3, reinstatement work shall be treated as a Change.

Reference: Reinstatement work statement

In accordance with clause 6.13.5.1, we attach a reinstatement work statement, in the amount of [*insert amount*] (exclusive of VAT) which relates to the following reinstatement work:

- the restoration of damaged work;
- the replacement or repair of any lost or damaged Site Materials; and
- the removal and disposal of any debris.

We also attach the basis on which this sum has been calculated.

We confirm that there has been no deduction of Retention in respect of this sum.

In accordance with clause 6.13.5.2, we have retained the amount of [*insert amount*] (exclusive of VAT), properly incurred by the Employer and notified to insurers in respect of professional fees [up to the aggregate amount of the percentage cover for those fees] / [up to the amount paid by insurers in respect of those fees].

Resource 216

Description	A notice of termination following material loss or damage to any of the Existing Structures.
Type	Notice: Contractor or Employer to the other Party
Provision references	clause 6.14
Drafting notes	Notices under this clause must be given in accordance with clause 1.7.4 (Notices and other communications). **Timing:** If either party wishes to terminate the Contractor's employment under the Contract, clause 6.14 requires them to give this notice within 28 days of the occurrence of the loss or damage. Clause 6.14 requires that it must be 'just and equitable' for the Contractor's employment under the Contract to be terminated, so the Parties are encouraged to give reasons for doing so. The Contractor should also first consider whether clause 7.8.2.2 (Rights for a Funder) applies, and whether any Funder Rights have been vested in a Funder as third party rights. This would make the Contractor's right to terminate its employment under the Contract (or to treat it as repudiated) subject to the provisions of paragraph 6 of Part 2 of Schedule 5 (Funder Rights) (known as 'step-in' rights).

Reference: Notice of termination following material loss or damage to the Existing Structures (clause 6.14)

In accordance with clause 6.14 (Loss or damage to Existing Structures – right of termination), and following the material loss or damage to the Existing Structures on [*insert date*], we write to notify you that we terminate [*Contractor*: our] / [*Employer*: your] employment under the Contract.

We consider that it is just and equitable to do so for the following reasons: [*insert reasons.*]

Resource 217

Description	The invocation of a dispute resolution procedure to determine whether the termination under clause 6.14 is just and equitable.
Type	Correspondence: Contractor or Employer to the other Party
Provision references	clause 6.14
Drafting notes	**Timing:** If either Party wishes to give this notice, clause 6.14.1. requires them to do so (i) within 7 days of receiving the initial notice of termination of the Contractor's employment under the Contract, or (ii) such longer period as the Parties may agree.

Reference: Notice of termination following material loss or damage to the Existing Structures (clause 6.14)

We write further to your notice dated [*insert date*], given in accordance with clause 6.14 (Loss or damage to Existing Structures – right of termination).

We do not consider that it is just and equitable to terminate [*Contractor*: our] / [*Employer*: your] employment under the Contract, for the following reasons: [*insert reasons.*]

In accordance with clause 6.14.1, we advise you that we will therefore invoke the following dispute resolution procedure under the Contract, to determine whether the termination is just and equitable: [*insert details of the relevant dispute resolution procedure, for example, negotiation (clause 9.1), mediation (clause 9.2), adjudication (Article 8), arbitration (Article 9), or litigation (Article 10).*]

Resource 218

Description	A request from the Employer to the Contractor for evidence of professional indemnity insurance.
Type	Request: Employer to Contractor
Provision references	clause 6.15.3
Drafting notes	The Employer is encouraged to make this request annually, following the renewal date of the Contractor's Professional Indemnity insurance policy.

Reference: Request for documentary evidence of Professional Indemnity insurance (clause 6.15.3)

In accordance with clause 6.15.3 (Professional Indemnity Insurance: Obligation to insure), we request that you produce for inspection documentary evidence to demonstrate that the Professional Indemnity insurance policy required under clause 6.15.1 and clause 6.15.2 has been effected and/or is being maintained.

Resource 219

Description	The provision by the Contractor to the Employer of evidence of professional indemnity insurance.
Type	Reply: Contractor to Employer
Provision references	clause 6.15.3

Reference: Request for documentary evidence of Professional Indemnity insurance (clause 6.15.3)

We write further to your request dated [*insert date*], made in accordance with clause 6.15.3 (Professional Indemnity Insurance: Obligation to insure).

We attach the documentary evidence to demonstrate that the Professional Indemnity policy required under clause 6.15.1 and clause 6.15.2 has been effected and/or is being maintained.

Resource 220

Description	A notice from the Contractor to the Employer of the increased cost or non-availability of professional indemnity insurance.
Type	Notice: Contractor to Employer
Provision references	clause 6.16
Drafting notes	**Timing:** Clause 6.16 requires the Contractor to give this notice 'immediately' upon the insurance ceasing to be available at commercially reasonable rates or terms. The Contractor is encouraged to provide a supporting letter from its insurance broker, to explain why and to what extent the insurance has ceased to be available at commercially reasonable rates or terms.

Reference: Notice of increased cost or non-availability of Professional Indemnity insurance (clause 6.16)

In accordance with clause 6.16 (Professional Indemnity Insurance: Increased cost and non-availability), we write to notify you that the Professional Indemnity insurance referred to in clause 6.15 (Professional Indemnity Insurance: Obligation to insure) has ceased to be available at commercially reasonable rates or terms.

[We attach a letter from our insurance broker to that effect.]

We propose to arrange a meeting as soon as possible, to discuss the means of best protecting our respective positions in the absence of such insurance. Please confirm your availability.

Resource 221

Description	A tracker template for insurances required under the Contract, and their renewal.
Type	Tracker: Contractor and Employer
Provision references	clause 6.1 to clause 6.16

Below is a tracker template for recording the insurances required under the Contract, and their renewal:

<u>Clause 6.4 (Contractor's insurance of his liability), to be provided by the Contractor</u>

Contractor's Public Liability insurance: injury to persons or property – the required level of cover is not less than [*insert amount*] for any one occurrence or series of occurrences arising out of one event.

Renewal date: [*insert date*]

<u>Clause 6.5 (Contractor's insurance of liability of Employer), to be provided by the Contractor</u>

Insurance – liability of Employer: the minimum amount of indemnity for any one occurrence or series of occurrences arising out of one event: [*insert amount*]

Renewal date: [*insert date*]

<u>Clause 6.7 (Insurance of the Works and Existing Structures: Insurance Options and period), to be provided by either the Contractor (where Insurance Option A applies) or the Employer (where Insurance Option B or Insurance Option C applies)</u>

Works insurance: Insurance Option applicable: [Insurance Option A] / [Insurance Option B] / [Insurance Option C]

Percentage to cover professional fees: [*insert percentage*]

Where Insurance Option A applies and cover is to be provided under the Contractor's annual policy (paragraph A.2), the annual renewal date is: [*insert date*]

Where Insurance Option C applies, paragraph C.1: [applies] / [is replaced by the provisions of the following C.1 Replacement Schedule: [*insert reference*]]

<u>Clause 6.10 (Terrorism Cover – policy extensions and premiums), to be provided by either the Contractor (where Insurance Option A applies) or the Employer (where Insurance Option B or Insurance Option C applies)</u>

Terrorism Cover: details of the required cover are set out in the following documents: [*insert reference*]

<u>Clause 6.15 (Professional Indemnity Insurance: Obligation to ensure), to be provided by the Contractor</u>

Level of cover: the amount of indemnity required [is for any one claim or series of claims arising out of one event] / [is the aggregate amount for any one period of insurance] and is: [*insert amount*].

Sub-limits within the overall level of cover: [*insert details of sub-limits*]

Specific exclusions listed in the relevant schedule(s) (or other policy document(s)) to the relevant policy: [*insert details of specific exclusions*]

Expiry of the required period of Professional Indemnity insurance is: [6 years] / [12 years] / [[*insert number*] years (not exceeding 12 years)]

Document link: [*insert link to documents*]

General

- Calendar reminder for renewal dates.

Resource 222

Description	The provision of a notice from the insurers of the Remedial Measures required, as a result of a breach of the Joint Fire Code.
Type	Correspondence from insurers: Employer / Contractor
Provision references	clause 6.19.1

Reference: Breach of Joint Fire Code; Remedial Measures (clause 6.19.1)

We write further to the breach of the Joint Fire Code, which occurred on [*insert date*].

The insurers under the Works Insurance Policy have specified by notice to us the Remedial Measures they require as a result of this breach.

In accordance with clause 6.19.1 (Breach of Joint Fire Code – Remedial Measures), we attach a copy of this notice.

[*Contractor*: We will ensure that the Remedial Measures are carried out.] /

[*Employer*: Please ensure that the Remedial Measures are carried out. In accordance with clause 6.19.2, if within 7 days of receipt of this notice, you do not begin to carry out or thereafter fail without reasonable cause regularly and diligently to proceed with the Remedial Measures, then the Employer may employ and pay other persons to carry out those Remedial Measures. You will be liable for all additional costs incurred by the Employer in connection with such employment, and an appropriate deduction will be made from the Contract Sum.]

Resource 223

Description	A reply from the Contractor to the Employer following receipt of the Remedial Measures required by the insurers, as a result of a breach of the Joint Fire Code.
Type	Reply: Contractor to Employer
Provision references	clause 6.19.1
Drafting notes	**Good practice** Clause 6.19.2 does not require the Contractor to give this reply to the Employer, but it is considered good practice, particularly in the context of a recent breach of the Joint Fire Code. **Timing:** Clause 6.19.2 requires the Contractor to begin to carry out and regularly and diligently proceed with the Remedial Measures within 7 days of receipt of the notice under clause 6.19.1 specifying those Remedial Measures. The Contractor is encouraged to begin carrying out the Remedial Works immediately.

Reference: Breach of Joint Fire Code; Remedial Measures (clause 6.19.1)

We write in reply to your correspondence dated [*insert date*], which attached the notice of Remedial Measures from the insurers under the Works Insurance Policy.

We will ensure that the Remedial Measures are carried out regularly and diligently.

We have carried out the following works in relation to the Remedial Measures to date: [*insert description of the extent to which the Contractor has carried out the Remedial Measures.*]

Please let us know if there is any further information we can provide in this regard.

Resource 224

Description	Confirmation from the Employer to the Contractor that the Employer will employ and pay others to carry out the Remedial Measures, following the Contractor's failure to proceed with the Remedial Measures.
Type	Notice: Employer to Contractor
Provision references	clause 6.19.2

Reference: Breach of Joint Fire Code; Remedial Measures (clause 6.19.2)

We write further to [our] / [your] correspondence dated [*insert date*], which attached the notice of Remedial Measures from the insurers under the Works Insurance Policy, in accordance with clause 6.19.1 (Breach of Joint Fire Code – Remedial Measures).

You have [not begun to carry out] / [failed without reasonable cause regularly and diligently to proceed with] the Remedial Measures, within 7 days of receipt of that notice of Remedial Measures.

Therefore, in accordance with clause 6.19.2, the Employer will employ and pay other persons to carry out those Remedial Measures. You will be liable for all additional costs incurred by the Employer in connection with such employment, and an appropriate deduction will be made from the Contract Sum.

Resource 225

Description	Confirmation from the Employer to the Contractor that the Employer has employed and paid others to carry out the Remedial Measures, following the Contractor's failure to proceed with the Remedial Measures.
Type	Notice: Employer to Contractor
Provision references	clause 6.19.2
Drafting notes	**Good practice** Clause 6.19.2 does not require the Employer to give this notice, but it is considered good practice to notify the Contractor of the costs incurred by the Employer.

Reference: Breach of Joint Fire Code; Remedial Measures (clause 6.19.2)

We write further to the following correspondence, given in accordance with clause 6.19 (Breach of Joint Fire Code – Remedial Measures):

- our correspondence dated [*insert date*], which attached the notice of Remedial Measures from the insurers under the Works Insurance Policy; and
- our correspondence dated [*insert date*], where we advised you that, given that you [did not begin to carry out] / [failed without reasonable cause regularly and diligently to proceed with] the Remedial Measures, the Employer would employ and pay other persons to carry out those Remedial Measures.

The Employer has now employed and paid other persons to carry out those Remedial Measures, and has incurred a total cost of [*insert amount*] (exclusive of VAT) in connection with such employment, particulars of which are attached.

A corresponding deduction will therefore be made from the Contract Sum.

Resource 226

Description	A request for consent from the other Party to assign the Contract or any rights thereunder.
Type	Consent: Contractor or Employer to the other Party
Provision references	clause 7.1

Reference: Consent to assign the Contract or any rights thereunder (clause 7.1)

In accordance with clause 7.1 (Assignment: General), we request your consent to [assign the Contract.] / [assign the following rights under the Contract: [*insert details of the proposed rights to be assigned.*]]

We propose to assign to: [*insert details of the assignee.*]

Please give your consent by [*insert date*].

Resource 227

Description	Consent given to assign the Contract or any rights thereunder.
Type	Consent given: Contractor or Employer to the other Party
Provision references	clause 7.1
Drafting notes	In accordance with clause 1.10 (Consents and approvals), the giving of consent under this clause 7.1 is at the 'sole discretion' from the Party from whom it is not. In other words, the general requirement that the provision of such consent should not be unreasonably delayed or withheld does not apply in this instance.

Reference: Consent to assign the Contract or any rights thereunder (clause 7.1)

We write in reply to your request dated [*insert date*], made in accordance with clause 7.1 (Assignment: General).

We give our consent for you to [assign the Contract] / [assign the rights under the Contract as stated in your request] to the assignee stated in your request.

Resource 228

Description	Consent withheld to assign the Contract or any rights thereunder.
Type	Consent withheld: Contractor or Employer to the other Party
Provision references	clause 7.1
Drafting notes	In accordance with clause 1.10 (Consents and approvals), the giving of consent under this clause 7.1 is at the 'sole discretion' from the Party from whom it is not. In other words, the general requirement that the provision of such consent should not be unreasonably delayed or withheld does not apply in this instance.

Reference: Consent to assign the Contract or any rights thereunder (clause 7.1)

We write in reply to your request dated [*insert date*], made in accordance with clause 7.1 (Assignment: General).

We withhold our consent for you to [assign this Contract] / [assign the rights under this Contract as stated in your request] to the assignee stated in your request [or otherwise].

Resource 229

Description	A request from the Contractor for the Employer's consent to provide a performance bond from a particular surety.
Type	Request: Contractor to Employer
Provision references	clause 7.3.1
Drafting notes	Footnote [25] of the Contract states that if a performance bond is required, the identity of the surety, as well as the operative terms of the bond, should be agreed prior to execution of the contract. This request should therefore only be required if this has not yet been agreed.

Reference: Performance Bond; approval of surety (clause 7.3.1)

In accordance with clause 7.3.1 (Performance Bonds and Guarantees), we request your approval to provide a performance bond or guarantee of our due performance of the Contract (in an amount equal to the percentage of the Contract Sum and for the period stated in the Contract Particulars) from the following surety: [*insert details of the proposed surety.*]

Please give your approval by [*insert date*].

Resource 230

Description	Approval given from the Employer for the Contractor to provide a performance bond from a particular surety.
Type	Approval given: Employer to Contractor
Provision references	clause 7.3.1

Reference: Performance Bond; approval of surety (clause 7.3.1)

We write in reply to your request dated [*insert date*], made in accordance with clause 7.3.1 (Performance Bonds and Guarantees).

We give our approval for you to provide the performance bond or guarantee from [*insert name of surety*] as surety.

Resource 231

Description	Approval withheld from the Employer for the Contractor to provide a performance bond from a particular surety.
Type	Approval withheld: Employer to Contractor
Provision references	clause 7.3.1
Drafting notes	Clause 1.10 requires that the Employer's approval must not be unreasonably delayed or withheld.

Reference: Performance Bond; approval of surety (clause 7.3.1)

We write in reply to your request dated [*insert date*], made in accordance with clause 7.3.1 (Performance Bonds and Guarantees).

We withhold our approval for you to provide the performance bond or guarantee from [*insert name of surety*] as surety, for the following reasons: [*insert reasons, to demonstrate why the approval was not unreasonably delayed or withheld.*]

Please propose an alternative surety for our approval, taking the reasons above into consideration.

Resource 232

Description	A request by the Contractor to the Employer to amend the form of performance bond or parent company guarantee.
Type	Request: Contractor to Employer
Provision references	clause 7.3
Drafting notes	Clause 7.3 requires that the performance bond or parent company guarantee should be 'substantially' in the form of the document identified by the Contract Particulars. This request can therefore be made where the Contractor wishes to make a more substantial amendment.

Reference: Request to amend form of [performance bond] / [parent company guarantee] (clause 7.3)

We write in relation to our obligation under clause 7.3 (Performance Bonds and Guarantees) to provide [a performance bond or guarantee of our due performance of the Contract from a bank or other surety approved by the Employer in an amount equal to the percentage of the Contract Sum and for the period stated in the Contract Particulars] / [a guarantee by our parent company identified in the Contract Particulars, namely [*insert name of parent company*]].

We request that such [performance bond] / [guarantee] be in the attached form, instead of the form identified by the Contract Particulars, namely [*insert reference*].

The reasons for the proposed amendment are as follows: [*insert reasons for the proposed amendment of the form of performance bond or parent company guarantee.*]

Please confirm your agreement by [*insert date*].

Resource 233

Description	An agreement by the Employer for the Contractor to change the proposed form of performance bond or parent company guarantee.
Type	Agreement: Employer to Contractor
Provision references	clause 7.3

Reference: Request to amend form of [performance bond] / [parent company guarantee] (clause 7.3)

We write in reply to your request dated [*insert date*], to change the proposed form of [performance bond] / [parent company guarantee] required under clause 7.3 (Performance Bonds and Guarantees).

We agree to you providing the [performance bond] / [parent company guarantee] in the form attached with your request.

Resource 234

Description	A refusal by the Employer for the Contractor to change the proposed form of performance bond or parent company guarantee.
Type	No agreement: Employer to Contractor
Provision references	clause 7.3
Drafting notes	Clause 1.10 requires that the Employer's consent must not be unreasonably delayed or withheld. It is not clear whether it applies in this instance, given that it is not a 'consent' or 'approval', but the Employer is encouraged to provide its reasons in any event.

Reference: Request to amend form of [performance bond] / [parent company guarantee] (clause 7.3)

We write in reply to your request dated [*insert date*], to change the proposed form of [performance bond] / [parent company guarantee] required under clause 7.3 (Performance Bonds and Guarantees).

We do not agree to you providing the [performance bond] / [parent company guarantee] in the form attached with your request, for the following reasons: [*insert reasons, to demonstrate why the consent was not unreasonably delayed or withheld.*]

Please provide the [performance bond] / [parent company guarantee] substantially in the form identified by the Contract Particulars, namely [*insert reference*].

Resource 235

Description	A tracker template for recording performance bonds and parent company guarantees, across multiple building contracts.
Type	Tracker: Contractor
Provision references	clause 7.3

Below is a tracker template to help the Contractor record the performance bonds and parent company guarantees it has provided under separate building contracts.

Performance Bonds

- Reference: [*insert reference*]
- Date: [*insert date*]
- Building contract: [*insert reference to applicable building contract*]
- Surety: [*insert name of surety*]
- Beneficiary: [*insert beneficiary, for example, the Employer*]
- Bond amount: [*insert amount*]
- Bond expiry: [*insert expiry*]

Parent Company Guarantees

- Reference: [*insert reference*]
- Date: [*insert date*]
- Building contract: [*insert reference to applicable building contract*]
- Guarantor: [*insert parent company providing the guarantee*]
- Beneficiary: [*insert beneficiary, for example, the Employer*]
- Expiry: [*insert expiry*]

Document link: [*insert link to documents*]

Resource 236

Description	A notice from the Employer to the Contractor of the vesting of P&T Rights as third party rights in a Purchaser or Tenant.
Type	Notice: Employer to Contractor
Provision references	clause 7.7.1
Drafting notes	Notices under this clause must be given in accordance with clause 1.7.4 (Notices and other communications). If paragraph 7.2 Part 1 of Schedule 5 (Third Party Rights for Purchasers and Tenants) applies, include an email address for the Purchaser or Tenant.

Reference: Rights for Purchasers and Tenants (clause 7.7.1)

In accordance with clause 7.7.1 (Rights for Purchasers and Tenants), and given that the Rights Particulars state that you will confer P&T Rights on a Purchaser or Tenant as third party rights, we write to notify you that P&T Rights will vest as third party rights in the following Purchaser or Tenant on the date of receipt of this notice:

- Name of Purchaser or Tenant: [*insert name*]
- Address: [*insert name*]
- [Registered company number: [*insert company number*]]
- Nature of their interest in the Works: [*insert the nature of the Purchaser or Tenant's interest in the Works.*]
- [Email address: [*insert email address*]]

Resource 237

Description	A request from the Contractor to a Purchaser or Tenant to amend or vary clause 7.7 or Part 1 of Schedule 5, relating to P&T Rights.
Type	Request: Contractor to Purchaser or Tenant
Provision references	clause 7.7.2

Reference: Rights for Purchasers and Tenants (clause 7.7.2)

We write in relation to the building contract (the 'Contract') dated [*insert date*] made between [*insert name of Contractor*] (the 'Contractor') and [*insert name of Employer*] (the 'Employer').

In accordance with clause 7.7 (Rights for Purchasers and Tenants) of the Contract, P&T Rights have vested in you as third party rights from [*insert date*], in your capacity as a [Purchaser] / [Tenant].

In accordance with clause 7.7.2, we request your consent to amend or vary the express provisions of clause 7.7 and/or Part 1 of Schedule 5 (Third Party Rights for Purchasers and Tenants) as follows:

[*Insert proposed amendments to clause 7.7 and/or Part 1 of Schedule 5.*]

We make this request for the following reasons: [*insert reasons.*]

Please give your consent by [*insert date*].

Resource 238

Description	A notice from the Employer to the Contractor of the vesting of Funder Rights as third party rights in a Funder.
Type	Notice: Employer to Contractor
Provision references	clause 7.8.1
Drafting notes	Notices under this clause must be given in accordance with clause 1.7.4 (Notices and other communications). If paragraph 11.2 Part 2 of Schedule 5 (Third Party Rights for a Funder) applies, include an email address for the Funder.

Reference: Rights for a Funder (clause 7.8.1)

In accordance with clause 7.8.1 (Rights for a Funder), and given that the Rights Particulars state that you will confer Funder Rights on a Funder as third party rights, we write to notify you that Funder Rights will vest as third party rights in the following Funder on the date of receipt of this notice:

- Name: [*insert name*]
- Address: [*insert name*]
- Registered company number: [*insert company number*]
- [Email address: [*insert email address*]]

Resource 239

Description	A request from the Contractor to the Funder to amend or vary clause 7.8 or Part 2 of Schedule 5, relating to Funder Rights.
Type	Request: Contractor to Funder
Provision references	clause 7.8.2

Reference: Rights for a Funder (clause 7.8.2)

We write in relation to the building contract (the ‘Contract’) dated [*insert date*] made between [*insert name of Contractor*] (the ‘Contractor’) and [*insert name of Employer*] (the ‘Employer’).

In accordance with clause 7.8 (Rights for a Funder) of the Contract, Funder Rights have vested in you as third party rights from [*insert date*].

In accordance with clause 7.8.2.1, we request your consent to amend or vary the express provisions of clause 7.8 and/or Part 2 of Schedule 5 (Third Party Rights for a Funder) as follows:

[*Insert proposed amendments to clause 7.8 and/or Part 2 of Schedule 5.*]

We make this request for the following reasons: [*insert reasons.*]

Please give your consent by [*insert date*].

Resource 240

Description	A notice from the Employer to the Contractor to enter into a Collateral Warranty with a Purchaser or Tenant.
Type	Notice: Employer to Contractor
Provision references	clause 7.9
Drafting notes	Notices under this clause must be given in accordance with clause 1.7.4 (Notices and other communications).

Reference: Contractor's Warranties – Purchasers and Tenants (clause 7.9)

In accordance with clause 7.9 (Contractor's Warranties – Purchasers and Tenants), and given that the Rights Particulars state that you will confer P&T Rights on a Purchaser or Tenant by way of collateral warranty, we write to notify you that we require you within 14 days from receipt of this notice to enter into a Collateral Warranty in the form CWa/P&T, completed in accordance with the relevant Rights Particulars, with the following Purchaser or Tenant:

- Name of Purchaser or Tenant: [*insert name, address, registered company number, etc.*]
- Nature of their interest in the Works: [*insert the nature of their interest in the Works.*]

Resource 241

Description	A notice from the Employer to the Contractor to enter into a Collateral Warranty with a Funder.
Type	Notice: Employer to Contractor
Provision references	clause 7.10
Drafting notes	Notices under this clause must be given in accordance with clause 1.7.4 (Notices and other communications).

Reference: Contractor's Warranty – Funder (clause 7.10)

In accordance with clause 7.10 (Contractor's Warranty – Funder), and given that the Rights Particulars state that you will confer Funder Rights on a Funder by way of collateral warranty, we write to notify you that we require you within 14 days from receipt of this notice to enter into a Collateral Warranty in the form CWa/F, completed in accordance with the relevant Rights Particulars, with the following Funder: [*insert name, address, registered company number, etc.*]

Resource 242

Description	A tracker template to record P&T Rights and Funder Rights, by way of third party rights or collateral warranty.
Type	Tracker: Contractor or Employer
Provision references	clause 7.7 to clause 7.10

Below is a tracker template to record those parties to whom the Contractor has conferred P&T Rights or Funder Rights by way of third party rights or collateral warranty:

P&T Rights: [Third party rights] / [Collateral warranty]

- Purchaser / Tenant: [*insert name, address, registered company number, etc.*]
- Nature of their interest in the Works: [*insert the nature of their interest in the Works.*]
- Date third party rights vested / date of collateral warranty: [*insert date*]

Funder Rights: [Third party rights] / [Collateral warranty]

- Funder: [*insert name, address, registered company number, etc.*]
- Date third party rights vested / date of collateral warranty: [*insert date*]

Document link: [*insert link to documents*]

Resource 243

Description	A notice from the Employer to the Contractor requesting third party rights or collateral warranties from a sub-contractor.
Type	Notice: Employer to Contractor
Provision references	clause 7.11.1.1
Drafting notes	Notices under this clause must be given in accordance with clause 1.7.4 (Notices and other communications).

Reference: Third Party Rights and Collateral Warranties from Sub-Contractors (clause 7.11.1)

In accordance with clause 7.11.1.1 (Third Party Rights and Collateral Warranties from Sub-Contractors), we write to notify you that we require the following:

- Type of right or warranty: [Third party rights] / [Collateral Warranty]
- From the following sub-contractor: [*insert name, address, registered company number, etc.*]
- In favour of the following beneficiary: [[*insert name of Purchaser*], a Purchaser] / [[*insert name of Tenant*], a Tenant] / [[*insert name of Funder*], a Funder] / [the Employer]

[*Third party rights*: Please promptly give notice to the sub-contractor of this request.] /

[*Collateral Warranty*: In accordance with clause 7.11.1.2, within 21 days of receipt of this notice, please take such steps as are required to obtain the Collateral Warranty, and promptly forward the executed document to the Employer, having also executed and delivered the document.]

Resource 244

Description	Confirmation or provision by the Contractor to the Employer of third party rights or a collateral warranty from a sub-contractor.
Type	Correspondence: Contractor to Employer
Provision references	clause 7.11.1.2
Drafting notes	**Timing:** In respect of third party rights, clause 7.11.1.1 requires the Contractor to 'promptly' notify the sub-contractor. **Timing:** In respect of Collateral Warranties, clause 7.11.1.2 requires the Contractor to obtain and forward the executed and delivered document from the sub-contractor to the Employer within 21 days from receipt of the Employer's notice.

Reference: Third Party Rights and Collateral Warranties from Sub-Contractors (clause 7.11.1)

We write in reply to your notice dated [*insert date*], given in accordance with clause 7.11.1 (Third Party Rights and Collateral Warranties from Sub-Contractors).

[*Third party rights*: We confirm that we notified the sub-contractor on [*insert date*] under clause [*insert reference*] of their sub-contract that third party rights have vested in the beneficiary identified in your notice.] /

[*Collateral Warranty*: We enclose a copy of the Collateral Warranty, executed and delivered by the sub-contractor, and by us.]

Resource 245

Description	Approval given/withheld to the sub-contractor to amend the form of third party rights or collateral warranty.
Type	Approval: Contractor or Employer to sub-contractor
Provision references	clause 7.11.2
Drafting notes	Clause 1.10 requires that the Employer's or Contractor's approval must not be unreasonably delayed or withheld.

Reference: Third Party Rights and Collateral Warranties from Sub-Contractors (clause 7.11.2)

We write in relation to the following contracts:

- the building contract (the 'Contract') dated [*insert date*] made between [*insert name of Contractor*] (the 'Contractor') and [*insert name of Employer*] (the 'Employer'); and
- the sub-contract (the 'Sub-contract') dated [*insert date*] made between the Contractor and [*insert name of Sub-Contractor*] (the 'Sub-Contractor').

We write in reply to your request dated [*insert date*], for our approval to amend the form of [third party rights] / [collateral warranty], as set out in that request.

In accordance with clause 7.11.2 (Third Party Rights and Collateral Warranties from Sub-Contractors) of the Contract, [we give our approval for this request.] / [we withhold our approval for this request, for the following reasons: [*insert reasons, to demonstrate why the approval was not unreasonably delayed or withheld.*]]

Resource 246

Description	A request from the Contractor to a beneficiary to amend, vary, or rescind the sub-contract.
Type	Request: Contractor to beneficiary
Provision references	clause 7.11.3
Drafting notes	Clause 7.11 assumes that the sub-contract will be based on the JCT Design and Build Sub-Contract, and makes reference to clause 2.26, clause 2.27, and Schedule 6 (Third Party Rights) of the JCT Design and Build Sub-Contract Conditions. Clause 7.11.3.2 only applies where the beneficiary is the Employer or a Funder.

Reference: Third Party Rights and Collateral Warranties from Sub-Contractors (clause 7.11.3)

We write in relation to the following contracts:

- the building contract (the 'Contract') dated [*insert date*] made between [*insert name of Contractor*] (the 'Contractor') and [*insert name of Employer*] (the 'Employer'); and
- the sub-contract (the 'Sub-Contract') dated [*insert date*] made between the Contractor and [*insert name of Sub-Contractor*] (the 'Sub-Contractor').

In accordance with clause 7.11 (Third Party Rights and Collateral Warranties from Sub-Contractors) of the Contract, third party rights have vested in you from [*insert date*], in your capacity as [a Purchaser] / [a Tenant] / [a Funder] / [the Employer].

[*All beneficiaries*: In accordance with clause 7.11.3, we request your consent for us to agree to amend or vary certain provisions of the Sub-Contract as follows: [*insert proposed amendments to the Sub-Contract, with references to the relevant provisions.*] /

[*Employer and Funder only*: In accordance with clause 7.11.3, we request your consent for us to rescind the Sub-Contract.]

We make this request [on the Sub-Contractor's behalf] for the following reasons: [*insert reasons.*]

Please give your consent by [*insert date*].

Resource 247

Description	Consent given/withheld by the Employer to the Contractor to amend the sub-contract.
Type	Consent: Employer to Contractor
Provision references	clause 7.11.3
Drafting notes	Clause 1.10 requires that the Employer's consent must not be unreasonably delayed or withheld.

Reference: Third Party Rights and Collateral Warranties from Sub-Contractors (clause 7.11.3)

We write in reply to your request dated [*insert date*], made in accordance with clause 7.11.3 (Third Party Rights and Collateral Warranties from Sub-Contractors), for our consent for you to agree to [amend or vary certain provisions of the Sub-Contract as stated in your request.] / [rescind the Sub-Contract.]

[We give our consent to this request.] /

[We withhold our consent to this request, for the following reasons: [*insert reasons, to demonstrate why the consent was not unreasonably delayed or withheld.*]]

Resource 248

Description	A tracker template to record third party rights and collateral warranties from sub-contractors.
Type	Tracker: Contractor and Employer
Provision references	clause 7.11

Below is a tracker template to record those sub-contractors who have conferred third party rights or executed and delivered a Collateral Warranty in favour of Purchasers, Tenants, Funders, or the Employer:

- Sub-contractor: [*insert name, address, registered company number, etc.*]
- Description of the sub-contract works: [*insert brief description of the sub-contract works*]
- Type of rights: [Third party rights] / [Collateral warranty]
- Date third party rights vested / date of collateral warranty: [*insert date*]

- Beneficiary: [*insert name, address, registered company number, etc. of the Purchaser, Tenant, Funder, or the Employer*]
- Date Collateral Warranty provided to Employer: [*insert date*]
- Document link: [*insert link to documents*]
- Status / action required: [*insert status, and any action required*]

Resource 249

Description	Confirmation of the reinstatement of the Contractor's employment under the Contract.
Type	Correspondence: Contractor or Employer to the other Party
Provision references	clause 8.3.2
Drafting notes	**Good practice** This confirmation is not required by clause 8.3.2, but it is considered good practice to record the date and terms of the reinstatement of the Contractor's employment under the Contract.

Reference: Reinstatement of Contractor's employment (clause 8.3.2)

In accordance with clause 8.3.2 (Other rights, reinstatement), and following the termination of [*Contractor*: our] / [*Employer*: your] employment under the Contract on [*insert date*], we write to confirm that [*Contractor*: our] / [*Employer*: your] employment was reinstated on [*insert date*], on the following terms: [*insert terms*].

Resource 250

Description	A notice from the Employer to the Contractor of a specified default.
Type	Notice: Employer to Contractor
Provision references	clause 8.4.1
Drafting notes	Notices under this clause must be given in accordance with clause 1.7.4 (Notices and other communications).

Reference: Termination; notice of specified default (clause 8.4.1)

In accordance with clause 8.4.1 (Termination by Employer: Default by Contractor), we write to notify you of the following specified default:

[*Insert the relevant specified default:*

- *Without reasonable cause, you have wholly or substantially suspended the carrying out of the Works. /*
- *You have failed to proceed regularly and diligently with the performance of your obligations under the Contract. /*

- *You have refused or neglected to comply with a notice or instruction from the Employer requiring you to remove any work, materials or goods not in accordance with the Contract, and by such refusal or neglect the Works have been materially affected. /*
- *You have failed to comply with clause 3.3 (Consent to sub-contracting). /*
- *You have failed to comply with clause 7.1 (Assignment: General). /*
- *You have failed to comply with clause 3.16 (CDM Regulations and Part 2A of the Building Regulations).*]

The details of the specified default are as follows: [*insert details of the specified default, including all relevant dates, parties involved, actions, correspondence, etc.*]

In accordance with clause 8.4.2, if this specified default continues for 14 days from receipt of this notice, we may on, or within 21 days from, the expiry of that 14-day period by a further notice to you terminate your employment under the Contract.

Resource 251

Description	A reply from the Contractor to the Employer's notice of a specified default.
Type	Reply: Contractor to Employer
Provision references	clause 8.4.1
Drafting notes	**Good practice** This reply is not required by clause 8.4.1, but it is considered good practice for the Contractor to promptly notify the Employer if the specified default is no longer continuing.

Reference: Termination; notice of specified default (clause 8.4.1)

We write in reply to your notice dated [*insert date*], given in accordance with clause 8.4.1 (Termination by Employer: Default by Contractor).

We can confirm that the specified default is no longer continuing, and we have promptly undertaken the following in that regard: [*insert steps taken to address the specified default.*]

[We attach the following documents to that effect: [*insert description of documents showing that the specified default is no longer continuing.*]]

Please let us know if there is anything further we can provide on this.

Resource 252

Description	A notice of termination from the Employer to the Contractor, following the continuation of a specified default.
Type	Notice: Employer to Contractor
Provision references	clause 8.4.2
Drafting notes	Notices under this clause must be given in accordance with clause 1.7.4 (Notices and other communications). **Timing:** If the Employer wishes to give this notice, clause 8.4.2 requires them to give it both (i) after the 14-day period under clause 8.4.2 has expired, and (ii) within 21 days from the expiry of that 14-day period.

Reference: Notice of termination (clause 8.4.2)

We write further to our notice dated [*insert date*], given in accordance with clause 8.4.1 (Termination by Employer: Default by Contractor).

The specified default has continued for 14 days from receipt of that notice.

Therefore, in accordance with clause 8.4.2, we write to notify you that your employment under the Contract is terminated.

Resource 253

Description	A notice of termination from the Employer to the Contractor, following the repetition of a specified default.
Type	Notice: Employer to Contractor
Provision references	clause 8.4.3
Drafting notes	Notices under this clause must be given in accordance with clause 1.7.4 (Notices and other communications). **Timing:** If the Employer wishes to give this notice, clause 8.4.3 requires them to do so within 'a reasonable time' after the repetition of a specified default. It is therefore recommended to give the notice within 7 days.

Reference: Notice of termination (clause 8.4.3)

We write further to our notice dated [*insert date*], given in accordance with clause 8.4.1 (Termination by Employer: Default by Contractor).

You have subsequently repeated the specified default, the details of which are as follows: [*insert details of the repeated specified default, including all relevant dates, parties involved, actions, correspondence, etc.*]

Therefore, in accordance with clause 8.4.3, we write to notify you that your employment under the Contract is terminated.

Resource 254

Description	A notice of termination from the Employer to the Contractor, following the Contractor becoming Insolvent.
Type	Notice: Employer to Contractor
Provision references	clause 8.5.1
Drafting notes	Notices under this clause must be given in accordance with clause 1.7.4 (Notices and other communications). Clause 8.5.1 permits the Employer to give this notice 'at any time', but the Employer is encouraged to give it as soon as it has received confirmation that the Contractor is Insolvent.

Reference: Notice of termination; Insolvency of Contractor (clause 8.5.1)

In accordance with clause 8.5.1 (Termination by Employer: Insolvency of Contractor), and given that you are Insolvent in accordance with clause 8.1 (Meaning of insolvency), we write to notify you that your employment under the Contract is terminated.

In accordance with clause 8.5.3.2, your obligations under Article 1 (Contractor's obligations) and the Conditions to carry out and complete the Works are suspended.

Resource 255

Description	A notice from the Contractor to the Employer relating to the Contractor's insolvency.
Type	Notice: Contractor to Employer
Provision references	clause 8.5.2
Drafting notes	Notices under this clause must be given in accordance with clause 1.7.4 (Notices and other communications). **Timing:** Clause 8.5.2 requires the Contractor to give this notice 'immediately' if it makes any such proposal, gives notice of any meeting, or becomes the subject of any such proceedings or appointment.

Reference: Insolvency of Contractor (clause 8.5.2)

In accordance with clause 8.5.2 (Termination by Employer: Insolvency of Contractor), we write to notify you that we have [made the following proposal] / [given notice of the following meeting] / [become the subject of the following proceedings or an appointment] relating to the matters referred to in clause 8.1 (Meaning of insolvency):

[*Insert details of the proposal, meeting, proceedings or appointment as applicable under clause 8.1.*]

Resource 256

Description	A notice of termination from the Employer to the Contractor, relating to corruption and regulation 73(1)(b) of the PC Regulations.
Type	Notice: Employer to Contractor
Provision references	clause 8.6
Drafting notes	Notices under this clause must be given in accordance with clause 1.7.4 (Notices and other communications). If the Employer wishes to give a notice under this clause 8.6, it is encouraged to do so as soon as possible after any of the events included in clause 8.6 have occurred.

Reference: Notice of termination; Corruption and regulation 73(1)(b) of the PC Regulations (clause 8.6)

We write in accordance with clause 8.6 (Termination by Employer: Corruption and regulation 73(1)(b) of the PC Regulations).

[You (or a person employed by you, or acting on your behalf) have committed an offence under the Bribery Act 2010.] /

[*Where the Employer is a Local or Public Authority*: You (or a person employed by you, or acting on your behalf) have given a fee or reward, the receipt of which is an offence under sub-section (2) of section 117 of the Local Government Act 1972.] /

[*Where the Contract is one to which regulation 73(1) of the PC Regulations applies*: The circumstances set out in regulation 73(1)(b) of the PC Regulations apply.]

The particulars are as follows: [*insert details of the offence or circumstances.*]

Therefore, in accordance with clause 8.6, we write to notify you that your employment under the Contract is terminated.

Resource 257

Description	A requirement by the Employer for the Contractor to remove from the Works temporary buildings, plant, tools, equipment, goods, and materials belonging to the Contractor.
Type	Requirement: Employer to Contractor
Provision references	clause 8.7.2.1

Reference: Consequences of termination under clauses 8.4 to 8.6 (clause 8.7.2.1)

In accordance with clause 8.7.2.1 (Termination by Employer: Consequences of termination under clauses 8.4 to 8.6), we require you to remove (or procure the removal) from the Works of the following temporary buildings, plant, tools, equipment, goods, and/or materials belonging to you or any Contractor's Persons: [*insert details of what is to be removed.*]

Please ensure that these are removed by no later than [*insert date*].

Resource 258

Description	The provision by the Contractor to the Employer of all Contractor's Design Documents then prepared.
Type	Correspondence: Contractor to Employer
Provision references	clause 8.7.2.2

Reference: Consequences of termination under clauses 8.4 to 8.6 (clause 8.7.2.2)

In accordance with clause 8.7.2.2 (Termination by Employer: Consequences of termination under clauses 8.4 to 8.6), we are required to provide the Employer with copies of all Contractor's Design Documents then prepared, whether or not previously provided.

All Contractor's Design Documents are available [to download at the following link: [*insert link*]] / [on our document management platform, to which you already have access].

Please let us know if there is any further assistance we can provide in this regard.

Resource 259

Description	A requirement by the Employer for the Contractor to assign agreements for the supply of materials or goods, and/or for the execution of any work, for the purposes of the Contract.
Type	Requirement: Employer to Contractor
Provision references	clause 8.7.2.3
Drafting notes	**Timing:** If this is required by the Employer, clause 8.7.2.3 requires them to make this request within 14 days of the date of termination. Footnote [58] of the Contract states that clause 8.7.2.3 may not be effectual, in cases of the Contractor's insolvency.

Reference: Consequences of termination under clauses 8.4 to 8.6 (clause 8.7.2.3)

In accordance with clause 8.7.2.3 (Termination by Employer: Consequences of termination under clauses 8.4 to 8.6), we require you to assign to the Employer (without charge) the benefit of the following agreements for the supply of materials or goods and/or for the execution of work for the purposes of the Contract: [*insert details of the agreements to be assigned.*]

Resource 260

Description	A notice from the Employer to the Contractor of the completion of the Works and the making good of defects in them
Type	Notice: Employer to Contractor
Provision references	clause 8.7.4
Drafting notes	Notices under this clause must be given in accordance with clause 1.7.4 (Notices and other communications). **Timing:** Clause 8.7.4 requires the Employer to give this notice 'forthwith' upon completion of the Works and the making good of defects in them (or of instructions otherwise, as referred to in clause 2.35).

Reference: Completion of the Works (clause 8.7.4)

In accordance with clause 8.7.4 (Consequences of termination under clauses 8.4 to 8.6), we write to notify you that completion of the Works and the making good of defects in them (or of instructions otherwise, as referred to in clause 2.35 (Schedules of defects and instructions)) took place on [*insert date*].

Resource 261

Description	A notice from the Employer to the Contractor of the Employer's decision not to complete the Works.
Type	Notice: Employer to Contractor
Provision references	clause 8.8.1
Drafting notes	Notices under this clause must be given in accordance with clause 1.7.4 (Notices and other communications). **Timing:** Clause 8.8.1 requires the Employer to notify the Contractor 'forthwith', if within the period of 6 months from the date of termination of the Contractor's employment the Employer decides not to have the Works carried out and completed.

Reference: Notice of Employer's decision not to complete the Works (clause 8.8.1)

In accordance with clause 8.8.1 (Termination by Employer: Employer's decision not to complete the Works), and following the termination of your employment under the Contract on [*insert date*], we write to notify you of our decision not to have the Works carried out and completed.

Resource 262

Description	A notice from the Contractor to the Employer of a specified default.
Type	Notice: Contractor to Employer
Provision references	clause 8.9.1
Drafting notes	Notices under this clause must be given in accordance with clause 1.7.4 (Notices and other communications).

Reference: Termination; notice of specified default (clause 8.9.1)

In accordance with clause 8.9.1 (Termination by Contractor: Default by Employer), we write to notify you of the following specified default:

[*Insert the relevant specified default:*

- *You have not paid by the final date for payment the amount due to us in accordance with clause 4.9 (Interim and final payments – final date and amount), and/ or any VAT properly chargeable on that amount. /*
- *You have failed to comply with clause 7.1 (Assignment: General). /*
- *You have failed to comply with clause 3.16 (CDM Regulations and Part 2A of the Building Regulations).]*

The particulars of the specified default are as follows: [*insert details of the specified default, including all relevant dates, parties involved, actions, correspondence, etc.*]

In accordance with clause 8.9.3, if this specified default continues for 14 days from receipt of this notice, we may on, or within 21 days from, the expiry of that 14-day period by a further notice to you terminate our employment under the Contract.

Resource 263

Description	A notice from the Contractor to the Employer of a specified suspension event.
Type	Notice: Contractor to Employer
Provision references	clause 8.9.2
Drafting notes	Notices under this clause must be given in accordance with clause 1.7.4 (Notices and other communications). The Contractor should also first consider whether clause 7.8.2.2 (Rights for a Funder) applies, and whether any Funder Rights have been vested in a Funder as third party rights. This would make the Contractor's right to terminate its employment under the Contract (or to treat it as repudiated) subject to the provisions of paragraph 6 of Part 2 of Schedule 5 (Funder Rights) (known as 'step-in' rights).

Reference: Termination; notice of specified suspension event (clause 8.9.2)

In accordance with clause 8.9.2 (Termination by Contractor: Default by Employer), we write to notify you of a specified suspension event.

We consider that the carrying out of the whole (or substantially the whole) of the uncompleted Works has been suspended for a continuous period of the length stated in the Contract Particulars (namely [*insert period*]), by reason of an impediment, prevention, or default, whether by act or omission, by the Employer, or an Employer's Person.

We further consider that this was not caused by our negligence or default, nor the negligence or default of any Contractor's Person.

The particulars of the specified suspension event are as follows: [*insert details of the specified suspension event, including all relevant dates, parties involved, actions, correspondence, etc.*]

In accordance with clause 8.9.3, if this specified suspension event continues for 14 days from receipt of this notice, we may on, or within 21 days from, the expiry of that 14-day period by a further notice to you terminate our employment under the Contract.

Resource 264

Description	A reply from the Employer to the Contractor's notice of a specified default or a specified suspension event.
Type	Reply: Employer to Contractor
Provision references	clause 8.9.1, clause 8.9.2
Drafting notes	**Good practice** This reply is not required by clause 8.9.1 or clause 8.9.2, but it is considered good practice for the Employer to promptly notify the Contractor if the specified default or specified suspension event is no longer continuing.

Reference: Reply to notice of [specified default (clause 8.9.1)] / [specified suspension event (clause 8.9.2)]

We write in reply to your notice dated [*insert date*], given in accordance with [*Specified default*: clause 8.9.1] / [*Specified suspension event*: clause 8.9.2] (Termination by Contractor: Default by Employer).

We can confirm that the [specified default] / [specified suspension event] is no longer continuing, and have promptly undertaken the following in that regard: [*insert steps taken to address the specified default or specified suspension event.*]

[We attach the following documents to that effect: [*insert description of documents showing that the specified default or specified suspension event is no longer continuing.*]]

Please let us know if there is anything further we can provide on this.

Resource 265

Description	A notice of termination from the Contractor to the Employer, following the continuation of a specified default or a specified suspension event.
Type	Notice: Contractor to Employer
Provision references	clause 8.9.3
Drafting notes	Notices under this clause must be given in accordance with clause 1.7.4 (Notices and other communications). **Timing:** If the Contractor wishes to give this Notice, clause 8.9.3 requires them to give it both (i) after the 14-day period under clause 8.9.3 has expired, and (ii) within 21 days from the expiry of that 14-day period. The Contractor should also first consider whether clause 7.8.2.2 (Rights for a Funder) applies, and whether any Funder Rights have been vested in a Funder as third party rights. This would make the Contractor's right to terminate its employment under the Contract (or to treat it as repudiated) subject to the provisions of paragraph 6 of Part 2 of Schedule 5 (Funder Rights) (known as 'step-in' rights).

Reference: Notice of termination (clause 8.9.3)

We write further to our notice dated [*insert date*], given in accordance with [*Specified default*: clause 8.9.1] / [*Specified suspension event*: clause 8.9.2] (Termination by Contractor: Default by Employer).

The [specified default] / [specified suspension event] has continued for 14 days from receipt of that notice.

Therefore, in accordance with clause 8.4.2, we write to notify you that our employment under the Contract is terminated.

Resource 266

Description	A notice of termination from the Contractor to the Employer, following the repetition of a specified default or a specified suspension event.
Type	Notice: Contractor to Employer
Provision references	clause 8.9.4
Drafting notes	Notices under this clause must be given in accordance with clause 1.7.4 (Notices and other communications). **Timing:** If the Contractor wishes to give this notice, clause 8.9.4 requires them to do so within 'a reasonable time' after the repetition of a specified default. It is therefore recommended to give the notice within 7 days. The Contractor should also first consider whether clause 7.8.2.2 (Rights for a Funder) applies, and whether any Funder Rights have been vested in a Funder as third party rights. This would make the Contractor's right to terminate its employment under the Contract (or to treat it as repudiated) subject to the provisions of paragraph 6 of Part 2 of Schedule 5 (Funder Rights) (known as 'step-in' rights).

Reference: Notice of termination (clause 8.9.4)

We write further to our notice dated [*insert date*], given in accordance with [*Specified default*: clause 8.9.1] / [*Specified suspension event*: clause 8.9.2] (Termination by Employer: Default by Contractor).

[*Specified default*: You have subsequently repeated the specified default, the particulars of which are as follows: [*insert details of the repeated specified default, including all relevant dates, parties involved, actions, correspondence, etc.*]] /

[*Specified suspension event*: The specified suspension event was repeated for a period of [*insert period*], such that the regular progress of the Works is, or is likely to be, materially affected. The particulars of this repeated specified suspension event are as follows: [*insert details of the repeated specified suspension event, including all relevant dates, parties involved, actions, correspondence, etc.*]]

Therefore, in accordance with clause 8.9.4, we write to notify you that our employment under the Contract is terminated.

Resource 267

Description	A notice of termination from the Contractor to the Employer, following the Employer becoming Insolvent.
Type	Notice: Contractor to Employer
Provision references	clause 8.10.1
Drafting notes	Notices under this clause must be given in accordance with clause 1.7.4 (Notices and other communications). Clause 8.10.1 does not specify by when this notice may be sent, but the Contractor is encouraged to give it as soon as it has received confirmation that the Employer is Insolvent. The Contractor should also first consider whether clause 7.8.2.2 (Rights for a Funder) applies, and whether any Funder Rights have been vested in a Funder as third party rights. This would make the Contractor's right to terminate its employment under the Contract (or to treat it as repudiated) subject to the provisions of paragraph 6 of Part 2 of Schedule 5 (Funder Rights) (known as 'step-in' rights).

Reference: Notice of termination; Insolvency of Employer (clause 8.10.1)

In accordance with clause 8.10.1 (Termination by Contractor: Insolvency of Employer), and given that you are Insolvent in accordance with clause 8.1 (Meaning of insolvency), we write to notify you that our employment under the Contract is terminated.

In accordance with clause 8.10.3, our obligations under Article 1 (Contractor's obligations) and the Conditions to carry out and complete the Works are suspended.

Resource 268

Description	A notice from the Employer to the Contractor relating to the Employer's insolvency.
Type	Notice: Employer to Contractor
Provision references	clause 8.10.2
Drafting notes	Notices under this clause must be given in accordance with clause 1.7.4 (Notices and other communications). **Timing:** Clause 8.10.2 requires the Employer to give this notice 'immediately' if it makes any such proposal, gives notice of any meeting, or becomes the subject of any such proceedings or appointment.

Reference: Insolvency of Employer (clause 8.10.2)

In accordance with clause 8.10.2 (Termination by Contractor: Insolvency of Employer), we write to notify you that we have [made the following proposal] / [given notice of the following meeting] / [become the subject of the following proceedings or an appointment] relating to the matters referred to in clause 8.1 (Meaning of insolvency):

[*Insert details of the proposal, meeting, proceedings, or appointment as applicable under clause 8.1.*]

Resource 269

Description	A notice of a potential termination of the Contractor's employment under the Contract, following a suspension of the carrying out of the uncompleted Works for a continuous period.
Type	Notice: Contractor or Employer to the other Party
Provision references	clause 8.11.1
Drafting notes	Notices under this clause must be given in accordance with clause 1.7.4 (Notices and other communications). **Timing:** If either Party wishes to give this notice, clause 8.11.1 permits them to do so upon the expiry of the relevant period of suspension stated in the Contract Particulars. In accordance with clause 8.11.2, the Contractor cannot give this notice in respect of clause 8.11.1.3 where the loss or damage to the Works was caused by the negligence or default of the Contractor or any Contractor's Person. The Contractor should also first consider whether clause 7.8.2.2 (Rights for a Funder) applies, and whether any Funder Rights have been vested in a Funder as third party rights. This would make the Contractor's right to terminate its employment under the Contract (or to treat it as repudiated) subject to the provisions of paragraph 6 of Part 2 of Schedule 5 (Funder Rights) (known as 'step-in' rights).

Reference: Termination by either Party and regulation 73(1)(a) of the PC Regulations (clause 8.11.1)

We write in accordance with clause 8.11.1 (Termination by either Party and regulation 73(1)(a) of the PC Regulations).

The carrying out of the whole, or substantially the whole, of the uncompleted Works has been suspended for the relevant continuous period of the length stated in the Contract Particulars (namely [*insert period*]) by reason of one or more of the following events:

[*Insert the applicable events:*

- *force majeure;*
- *Employer's instructions under clause 2.13 (Notice of discrepancies etc.), clause 3.9 (Instructions requiring Changes), or clause 3.10 (Postponement of work) issued as a result of the negligence or default of any Statutory Provider;*
- *loss or damage to the Works occasioned by any risk covered by the Works Insurance Policy or by an Excepted Risk;*
- *civil commotion or the use or threat of terrorism and/or the activities of the relevant authorities in dealing with such event or threat;*
- *an epidemic:*
  - *first occurring after the Base Date which affects the execution of the Works; or*
  - *first occurring before the Base Date whose effects change after the Base Date and any such changes affect the execution of the Works;*

  by limiting the availability or use of labour or the availability of persons engaged in providing services for the Works where such labour and/or persons are necessary for the proper carrying out of the Works, or preventing the Contractor from, or delaying the Contractor in, securing such goods or materials or such services as are necessary for the proper carrying out of the Works;

- *the occurrence after the Base Date of any of the following which affects the execution of the Works:*
  - *the passing into law of any statute, statutory instrument or other subordinate legislation, regulation or bye-law, whether to make a new law or change or repeal an existing law;*
  - *the exercise of any statutory power, except to the extent caused or contributed to by any default, whether by act or omission, of the Contractor or any Contractor's Person; or*
  - *the publication of any guidance,*

  by the United Kingdom Government or any of the devolved administrations, or any Local or Public Authority or any equivalent authority governed by public law in any of the devolved administrations, or in the case of guidance by the Construction Leadership Council or its successor;

- *delay in receipt of any permission or approval for the purposes of Development Control Requirements necessary for the Works to be carried out or proceed, which delay the Contractor has taken all practicable steps to avoid or reduce.]*

The particulars of the suspension are as follows: [*insert details of the suspension.*]

We notify you that, unless the suspension ceases within 7 days after the date of receipt of this notice, we may terminate [*Contractor*: our] / [*Employer*: your] employment under the Contract.

Resource 270

Description	A notice of termination of the Contractor's employment under the Contract, following a suspension of the carrying out of the uncompleted Works for a continuous period.
Type	Notice: Contractor or Employer to the other Party
Provision references	clause 8.11.1
Drafting notes	Notices under this clause must be given in accordance with clause 1.7.4 (Notices and other communications). **Timing:** If either Party wishes to give this notice, clause 8.11.1 permits them to do so if the suspension has not ceased after the expiry of the 7-day period referred to in clause 8.11.1. In accordance with clause 8.11.2, the Contractor cannot give this notice in respect of clause 8.11.1.3 where the loss or damage to the Works was caused by the negligence or default of the Contractor or any Contractor's Person.

Reference: Notice of termination (clause 8.11.1)

We write further to our notice dated [*insert date*], given in accordance with clause 8.11.1 (Termination by either Party and regulation 73(1)(a) of the PC Regulations).

The suspension has not ceased within 7 days after the date of receipt of that notice.

Therefore, in accordance with clause 8.11.1, we write to notify you that [*Contractor*: our] / [*Employer*: your] employment under the Contract is terminated.

Resource 271

Description	A notice of termination from the Employer to the Contractor, relating to regulation 73(1) of the PC Regulations.
Type	Notice: Employer to Contractor
Provision references	clause 8.11.3
Drafting notes	This notice may only be sent where the Contract is one to which regulation 73(1)(a) of the PC Regulations applies. Notices under this clause must be given in accordance with clause 1.7.4 (Notices and other communications).

Reference: Notice of termination (clause 8.11.3)

We write in accordance with clause 8.11.3 (Termination by either Party and regulation 73(1)(a) of the PC Regulations).

We consider that the grounds set out in regulation 73(1)(a) of the PC Regulations apply as follows: [*insert explanation of how the grounds in regulation 73(1)(a) of the PC Regulations apply.*]

Therefore, we write to notify you that your employment under the Contract is terminated.

Resource 272

Description	The provision of documents relating to clause 2.37 (As-built Drawings) from the Contractor to the Employer.
Type	Correspondence: Contractor to Employer
Provision references	clause 8.12.2

Reference: Consequences of termination under clauses 8.9 to 8.11; provision of documents (clause 8.12.2)

In accordance with clause 8.12.2 (Consequences of Termination under clauses 8.9 to 8.11, etc.), we are required to provide the Employer with copies of all documents referred to in clause 2.37 (As-built Drawings) then prepared.

These documents are available [to download at the following link: [*insert link*]] / [on our document management platform, to which you already have access].

Please let us know if there is any further assistance we can provide in this regard.

Resource 273

Description	The provision of all documents necessary for calculating the amounts referred to in clause 8.12.5, from the Contractor to the Employer.
Type	Correspondence: Contractor to Employer
Provision references	clause 8.12.3
Drafting notes	The amount referred to in clause 8.12.5.5 may only be included where the Contractor's employment is terminated either under (i) clause 8.9, (ii) clause 8.10, or (iii) clause 8.11.1.3, if the loss or damage to the Works was caused by the negligence or default of the Employer or any Employer's Person. **Timing:** Clause 8.12.3 requires the Contractor to provide these documents 'not later than 2 months after the date of termination'.

Reference: Consequences of termination under clauses 8.9 to 8.11; provision of documents (clause 8.12.3)

In accordance with clause 8.12.3 (Consequences of Termination under clauses 8.9 to 8.11, etc.), we are required to provide you with all documents necessary for calculating the amounts referred to in clause 8.12.5, namely the following:

- in accordance with clause 8.12.5.1, the total value of work properly executed at, and of any design work properly carried out before, the date of termination of our employment, ascertained in accordance with these Conditions as if the employment had not been terminated, together with any other amounts due to us under these Conditions;
- in accordance with clause 8.12.5.2, any sums ascertained in respect of direct loss and/or expense under clause 4.20 (whether ascertained before or after the date of termination);
- in accordance with clause 8.12.5.3, the reasonable cost of removal under clause 8.12.2;
- in accordance with clause 8.12.5.4, the cost of materials or goods (including Site Materials) properly ordered for the Works for which we then have paid or are legally bound to pay; and
- [in accordance with clause 8.12.5.5, any direct loss and/or damage caused to us by the termination.]

These documents are [attached] / [available to download at the following link: [*insert link*]].

Resource 274

Description	The provision of an account of the amounts referred to in clause 8.12.5, from the Employer to the Contractor.
Type	Account: Employer to Contractor
Provision references	clause 8.12.3
Drafting notes	The amount referred to in clause 8.12.5.5 may only be included where the Contractor's employment is terminated either under (i) clause 8.9, (ii) clause 8.10, or (iii) clause 8.11.1.3, if the loss or damage to the Works was caused by the negligence or default of the Employer or any Employer's Person. **Timing:** Clause 8.12.3 requires the Employer to prepare and send to the Contractor this account 'not later than 3 months' after receipt of the documents referred to in clause 8.12.3.

Reference: Consequences of termination under clauses 8.9 to 8.11; statement of account (clause 8.12.3)

In accordance with clause 8.12.3 (Consequences of Termination under clauses 8.9 to 8.11, etc.), and following the termination of your employment under the Contract under [clause 8.9 (Termination by Contractor: Default by Employer)] /

[clause 8.10 (Termination by Contractor: Insolvency of Employer)] / [clause 8.11 (Termination by either Party and regulation 73(1)(a) of the PC Regulations)] / [clause 6.11.2.2 (Terrorism Cover – non-availability – Employer's options)] / [clause 6.14 (Loss or damage to Existing Structures – right of termination)], we attach an account setting out the amounts referred to in clause 8.12.5, namely the following:

- in accordance with clause 8.12.5.1, the total value of work properly executed at, and of any design work properly carried out before, the date of termination of your employment, ascertained in accordance with these Conditions as if the employment had not been terminated, together with any other amounts due to you under these Conditions;
- in accordance with clause 8.12.5.2, any sums ascertained in respect of direct loss and/or expense under clause 4.20 (whether ascertained before or after the date of termination);
- in accordance with clause 8.12.5.3, the reasonable cost of removal under clause 8.12.2;
- in accordance with clause 8.12.5.4, the cost of materials or goods (including Site Materials) properly ordered for the Works for which you then have paid or are legally bound to pay; and
- [in accordance with clause 8.12.5.5, any direct loss and/or damage caused to you by the termination.]

(Note: All amounts stated are exclusive of VAT.)

Resource 275

Description	A template Employer's Termination Payment Notice, from the Employer to the Contractor.
Type	Template: Employer's Termination Payment Notice
Provision references	clause 8.13.2
Drafting notes	**Timing:** Clause 8.13.2 requires the Employer to give the Employer's Termination Payment Notice not later than 5 days after the due date for the Termination Payment, determined by reference to either (i) clause 8.7.5, (ii) clause 8.8.2, or (iii) clause 8.12.4.

Reference: Employer's Termination Payment Notice (clause 8.13.2)

In accordance with clause 8.13.2 (Termination Payment – final date, notices and amount), we attach the Employer's Termination Payment Notice, which states the sum we consider to be due to you at the due date, and the basis on which that sum has been calculated.

The Employer's Termination Payment Notice has been calculated in accordance with [clause 8.7.6 (Consequences of termination under clauses 8.4 to 8.6)] / [clause 8.8.3 (Employer's decision not to complete the Works)] / [clause 8.12.5 (Consequences of Termination under clauses 8.9 to 8.11, etc.)].

Employer's Termination Payment Notice

- Employer's Termination Payment Notice date: [*insert date*]
- Due date: [*insert due date, by reference to either (i) clause 8.7.5, (ii) clause 8.8.2, or (iii) clause 8.12.4*]
- Final date for payment: [*insert final date for payment, by reference to clause 8.13.1*]

At the due date stated above, the sum due in this Employer's Termination Payment Notice is: [*insert amount in words*], as a balance due from [the Employer to the Contractor] / [the Contractor to the Employer].

We attach the basis on which this sum was calculated, which includes the following amounts:

[clause 8.7.6:

'the amount due on termination shall be calculated in accordance with the following amounts:

8.7.6.1 the amount of expenses properly incurred by the Employer, including those incurred pursuant to clause 8.7.1 and, where applicable, clause 8.5.3.3, and of any direct loss and/or damage caused to the Employer and for which the Contractor is liable, whether arising as a result of the termination or otherwise;

8.7.6.2 the amount of payments made to the Contractor; and

8.7.6.3 the total amount which would have been payable for the Works in accordance with this Contract,

and if the sum of the amounts referred to in clauses 8.7.6.1 and 8.7.6.2 exceeds the amount referred to in clause 8.7.6.3, the difference shall be an amount payable by the Contractor to the Employer or, if that sum is less, by the Employer to the Contractor (the 'Termination Payment').'] /

[clause 8.8.3:

'The amount due on termination shall be calculated in accordance with the following amounts:

8.8.3.1 the total value of work properly executed at the date of termination or date on which the Contractor became Insolvent, ascertained in accordance with these Conditions as if that employment had not been terminated, together with any amounts due to the Contractor under these Conditions not included in such total value; and

8.8.3.2 the aggregate amount of any expenses properly incurred by the Employer and of any direct loss and/or damage caused to the Employer and for which the Contractor is liable, whether arising as a result of the termination or otherwise,

and after taking into account amounts previously paid to the Contractor under this Contract, if the amount referred to in clause 8.8.3.2 exceeds the amount referred to in clause 8.8.3.1, the difference shall be an amount payable by the Contractor to the Employer or, if the clause 8.8.3.2 amount is less, by the Employer to the Contractor (the 'Termination Payment').'] /

[clause 8.12.5:

'the amount due on termination from the Employer to the Contractor or (if a negative amount) from the Contractor to the Employer shall be the aggregate of:

8.12.5.1 *the total value of work properly executed at, and of any design work properly carried out before, the date of termination of the Contractor's employment, ascertained in accordance with these Conditions as if the employment had not been terminated, together with any other amounts due to the Contractor under these Conditions;*

8.12.5.2 *any sums ascertained in respect of direct loss and/or expense under clause 4.20 (whether ascertained before or after the date of termination);*

8.12.5.3 *the reasonable cost of removal under clause 8.12.2;*

8.12.5.4 *the cost of materials or goods (including Site Materials) properly ordered for the Works for which the Contractor then has paid or is legally bound to pay;*

8.12.5.5 *(only where the Contractor's employment is terminated either under clause 8.9 or 8.10, or under clause 8.11.1.3 if the loss or damage to the Works was caused by the negligence or default of the Employer or any Employer's Person) any direct loss and/or damage caused to the Contractor by the termination,*

less amounts previously paid to the Contractor under this Contract, but without deduction of any Retention (the 'Termination Payment'). Payment by the Employer for any such materials and goods as are referred to in clause 8.12.5.4 shall be subject to those materials and goods thereupon becoming the Employer's property.']

Note: All sums stated are exclusive of VAT.

Resource 276

Description	A template Contractor's Termination Payment Notice, from the Contractor to the Employer.
Type	Template: Contractor's Termination Payment Notice
Provision references	clause 8.13.5.1
Drafting notes	**Timing:** Clause 8.13.5.1 entitles the Contractor to give the Contractor's Termination Payment Notice at any time after the 5-day period referred to in clause 8.13.2, namely 5 days after the due date for the Termination Payment, determined by reference to either (i) clause 8.7.5, (ii) clause 8.8.2, or (iii) clause 8.12.4.

Reference: Contractor's Termination Payment Notice (clause 8.13.5.1)

In accordance with clause 8.13.5.1 (Termination Payment – final date, notices and amount), we attach the Contractor's Termination Payment Notice, which states the sum we consider to be due to [us] / [you] at the due date, and the basis on which that sum has been calculated.

The Contractor's Termination Payment Notice has been calculated in accordance with [clause 8.7.6 (Consequences of termination under clauses 8.4 to 8.6)] / [clause 8.8.3 (Employer's decision not to complete the Works)] / [clause 8.12.5 (Consequences of Termination under clauses 8.9 to 8.11, etc.)].

Contractor's Termination Payment Notice

- Contractor's Termination Payment Notice date: [*insert date*]
- Due date: [*insert due date, by reference to either (i) clause 8.7.5, (ii) clause 8.8.2, or (iii) clause 8.12.4*]
- Final date for payment: [*insert final date for payment, by reference to clause 8.13.1 and clause 8.13.5.2*]

At the due date stated above, the sum due in this Contractor's Termination Payment Notice is: [*insert amount in words*], as a balance due from [the Employer to the Contractor] / [the Contractor to the Employer].

We attach the basis on which this sum was calculated, which includes the following amounts:

[clause 8.7.6:

the amount due on termination shall be calculated in accordance with the following amounts:

8.7.6.1 *the amount of expenses properly incurred by the Employer, including those incurred pursuant to clause 8.7.1 and, where applicable, clause 8.5.3.3, and of any direct loss and/or damage caused to the Employer and for which the Contractor is liable, whether arising as a result of the termination or otherwise;*

8.7.6.2 *the amount of payments made to the Contractor; and*

8.7.6.3 *the total amount which would have been payable for the Works in accordance with this Contract,*

and if the sum of the amounts referred to in clauses 8.7.6.1 and 8.7.6.2 exceeds the amount referred to in clause 8.7.6.3, the difference shall be an amount payable by the Contractor to the Employer or, if that sum is less, by the Employer to the Contractor (the 'Termination Payment').] /

[clause 8.8.3:

The amount due on termination shall be calculated in accordance with the following amounts:

8.8.3.1 *the total value of work properly executed at the date of termination or date on which the Contractor became Insolvent, ascertained in accordance with these Conditions as if that employment had not been terminated, together with any amounts due to the Contractor under these Conditions not included in such total value; and*

8.8.3.2 *the aggregate amount of any expenses properly incurred by the Employer and of any direct loss and/or damage caused to the Employer and for which the Contractor is liable, whether arising as a result of the termination or otherwise, and after taking into account amounts previously paid to the Contractor under this Contract, if the amount referred to in clause 8.8.3.2 exceeds the amount referred to in clause 8.8.3.1, the difference shall be an amount payable by the Contractor to the Employer or, if the clause 8.8.3.2 amount is less, by the Employer to the Contractor (the 'Termination Payment').*] /

[clause 8.12.5:

the amount due on termination from the Employer to the Contractor or (if a negative amount) from the Contractor to the Employer shall be the aggregate of:

8.12.5.1 *the total value of work properly executed at, and of any design work properly carried out before, the date of termination of the Contractor's employment, ascertained in accordance with these Conditions as if the employment had not been terminated, together with any other amounts due to the Contractor under these Conditions;*

8.12.5.2 *any sums ascertained in respect of direct loss and/or expense under clause 4.20 (whether ascertained before or after the date of termination);*

8.12.5.3 *the reasonable cost of removal under clause 8.12.2;*

8.12.5.4 *the cost of materials or goods (including Site Materials) properly ordered for the Works for which the Contractor then has paid or is legally bound to pay;*

8.12.5.5 *(only where the Contractor's employment is terminated either under clause 8.9 or 8.10, or under clause 8.11.1.3 if the loss or damage to the Works was caused by the negligence or default of the Employer or any Employer's Person) any direct loss and/or damage caused to the Contractor by the termination,*

less amounts previously paid to the Contractor under this Contract, but without deduction of any Retention (the 'Termination Payment'). Payment by the Employer for any such materials and goods as are referred to in clause 8.12.5.4 shall be subject to those materials and goods thereupon becoming the Employer's property.]

Note: All sums stated are exclusive of VAT.

Resource 277

Description	A notice of a matter which appears likely to give rise to a dispute or difference.
Type	Notice: Contractor or Employer to the other Party
Provision references	clause 9.1
Drafting notes	**Timing:** Clause 9.1 requires the Parties to notify the other Party 'promptly' of any matter that appears likely to give rise to a dispute or difference.

Reference: Notification and negotiation of disputes (clause 9.1)

In accordance with clause 9.1 (Notification and negotiation of disputes), and with a view to the avoidance or early resolution of disputes or differences, we write to notify you of the following matter which appears likely to give rise to a dispute or difference: [*insert details of the potential dispute or difference.*]

We propose to arrange a meeting between the senior executives nominated in the Contract Particulars, namely [*insert name*] (on behalf of the Contractor) and [*insert name*] (on behalf of the Employer), as soon as possible, for direct, good faith negotiations to resolve the matter. Please confirm the availability of your nominated senior executive to attend this meeting.

Resource 278

Description	A request to mediate a difference or dispute between the Parties.
Type	Request: Contractor or Employer to the other Party
Provision references	clause 9.2

Reference: Request to refer difference or dispute to mediation (clause 9.2)

We refer to the following dispute or difference between the Parties:

[*Insert details of the dispute or difference.*]

Given that, to date, we have been unable to resolve this dispute or difference by direct negotiations, we request that the matter be referred to mediation.

In accordance with clause 9.2 (Mediation), please give serious consideration to this request, and notify us of your decision by [*insert date*].

Resource 279

Description	The submission by the Contractor of Contractor's Design Documents.
Type	Submission: Contractor to Employer
Provision references	Schedule 1, paragraph 1
Drafting notes	Paragraph 1 assumes that the Contractor will submit two paper copies of each Contractor's Design Document. In practice, it is more likely that the Design Submission Procedure will be implemented by way of a document management platform. The Contractor must ensure that each Contractor's Design Document is prepared and submitted by the means and in the format stated in the Employer's Requirements or in the Contractor's Proposals. **Timing:** The Contractor must also submit each Contractor's Design Document in sufficient time to allow any comments of the Employer to be incorporated prior to the relevant Contractor's Design Document being used for procurement, and/or in the carrying out of the Works.

Reference: Design Submission Procedure (Schedule 1, paragraph 1)

In accordance with paragraph 1 of Schedule 1 (Design Submission Procedure), we [enclose two copies] / [attach a copy] of the following Contractor's Design Documents: [*insert details of the Contractor's Design Documents.*]

Resource 280

Description	The return by the Employer of Contractor's Design Documents.
Type	Return: Employer to Contractor
Provision references	Schedule 1, paragraph 2
Drafting notes	Paragraph 2 assumes that the Contractor will return one paper copy of each Contractor's Design Document. In practice, it is more likely that the Design Submission Procedure will be implemented by way of a document management platform. **Timing:** The Employer must return one copy of each Contractor's Design Document either (i) within 14 days from the date of receipt, or (ii) if later (and if applicable), 14 days from either the date or expiry of the period for submission of the same stated in the Contract Documents.

Reference: Design Submission Procedure (Schedule 1, paragraph 2)

In accordance with paragraph 2 of Schedule 1 (Design Submission Procedure), we [enclose one copy] / [attach a copy] of the following marked Contractor's Design Documents:

[*Insert details of the returned Contractor's Design Documents, and whether they have been marked 'A', 'B', or 'C'.*]

[*For Contractor's Design Documents marked 'B' or 'C'*: In accordance with paragraph 4, where we have marked a Contractor's Design Document 'B' or 'C', we have identified by means of a written comment why we consider that it is not in accordance with the Contract.]

[*For Contractor's Design Documents marked 'C'*: In accordance with paragraph 6, you must not carry out any work in accordance with any Contractor's Design Document marked 'C', and the Employer will not be liable to pay for any work within the Works executed otherwise than in accordance with Contractor's Design Documents marked 'A' or 'B'.]

Resource 281

Description	Confirmation by the Contractor to the Employer of Contractor's Design Documents regarded as marked 'A'.
Type	Confirmation: Contractor to Employer
Provision references	Schedule 1, paragraph 3
Drafting notes	**Good practice** This correspondence is not required by paragraph 3, but it is considered good practice to notify the Employer, in case it has inadvertently failed to comply with its obligations under paragraph 2.

Reference: Design Submission Procedure (Schedule 1, paragraph 3)

We refer to the following Contractor's Design Documents: [*insert details of the Contractor's Design Documents*.]

You have not responded to these Contractor's Design Documents in the time stated in paragraph 2 of Schedule 1 (Design Submission Procedure), namely [*insert period. For example, 14 days from the date of receipt*].

In accordance with paragraph 3, these Contractor's Design Documents will therefore be regarded as marked 'A'.

In accordance with paragraph 5.1, we will carry out the Works in strict accordance with these Contractor's Design Documents.

Resource 282

Description	A reply from the Contractor to the Employer to marked Contractor's Design Documents.
Type	Reply: Contractor to Employer
Provision references	Schedule 1, paragraph 5
Drafting notes	**Good practice** This correspondence is not required by paragraph 5, but it is considered good practice to notify the Employer of the Contractor's intentions in respect of each Contractor's Design Document.

Reference: Design Submission Procedure (Schedule 1, paragraph 5)

We refer to the following Contractor's Design Documents:

[*Insert details of the Contractor's Design Documents, and whether they have been marked 'A', 'B', or 'C'.*]

In accordance with paragraph 5 of Schedule 1 (Design Submission Procedure):

- for those Contractor's Design Documents marked 'A', we will carry out the Works in strict accordance with those documents;
- for those Contractor's Design Documents marked 'B', we will incorporate the Employer's comments into them and promptly submit amended copies to the Employer, and we will carry out the Works in accordance with the documents so amended; and
- for those Contractor's Design Documents marked 'C', we will take due account of the Employer's comments on them, and will either resubmit them to the Employer in amended form for comment, or notify the Employer under paragraph 7.

Resource 283

Description	The resubmission by the Contractor of a Contractor's Design Document marked 'B'.
Type	Resubmission: Contractor to Employer
Provision references	Schedule 1, paragraph 5.2
Drafting notes	**Timing:** Paragraph 5.2 requires the Contractor to submit the amended Contractor's Design Document 'promptly' to the Employer.

Reference: Design Submission Procedure (Schedule 1, paragraph 5.2)

We refer to the following Contractor's Design Document, which has been marked 'B': [*insert details of the Contractor's Design Document.*]

In accordance with paragraph 5.2 of Schedule 1 (Design Submission Procedure), we [enclose two copies] / [attach a copy] of the amended Contractor's Design Document, with the Employer's comments incorporated.

We confirm that we will carry out the Works in accordance with this amended Contractor's Design Document.

Resource 284

Description	The resubmission by the Contractor of Contractor's Design Document marked 'C'.
Type	Resubmission: Contractor to Employer
Provision references	Schedule 1, paragraph 5.3
Drafting notes	**Timing:** Paragraph 5.3 requires that the Contractor's Design Document should be resubmitted 'forthwith'.

Reference: Design Submission Procedure (Schedule 1, paragraph 5.3)

We refer to the following Contractor's Design Document, which has been marked 'C': [*insert details of the Contractor's Design Document.*]

In accordance with paragraph 5.3 of Schedule 1 (Design Submission Procedure), we have taken due account of the Employer's comments on it, and resubmit [two copies] / [a copy] of the Contractor's Design Document in amended form for comment, in accordance with paragraph 1.

Resource 285

Description	A notice from the Contractor to the Employer of a disagreement with the Employer's comment on a Contractor's Design Document.
Type	Notice: Contractor to Employer
Provision references	Schedule 1, paragraph 7
Drafting notes	**Timing:** If the Contractor disagrees with the Employer's comment, paragraph 7 requires them to give this notice within 7 days of receipt of that comment. Refer to clause 5.1 (Definition of Changes) to help determine whether compliance with the Employer's comment would give rise to a Change.

Reference: Design Submission Procedure (Schedule 1, paragraph 7)

We refer to the following Contractor's Design Document, which has been marked 'C': [*insert details of the Contractor's Design Document.*]

In accordance with paragraph 7 of Schedule 1 (Design Submission Procedure), we write to notify you that:

- we consider that the Contractor's Design Document in question is in accordance with the Contract, and
- we disagree with the following comment, compliance with which would give rise to a Change: [*insert Employer's comment.*]

We consider that compliance with the comment would give rise to a Change for the following reasons: [*insert reasons why compliance with the comment would constitute a Change.*]

Within 7 days of receipt of this notice, please either confirm or withdraw the comment.

Resource 286

Description	A reply from the Employer to the Contractor's notice of disagreement with the Employer's comment on a Contractor's Design Document.
Type	Confirmation: Employer to Contractor
Provision references	Schedule 1, paragraph 7
Drafting notes	**Timing:** Paragraph 7 requires the Employer to either confirm or withdraw the comment within 7 days of receipt of the Contractor's notice under paragraph 7. Firstly, the Employer needs to decide whether to confirm or withdraw the comment. If the Employer confirms the comment, it may also wish to advise whether or not it considers that compliance with the comment would give rise to a Change.

Reference: Design Submission Procedure (Schedule 1, paragraph 7)

We write in reply to your notice dated [*insert date*], given in accordance with paragraph 7 of Schedule 1 (Design Submission Procedure), and in relation to the following Contractor's Design Document, which has been marked 'C': [*insert details of the Contractor's Design Document.*]

[*Employer's comment confirmed*: In accordance with paragraph 7, we confirm the comment. Please amend and resubmit the Contractor's Design Document accordingly.]

[We accept that compliance with the comment would give rise to a Change, and attach an instruction to that effect.] / [We do not consider that compliance with the comment would give rise to a Change, for the following reasons: [*insert reasons.*] In accordance with paragraph 8.3, confirmation of this comment does not signify acceptance by us that compliance with the comment will give rise to a Change.] /

[*Employer's comment withdrawn*: In accordance with paragraph 7, we withdraw the comment.]

Resource 287

Description	A tracker template for recording Contractor's Design Documents.
Type	Tracker: Contractor and Employer
Provision references	Schedule 1

Below is a tracker template for recording the submission of Contractor's Design Documents in accordance with the Design Submission Procedure:

- Contractor's Design Document reference: [*insert reference*]
- Description / file name: [*insert description or electronic file name of Contractor's Design Document*]

- Date for submission by the Contractor: [*insert the date by which the Contractor was required to have submitted the Contractor's Design Document*]
- Date submitted by the Contractor: [*insert date submitted by the Contractor*]
- Date for return by the Employer: [*insert the date by which the Employer was required to have returned the Contractor's Design Document*]
- Date returned by the Employer: [*insert date returned by the Employer*]
- Status / Mark: [*insert either 'A', 'B', or 'C'*]
- Date resubmitted by the Contractor (if applicable): [*insert date resubmitted by the Contractor*]
- Date of Contractor's disagreement with Employer's comment under paragraph 7 (if applicable): [*insert date*]
- Comment confirmed or withdrawn by the Employer (if applicable): [*confirmed*] / [*withdrawn*]
- Date comment confirmed or withdrawn by the Employer (if applicable): [*insert date*]
- Document link: [*insert link to documents*]
- Status / action required: [*insert status, and any action required*]

Resource 288

Description	A notice from the Contractor to the Employer of the Contractor's inability to enter into a sub-contract with a Named Sub-Contractor.
Type	Notice: Contractor to Employer
Provision references	Supplemental Provision 1, paragraph 1.1.2
Drafting notes	**Timing:** Paragraph 1.1.2 requires the Contractor to give this notice 'immediately', if for any reason it is unable to enter into a sub-contract with a Named Sub-Contractor.

Reference: Named Sub-Contractors (Supplemental Provision 1, Schedule 2)

In accordance with paragraph 1.1.2 of Supplemental Provision 1 (Named Sub-Contractors), we write to notify you that we have been unable to enter into a sub-contract with the following Named Sub-Contractor: [*insert details of Named Sub-Contractor.*]

We have been unable to enter into this sub-contract for the following reasons: [*insert reasons why the parties were unable to enter into the sub-contract.*]

Therefore, please issue a Change instruction, either:

(a) removing the grounds for us being unable to enter into this sub-contract;
(b) directing us to carry out that work using either our own resources or, at our option, a sub-contractor selected by us and approved by the Employer; or
(c) omitting the Named Sub-Contract Work from the Contract.

[Our preference is for option [*insert option (a), (b), or (c)*].]

Resource 289

Description	An instruction from the Employer following a notice from the Contractor of the Contractor's inability to enter into a sub-contract with a Named Sub-Contractor.
Type	Instruction: Employer to Contractor
Provision references	Supplemental Provision 1, paragraph 1.1.2
Drafting notes	Paragraph 1.1.2 states that the Employer cannot insist that the relevant work be executed by another Named Sub-Contractor.

Reference: Named Sub-Contractors; Employer's instruction [*insert reference*] (Supplemental Provision 1, Schedule 2)

We write in reply to your notice dated [*insert date*], given in accordance with paragraph 1.1.2 of Supplemental Provision 1 (Named Sub-Contractors), and in relation to the following Named Sub-Contractor: [*insert details of Named Sub-Contractor.*]

We instruct you as follows:

[*Insert details of the instruction, either:*

(a) removing the grounds for the Contractor being unable to enter into the sub-contract;
(b) directing the Contractor to carry out that work using either its own resources or, at the Contractor's option, a sub-contractor selected by them and approved by the Employer; or
(c) omitting the Named Sub-Contract Work from the Contract.]

[In accordance with paragraph 1.2, we confirm that the provisions of clause 5.2 (Valuation of Changes), clause 2.23 to clause 2.26 (Adjustment of Completion Date), and clause 4.19 to clause 4.23 (Loss and Expense) will, as relevant, apply to this Change instruction.]

Resource 290

Description	A notice from the Contractor to the Employer of the potential termination of a Named Sub-Contractor's employment under their sub-contract.
Type	Notice: Contractor to Employer
Provision references	Supplemental Provision 1, paragraph 1.3.1
Drafting notes	**Timing:** Paragraph 1.1.3 requires the Contractor to give this notice 'promptly', if any of the three circumstances are met.

Reference: Named Sub-Contractors (Supplemental Provision 1, Schedule 2)

In accordance with paragraph 1.3.1 of Supplemental Provision 1 (Named Sub-Contractors), we write in respect of the following Named Sub-Contractor: [*insert details of Named Sub-Contractor.*]

We write to notify you that we have become entitled to [terminate its employment under the sub-contract] / [give notice to them of a specified default which, if continued, would be grounds for termination of its employment under the sub-contract] / [treat the sub-contract as repudiated], for the following reasons: [*insert reasons.*]

Please let us know if you would like to arrange a consultation to discuss this further.

Resource 291

Description	A request from the Contractor for the Employer's consent to terminate a Named Sub-Contractor's employment under their sub-contract.
Type	Request for consent: Contractor to Employer
Provision references	Supplemental Provision 1, paragraph 1.3.2
Drafting notes	Paragraph 1.3.2 does not apply where the Named Sub-Contractor is or becomes Insolvent.

Reference: Named Sub-Contractors (Supplemental Provision 1, Schedule 2)

We write further to our notice dated [*insert date*], given in accordance with paragraph 1.3.1 of Supplemental Provision 1 (Named Sub-Contractors), and in respect of following Named Sub-Contractor: [*insert details of Named Sub-Contractor.*]

In accordance with paragraph 1.3.2, we request the Employer's consent to [give notice of termination to the Named Sub-Contractor] / [give notice to the Named Sub-Contractor that we are treating the sub-contract as repudiated].

Please give your consent by [*insert date*].

Resource 292

Description	Consent given by the Employer for the Contractor to terminate a Named Sub-Contractor's employment under their sub-contract.
Type	Consent given: Employer to Contractor
Provision references	Supplemental Provision 1, paragraph 1.3.2

Reference: Named Sub-Contractors (Supplemental Provision 1, Schedule 2)

We write further to your notice dated [*insert date*], given in accordance with paragraph 1.3.2 of Supplemental Provision 1 (Named Sub-Contractors), and in respect of following Named Sub-Contractor: [*insert details of Named Sub-Contractor.*]

We give our consent for you to [give notice of termination to the Named Sub-Contractor] / [give notice to the Named Sub-Contractor that you are treating the sub-contract as repudiated].

In accordance with paragraph 1.3.3, please send the Employer a copy of any such notice that you give to the Named Sub-Contractor.

Resource 293

Description	Consent withheld by the Employer for the Contractor to terminate a Named Sub-Contractor's employment under their sub-contract.
Type	Consent withheld: Employer to Contractor
Provision references	Supplemental Provision 1, paragraph 1.3.2
Drafting notes	Clause 1.10 requires that the Employer's consent must not be unreasonably delayed or withheld.

Reference: Named Sub-Contractors (Supplemental Provision 1, Schedule 2)

We write further to your notice dated [*insert date*], given in accordance with paragraph 1.3.2 of Supplemental Provision 1 (Named Sub-Contractors), and in respect of following Named Sub-Contractor: [*insert details of Named Sub-Contractor*.]

We withhold our consent for you to [give notice of termination to the Named Sub-Contractor] / [give notice to the Named Sub-Contractor that you are treating the sub-contract as repudiated], for the following reasons: [*insert reasons, to demonstrate why the consent was not unreasonably delayed or withheld*.]

Resource 294

Description	The provision of estimates by the Contractor to the Employer of the effects of a Change instruction.
Type	Estimates: Contractor to Employer
Provision references	Supplemental Provision 2, paragraph 2.2
Drafting notes	**Timing:** Paragraph 2.2 requires the Contractor to submit the estimates either (i) within 14 days of the date of the relevant instruction, or (ii) within such other period as may be agreed, or (iii) failing agreement, within such other period as may be reasonable in all the circumstances.

Reference: Valuation of Changes – Contractor's estimates (Supplemental Provision 2, Schedule 2)

We write further to the Employer's instruction dated [*insert date*], with reference [*insert reference*], and issued in accordance with clause 3.9 (Instructions requiring Changes).

In our opinion, compliance with this instruction will entail:

- a Valuation under clause 5.2 (Valuation of Changes and provisional sum work); and/or
- the making of an adjustment of time in respect of the Relevant Event in clause 2.26 (Relevant Events); and/or
- the ascertainment of direct loss and/or expense under clause 4.20 (Loss and Expense: Notification and ascertainment).

In accordance with paragraphs 2.2 and 2.3 of Supplemental Provision 2 (Valuation of Changes – Contractor's estimates), and before complying with this instruction, we attach estimates of the following:

- the value of the adjustment to the Contract Sum, supported by all necessary calculations, by reference to the values in the Contract Sum Analysis;
- the additional resources (if any) required to comply with the instruction;
- a method statement for compliance with the instruction;
- the length of any extension of time required, and the resultant change in the Completion Date(s); and
- the amount of any direct loss and/or expense, not included in any other estimate, which results from the regular progress of the Works or any part of them being materially affected by compliance with the instruction under clause 3.9.

In accordance with paragraph 2.4, we propose to meet as soon as possible to agree these estimates, following which they will be binding on the Parties (and in substitution for any Valuation under clause 5.2 and/or any ascertainment under clause 4.20).

Resource 295

Description	Confirmation from the Employer to the Contractor that estimates are not required.
Type	Correspondence: Employer to Contractor
Provision references	Supplemental Provision 2, paragraph 2.2.1
Drafting notes	**Timing:** Paragraph 2.2.1 requires the Employer to state that the estimates are not required either (i) with the instruction, or (ii) within 14 days after issuing the instruction.

Reference: Valuation of Changes – Contractor's estimates (Supplemental Provision 2, Schedule 2)

We write further to the Employer's instruction dated [*insert date*], with reference [*insert reference*], and issued in accordance with clause 3.9 (Instructions requiring Changes).

In accordance with paragraph 2.2.1 of Supplemental Provision 2 (Valuation of Changes – Contractor's estimates), we write to confirm that the estimates referred to in paragraph 2.2 are not required.

Resource 296

Description	An objection from the Contractor to the Employer to the provision of estimates.
Type	Objection: Contractor to Employer
Provision references	Supplemental Provision 2, paragraph 2.2.2
Drafting notes	**Timing:** If the Contractor wishes to raise any reasonable objections to the provision of all or any such estimates, paragraph 2.2.2 requires them to do so within 10 days of receipt of the relevant instruction.

Reference: Valuation of Changes – Contractor's estimates (Supplemental Provision 2, Schedule 2)

We write further to the Employer's instruction dated [*insert date*], with reference [*insert reference*], and issued in accordance with clause 3.9 (Instructions requiring Changes).

In accordance with paragraph 2.2.2 of Supplemental Provision 2 (Valuation of Changes – Contractor's estimates), we write to raise the following objection to the provision of the estimates referred to in paragraph 2.2: [*insert objections, whether to the provision of some or all of the estimates, and whether on the Contractor's behalf, or on behalf of any sub-contractor.*]

Resource 297

Description	Confirmation or withdrawal by the Employer of an instruction to the Contractor.
Type	Confirmation / withdrawal of instruction: Employer to Contractor
Provision references	Supplemental Provision 2, paragraph 2.5
Drafting notes	**Timing:** Paragraph 2.5 requires the Employer to either (i) instruct compliance or (ii) withdraw the instruction, within 10 days of receipt of the Contractor's estimates.

Reference: Valuation of Changes – Contractor's estimates (Supplemental Provision 2, Schedule 2)

We write further to the Employer's instruction dated [*insert date*], with reference [*insert reference*], and issued in accordance with clause 3.9 (Instructions requiring Changes).

The Parties have been unable to reach agreement on the estimates you provided.

Therefore, in accordance with paragraph 2.5 of Supplemental Provision 2 (Valuation of Changes – Contractor's estimates), we [instruct you to comply with the instruction, and Supplemental Provision 2 will not apply in respect of this instruction.] / [withdraw the instruction. This withdrawal will be at no cost to the Employer, except where the preparation of the estimates involved you in any additional design work, solely and necessarily carried out for the purpose of preparing your estimates, in which case such design work will be treated as a Change.]

Resource 298

Description	A tracker template for instructions and estimates relating to Supplemental Provision 2.
Type	Tracker: Contractor and Employer
Provision references	Supplemental Provision 2

Below is a tracker template for recording instructions and estimates in relation to Supplemental Provision 2 (Valuation of Changes – Contractor's estimates):

- Instruction date: [*insert date*]
- Instruction reference: [*insert reference*]
- Estimates required by the Employer: [*yes / no*]
- Contractor's objection to providing estimates: [*yes / no*]
- Date estimates provided:
  - Adjustment to Contract Sum: [*insert date*]
  - Additional resources required: [*insert date*]
  - Method statement for compliance: [*insert date*]
  - Extension of time required: [*insert date*]
  - Amount of direct loss and/or expense: [*insert date*]
- Date of meeting to agree estimates: [*insert date*]
- Estimates agreed:
  - Adjustment to Contract Sum: [*yes*] / [*no*]
  - Additional resources required: [*yes*] / [*no*]
  - Method statement for compliance: [*yes*] / [*no*]
  - Extension of time required: [*yes*] / [*no*]
  - Amount of direct loss and/or expense: [*yes*] / [*no*]
- Adjustment to Contract Sum: [*insert adjustment*]
- Adjustment to Completion Date(s): [*insert adjustments*]
- Document link: [*insert link to documents*]
- Status / action required: [*insert status, and any action required*]

Resource 299

Description	The provision by the Contractor to the Employer of an estimate of the direct loss and expense incurred by the Contractor.
Type	Estimate: Contractor to Employer
Provision references	Supplemental Provision 3, paragraph 3.2
Drafting notes	**Timing:** Paragraph 3.2 requires the Contractor to submit the estimate 'on presentation of the next Interim Payment Application'.

Reference: Loss and expense – Contractor's estimates (Supplemental Provision 3, Schedule 2)

In accordance with clause 4.19 (Loss and Expense: Matters materially affecting regular progress), we consider that we are entitled to an amount in respect of direct loss and/or expense, to be added to the Contract Sum.

In accordance with paragraph 3.2 of Supplemental Provision 3 (Loss and expense – Contractor's estimates), we attach an estimate of the addition to the Contract Sum which we require in respect such loss and/or expense, which we have incurred in the period immediately preceding that for which the corresponding Interim Payment Application is made (namely from [*insert date*] to [*insert date*]).

Resource 300

Description	The provision by the Contractor to the Employer of further estimates of the direct loss and expense incurred by the Contractor.
Type	Further estimates: Contractor to Employer
Provision references	Supplemental Provision 3, paragraph 3.3
Drafting notes	Following the submission of an estimate under paragraph 3.2, and for so long as it has incurred direct loss and/or expense to which clause 4.19 refers, paragraph 3.3 requires the Contractor to submit further estimates for each period immediately preceding that for which each Interim Payment Application is made.

Reference: Loss and expense – Contractor's estimates (Supplemental Provision 3, Schedule 2)

We write further to our estimate submitted on presentation of the Interim Payment Application made on [*insert date*], with reference [*insert reference*], in accordance with paragraph 3.2 of Supplemental Provision 3 (Loss and expense – Contractor's estimates).

In accordance with paragraph 3.3, we attach a further estimate of the addition to the Contract Sum which we require in respect such loss and/or expense, which we have incurred in the period immediately preceding that for which the corresponding Interim Payment Application is made (namely from [*insert date*] to [*insert date*]).

Resource 301

Description	A request by the Employer to the Contractor for further information and details relating to the Contractor's estimate of the direct loss and expense incurred by the Contractor.
Type	Request: Employer to Contractor
Provision references	Supplemental Provision 3, paragraph 3.4
Drafting notes	**Timing:** If the Employer wishes to make this request, paragraph 3.4 requires them to do so within 21 days of receipt of the Contractor's estimate. They are therefore encouraged to make any such request as soon as possible.

Reference: Loss and expense – Contractor's estimates (Supplemental Provision 3, Schedule 2)

We write further to your estimate, submitted in accordance with paragraph [3.2] / [3.3] of Supplemental Provision 3 (Loss and expense – Contractor's estimates), and received with Interim Payment Application dated [*insert date*], with reference [*insert reference*].

In accordance with paragraph 3.4, we require the following further information and details, in support of your estimate: [*insert description of further information or details required.*]

Resource 302

Description	A notice from the Employer to the Contractor relating to the Contractor's estimate of the direct loss and expense incurred by the Contractor.
Type	Notice: Employer to Contractor
Provision references	Supplemental Provision 3, paragraph 3.4
Drafting notes	**Timing:** Paragraph 3.4 requires the Employer to give this notice within 21 days of receipt of the Contractor's estimate (irrespective of when the Employer may have requested further information or details from the Contractor, also in accordance with paragraph 3.4).

Reference: Loss and expense – Contractor's estimates (Supplemental Provision 3, Schedule 2)

We write further to your estimate, submitted in accordance with paragraph [3.2] / [3.3] of Supplemental Provision 3 (Loss and expense – Contractor's estimates), and received with Interim Payment Application dated [*insert date*], with reference [*insert reference*].

In accordance with paragraph 3.4, we write to notify you that [we accept the estimate.] / [we wish to negotiate on the amount of the addition to the Contract Sum.] / [the provisions of clause 4.19 (Loss and Expense: Matters materially affecting regular progress) and clause 4.20 (Loss and Expense: Notification and ascertainment) will apply in respect of the loss and/or expense to which the estimate relates.]

Resource 303

Description	A tracker template for estimates of direct loss and expense incurred by the Contractor, relating to Supplemental Provision 3.
Type	Tracker: Contractor and Employer
Provision references	Supplemental Provision 3

Below is a tracker template for recording estimates in relation to Supplemental Provision 3 (Loss and expense – Contractor's estimates).

- Date of Contractor's notice under clause 4.20: [*insert date*]
- Deferment of possession / Relevant Matter: [*insert details*]
- Interim Payment Application date: [*insert date*]
- Interim Payment Application reference: [*insert reference*]
- Estimate reference: [*insert reference*]
- Estimate amount (exclusive of VAT): [*insert amount*]
- Estimate agreed (exclusive of VAT): [*insert amount*]
- Document link: [*insert link to documents*]
- Status / action required: [*insert status, and any action required*]

Resource 304

Description	A request by the Employer for the Contractor to provide an Acceleration Quotation.
Type	Request: Employer to Contractor
Provision references	Supplemental Provision 4, paragraph 4.1.1

Reference: Acceleration Quotation (Supplemental Provision 4, Schedule 2)

In accordance with paragraph 4.1.1 of Supplemental Provision 4 (Acceleration Quotation), we wish to investigate the possibility of achieving practical completion before the Completion Date for [the Works] / [Section [*insert reference*]].

Therefore, please either:

- provide an Acceleration Quotation accordingly, identifying the following:
  - the time that can be saved;
  - the amount of the adjustment to the Contract Sum (including the amount to be paid in respect of any direct loss and/or expense that is not included in any other Confirmed Acceptance or in any ascertainment under clause 4.20

(Loss and Expense: Notification and ascertainment), together with a fair and reasonable amount in respect of the cost of preparing the quotation); and
  - any other conditions attached; or
- explain why it would be impracticable to achieve practical completion earlier than the Completion Date.

Resource 305

Description	The provision by the Contractor to the Employer of an Acceleration Quotation, or an explanation why it would be impracticable to achieve practical completion earlier than the Completion Date.
Type	Acceleration Quotation: Contractor to Employer
Provision references	Supplemental Provision 4, paragraph 4.1.1
Drafting notes	**Timing:** Unless otherwise agreed between the Parties, Paragraph 4.2.1 requires the Contractor to submit the Acceleration Quotation in compliance with the Employer's invitation not later than 21 days from the later of (i) the date of receipt of the Employer's invitation, or (ii) the date of receipt by the Contractor of information sufficient to enable them to prepare the quotation.

Reference: Acceleration Quotation (Supplemental Provision 4, Schedule 2)

We write in reply to your invitation dated [*insert date*], for us to provide an Acceleration Quotation, in accordance with paragraph 4.1.1 of Supplemental Provision 4 (Acceleration Quotation).

[We attach an Acceleration Quotation, identifying the following:

- the time that can be saved;
- the amount of the adjustment to the Contract Sum (including the amount to be paid in respect of any direct loss and/or expense that is not included in any other Confirmed Acceptance or in any ascertainment under clause 4.20 (Loss and Expense: Notification and ascertainment), together with a fair and reasonable amount in respect of the cost of preparing the quotation); and
- any other conditions attached.] /

[We consider that it would be impracticable to achieve practical completion earlier than the Completion Date, for the following reasons: [*insert reasons*.]]

Resource 306

Description	An Acceleration Quotation template, from the Contractor to the Employer.
Type	Template: Acceleration Quotation
Provision references	Supplemental Provision 4, paragraph 4.1.1

Acceleration Quotation (Supplemental Provision 4, Schedule 2)

This Acceleration Quotation is provided in reply to the Employer's invitation dated [*insert date*], and in respect of [the Works] / [Section [*insert reference*]].

Proposal Summary

[*Insert a summary of how practical completion can be achieved before the Completion Date.*]

Programme implications

Time that can be saved: [*insert duration*]

Corresponding adjustment to the Completion Date(s): [*insert revised Completion Dates*]

Cost implications

Amount to be paid in respect of direct loss and/or expense: [*insert amount*]

Cost of preparing the quotation: [*insert amount*]

Total adjustment to the Contract Sum: [*insert amount*]

(Note: All amounts stated are exclusive of VAT.)

Any other conditions

This Acceleration Quotation is subject to the following further conditions:

[*Insert any other relevant conditions, for example, access to the site, working space, working hours, the execution or completion of the work in a specific order, etc.*]

Resource 307

Description	A request by the Employer for the Contractor to give revised proposals, in respect of an Acceleration Quotation.
Type	Request: Employer to Contractor
Provision references	Supplemental Provision 4, paragraph 4.1.3

Reference: Acceleration Quotation (Supplemental Provision 4, Schedule 2)

We write [in reply to your Acceleration Quotation dated [*insert date*], with reference [*insert reference*]] / [further to our invitation dated [*insert date*] for you to provide an Acceleration Quotation].

In accordance with paragraph 4.1.3 of Supplemental Provision 4 (Acceleration Quotation), we request that you provide revised proposals as follows: [*insert details of how the Employer wishes for the proposals to be revised.*]

Please provide the revised proposals by [*insert date*].

Resource 308

Description	A notice from the Employer to the Contractor of an adjustment to the periods referred to in Supplemental Provision 4.
Type	Notice: Employer to Contractor
Provision references	Supplemental Provision 4, paragraph 4.2.3
Drafting notes	Paragraph 4.2.3 permits the Parties to either increase or reduce any of the periods referred to in Supplemental Provision 4.

Reference: Acceleration Quotation (Supplemental Provision 4, Schedule 2)

In accordance with paragraph 4.2.3 of Supplemental Provision 4 (Acceleration Quotation), we write to confirm the adjustment to the following periods referred to in paragraph 4, to which the Parties have agreed:

[*Insert details of the increase or reduction of any of the periods referred to in paragraph 4.*]

Resource 309

Description	A Confirmed Acceptance from the Employer to the Contractor, by way of an instruction.
Type	Instruction: Employer to Contractor
Provision references	Supplemental Provision 4, paragraph 4.3
Drafting notes	**Timing:** Paragraph 4.2.2 provides that the Acceleration Quotation will remain open for acceptance by the Employer for not less than 7 days from its receipt.

Reference: Confirmed Acceptance; Employer's instruction [*insert reference*] (Supplemental Provision 4, Schedule 2)

In accordance with paragraph 4.3 of Supplemental Provision 4 (Acceleration Quotation), we wish to accept the Acceleration Quotation dated [*insert date*], with reference [*insert reference*], and attach an instruction to that effect, which includes the following:

- the adjustment of the Contract Sum (including the amount to be paid in respect of any direct loss and/or expense that is not included in any other Confirmed Acceptance or in any ascertainment under clause 4.20 (Loss and Expense: Notification and ascertainment), together with a fair and reasonable amount in respect of the cost of preparing the quotation);
- the adjustment to the time required by you for completion of [the Works] / [Section [*insert reference*]], and the resultant revised Completion Date(s); and
- any other conditions referred to in your Acceleration Quotation.

Resource 310

Description	A checklist of the Contractor's duties in relation to health and safety.
Type	Checklist: Contractor
Provision references	Supplemental Provision 5

In accordance with paragraph 5.1 of Supplemental Provision 5 (Health and safety), the Parties must endeavour to establish and maintain a culture and working environment in which health and safety is of paramount concern to everybody involved with the project.

Below is a checklist of the Contractor's other obligations under that provision, in addition to the specific health and safety requirements of the Contract:

- comply with any and all approved codes of practice produced or promulgated by the Health and Safety Executive;
- ensure that all personnel engaged by the Contractor and members of the Contractor's supply chain on site receive appropriate site-specific health and safety induction training, and regular refresher training;
- ensure that all such personnel have access at all times to competent health and safety advice, in accordance with regulation 7 of the Management of Health and Safety at Work Regulations 1999; and
- ensure that there is full and proper health and safety consultation with all such personnel, in accordance with the Health and Safety (Consultation with Employees) Regulations 1996.

Resource 311

Description	A proposed change to the Works by the Contractor, for the Employer's benefit.
Type	Proposed change: Contractor to Employer
Provision references	Supplemental Provision 6, paragraph 6.1

Reference: Cost savings and value improvements (Supplemental Provision 6, Schedule 2)

In accordance with paragraph 6.1 of Supplemental Provision 6 (Cost savings and value improvements), we are encouraged to propose changes to designs and specifications for the Works and/or to the programme for their execution that may benefit the Employer, whether in the form of a reduction in the cost of the Works or their associated life cycle costs, through practical completion at a date earlier than the Completion Date, or otherwise.

In accordance with paragraph 6.2, we attach details of such a proposed change, together with a quotation setting out our assessment of the benefit we believe the Employer may obtain, in financial terms.

Resource 312

Description	An instruction by the Employer of a change proposed by the Contractor.
Type	Instruction: Employer to Contractor
Provision references	Supplemental Provision 6, paragraph 6.3

Reference: Cost savings and value improvements; Employer's instruction [*insert reference*] (Supplemental Provision 6, Schedule 2)

We write further to your correspondence dated [*insert date*], given in accordance with paragraph 6.2 of Supplemental Provision 6 (Cost savings and value improvements).

In accordance with paragraph 6.3, we attach an instruction confirming:

- the change;
- the amount of the adjustment to the Contract Sum;
- the share of the financial benefit to be paid to you; and
- any adjustment to the Completion Date.

Resource 313

Description	A tracker template for proposals made by the Contractor, relating to Supplemental Provision 6.
Type	Tracker: Contractor and Employer
Provision references	Supplemental Provision 6

Below is a tracker template for recording proposals, in relation to Supplemental Provision 6 (Cost savings and value improvements):

Contractor's proposed change: [*insert description*]

- Proposal date: [*insert date*]
- Agreement reached on value, financial benefit, and adjustment to Completion Date: [*yes / no*]
- Employer's instruction date: [*insert date*]
- Employer's instruction reference: [*insert reference*]
- Adjustment to the Contract Sum: [*insert amount*] (exclusive of VAT)
- Share of the financial benefit to be paid to the Contractor: [*insert amount*] (exclusive of VAT)
- Adjustment to Completion Date: [*insert adjustment*]
- Document link: [*insert link to documents*]
- Status / action required: [*insert status, and any action required*]

Resource 314

Description	The provision by the Contractor to the Employer of information for performance indicators and monitoring.
Type	Provision of information: Contractor to Employer
Provision references	Supplemental Provision 7, paragraph 7.2

Reference: Performance indicators and monitoring (Supplemental Provision 7, Schedule 2)

In accordance with paragraph 7.2 of Supplemental Provision 7 (Performance Indicators and monitoring), we attach the following information that the Employer requires to monitor and assess our performance against the targets for the following performance indicators stated or identified in the Contract Documents: [*insert details of the information provided, and their associated performance indicators.*]

Resource 315

Description	Informing the Contractor that targets for performance indicators may not be met.
Type	Correspondence: Employer to Contractor
Provision references	Supplemental Provision 7, paragraph 7.3

Reference: Performance indicators and monitoring (Supplemental Provision 7, Schedule 2)

In accordance with paragraph 7.3 of Supplemental Provision 7 (Performance Indicators and monitoring), we write to inform you that we consider that the following target for the following performance indicators stated or identified in the Contract Documents may not be met: [*insert details of the target, and the associated performance indicators.*]

Therefore, please submit your proposals for improving your performance against this target.

Resource 316

Description	Informing the Contractor of a request made in accordance with the Freedom of Information Act 2000, in relation to the Contract.
Type	Correspondence: Employer to Contractor
Provision references	Supplemental Provision 8, paragraph 8.2
Drafting notes	**Timing:** Paragraph 8.2 requires the Employer to inform the Contractor 'promptly' of any request for disclosure that it receives in relation to the Contract.

Reference: Transparency; Request for disclosure under the Freedom of Information Act 2000 (Supplemental Provision 8, Schedule 2)

In accordance with paragraph 8.2 of Supplemental Provision 8 (Transparency), we write to inform you of the following request for disclosure, made in accordance with the Freedom of Information Act 2000, that the Employer has received in relation to the Contract: [*insert details of the request.*]

Resource 317

Description	The provision by the Contractor to the Employer of the Joint Names Policy.
Type	Correspondence: Contractor to Employer
Provision references	Schedule 3, paragraph A.1
Drafting notes	**Good practice** This correspondence is not required by paragraph A.1, but it is considered good practice to share the Joint Names Policy with the Employer. Check the entry against the references to clause 6.7 and Schedule 3 in the Contract Particulars for the percentage to cover professional fees.

Reference: Insurance Option A (New Buildings – All Risks Insurance of the Works by the Contractor); Joint Names Policy

In accordance with paragraph A.1 (Contractor to effect and maintain a Joint Names Policy) of Schedule 3 (Insurance Options), we have effected (and for the period specified in clause 6.7.2 (Insurance Options and period) will maintain with insurers approved by the Employer) a Joint Names Policy for All Risks Insurance, with cover no less than that specified in clause 6.8 (Insurance of the Works and Existing Structures: Related Definitions) for the full reinstatement value of [the Works] / [the Sections] [(plus [*insert percentage*] to cover professional fees, as stated in the Contract Particulars)].

We attach a copy of the Joint Names Policy.

Resource 318

Description	Confirmation by the Contractor to the Employer of the use of the Contractor's annual policy, as an alternative to a Joint Names Policy.
Type	Correspondence: Contractor to Employer
Provision references	Schedule 3, paragraph A.2
Drafting notes	**Good practice** This correspondence is not required by paragraph A.2, but it is considered good practice for the Contractor to share details of its annual insurance policy with the Employer.

Reference: Insurance Option A (New Buildings – All Risks Insurance of the Works by the Contractor); use of Contractor's annual policy as alternative

In accordance with paragraph A.2 (Use of Contractor's annual policy – as alternative) of Schedule 3 (Insurance Options), if and so long as we, independently of the Contract, maintain an annual insurance policy which in respect of the Works:

- provides (inter alia) All Risks Insurance with cover and in amounts no less than those stated in paragraph A.1 (Contractor to effect and maintain a Joint Names Policy); and
- is a Joint Names Policy,

that policy will satisfy our obligations under paragraph A.1.

We confirm that our annual insurance policy meets these requirements, and attach a copy.

The annual renewal date of the policy is [*insert date*], as stated in the Contract Particulars.

Resource 319

Description	The provision by the Employer to the Contractor of the Joint Names Policy.
Type	Correspondence: Employer to Contractor
Provision references	Schedule 3, paragraph B.1
Drafting notes	**Good practice** This correspondence is not required by paragraph B.1, but it is considered good practice to share the Joint Names Policy with the Contractor. Check the entry against clause 6.7 and Schedule 3 in the Contract Particulars for the percentage to cover professional fees.

Reference: Insurance Option B (New Buildings – All Risks Insurance of the Works by the Employer); Joint Names Policy

In accordance with paragraph B.1 (Employer to effect and maintain a Joint Names Policy) of Schedule 3 (Insurance Options), the Employer has effected (and for the period specified in clause 6.7.2 (Insurance Options and period) will maintain) a Joint Names Policy for All Risks Insurance, with cover no less than that specified in clause 6.8 (Insurance of the Works and Existing Structures: Related Definitions) for the full reinstatement value of [the Works] / [the Sections] [(plus [*insert percentage*] to cover professional fees, as stated in the Contract Particulars)].

We attach a copy of the Joint Names Policy.

Resource 320

Description	The provision by the Employer to the Contractor of the Joint Names Policy for Specified Perils.
Type	Correspondence: Employer to Contractor
Provision references	Schedule 3, paragraph C.1
Drafting notes	**Good practice** This correspondence is not required by paragraph C.1, but it is considered good practice to share the Joint Names Policy for Specified Perils with the Contractor. Refer to clause 6.8 (Insurance of the Works and Existing Structures: Related definitions) for the definition of '*Specified Perils*'.

Reference: Insurance Option C (Joint Names Insurance by the Employer of Existing Structures and Works in or Extensions to them); Joint Names Policy for Specified Perils

In accordance with paragraph C.1 (Existing Structures and contents – Joint Names Policy for Specified Perils) of Schedule 3 (Insurance Options), we have effected (and for the period specified in clause 6.7.2 (Insurance Options and period) will maintain) a Joint Names Policy in respect of the Existing Structures, together with the contents of them owned by the Employer or for which the Employer is responsible, for the full cost of reinstatement, repair, or replacement of loss or damage due to any of the Specified Perils.

We attach a copy of the Joint Names Policy.

Resource 321

Description	The provision by the Employer to the Contractor of the Joint Names Policy for All Risks.
Type	Correspondence: Employer to Contractor
Provision references	Schedule 3, paragraph C.2
Drafting notes	**Good practice** This correspondence is not required by paragraph C.2, but it is considered good practice to share the Joint Names Policy for All Risks with the Contractor. Check the entry for clause 6.7 and Schedule 3 in the Contract Particulars for the percentage to cover professional fees.

Reference: Insurance Option C (Joint Names Insurance by the Employer of Existing Structures and Works in or Extensions to them); Joint Names Policy for All Risks

In accordance with paragraph C.2 (The Works – Joint Names Policy for All Risks) of Schedule 3 (Insurance Options), the Employer has effected (and for the period specified in clause 6.7.2 (Insurance Options and period) will maintain) a Joint Names Policy for All Risks Insurance, with cover no less than that specified in clause 6.8 (Related Definitions) for the full reinstatement value of [the Works] / [the Sections] [(plus [*insert percentage*] to cover professional fees, as stated in the Contract Particulars)].

We attach a copy of the Joint Names Policy.

Resource 322

Description	A notice from the Contractor to a Purchaser or Tenant, if Professional Indemnity insurance ceases to be available.
Type	Notice: Contractor to Purchaser or Tenant
Provision references	Part 1 of Schedule 5, paragraph 5
Drafting notes	Paragraph 7 of the P&T Rights requires that notices under this paragraph must be delivered by hand, or sent by Signed For 1st class or Special Delivery Guaranteed post, or where an email address has been stated for the Purchaser or Tenant in the Employer's notice under clause 7.7.1, by email to the Purchaser or Tenant at such address or email address as they may from time to time notify the Contractor or (if no such address is then current) their last known principal business address or (where a body corporate) their registered or principal office. **Timing:** Paragraph 5 of the P&T Rights requires the Contractor to give this notice 'immediately' upon such insurance ceasing to be available at commercially reasonable rates or terms. It may be helpful to attach a copy of the P&T Rights, to assist the Purchaser or Tenant.

Reference: P&T Rights; Professional Indemnity insurance (Part 1 of Schedule 5)

We refer to the P&T Rights, included in Part 1 (Third Party Rights for Purchasers and Tenants) of Schedule 5 (Third Party Rights) of the building contract (the 'Contract') dated [*insert date*] made between us (the 'Contractor') and [*insert name of Employer*] (the 'Employer').

[A copy of the P&T Rights is enclosed, for reference.]

In accordance with paragraph 5 of the P&T Rights, we write to notify you that the Professional Indemnity insurance required under the Contract (on the terms and for the period referred to in clause 6.15 (Professional Indemnity Insurance:

Obligation to insure) and the related Contract Particulars) has ceased to be available at commercially reasonable rates or terms.

[We enclose a letter from our insurance broker to that effect.]

We propose to arrange a meeting as soon as possible, to discuss the means of best protecting our respective positions in the absence of such insurance.

Resource 323

Description	The provision by the Contractor to a Purchaser or Tenant of documentary evidence of Professional Indemnity insurance.
Type	Correspondence: Contractor to Purchaser or Tenant
Provision references	Part 1 of Schedule 5, paragraph 5

Reference: P&T Rights; Professional Indemnity insurance (Part 1 of Schedule 5)

We write further to your request dated [*insert date*], made in accordance with paragraph 5 of the P&T Rights, included in Part 1 (Third Party Rights for Purchasers and Tenants) of Schedule 5 (Third Party Rights) of the building contract (the 'Contract') dated [*insert date*] made between us (the 'Contractor') and [*insert name of Employer*] (the 'Employer').

We enclose documentary evidence to demonstrate that the Professional Indemnity policy required under paragraph 5 of the P&T Rights is being maintained.

Resource 324

Description	A notice from the Contractor to a Funder, relating to step-in rights, and prior to the Contractor being entitled to terminate its employment under the Contract.
Type	Notice: Contractor to Funder
Provision references	Part 2 of Schedule 5, paragraph 6.1.1
Drafting notes	Paragraph 11 of the Funder Rights requires that notices under this paragraph must be delivered by hand, or sent by Signed For 1st class or Special Delivery Guaranteed post, or where an email address has been stated for the Funder in the Employer's notice under clause 7.8.1, by email to the Funder at such address or email address as they may from time to time notify the Contractor or (if no such address is then current) their last known principal business address or (where a body corporate) their registered or principal office. It may be helpful to provide a copy of the Funder Rights, to assist the Funder.

Reference: Funder Rights (Part 2 of Schedule 5)

We refer to the Funder Rights, included in Part 2 (Third Party Rights for a Funder) of Schedule 5 (Third Party Rights) of the building contract (the 'Contract') dated [*insert date*] made between us (the 'Contractor') and [*insert name of Employer*] (the 'Employer').

[A copy of the Funder Rights is enclosed, for reference.]

In accordance with paragraph 6.1.1 of the Funder Rights, we enclose copies of the following notices required by the Contract to be sent to the Employer, prior to us being entitled to give notice under the Contract that our employment under the Contract is terminated: [*insert description of notices, along with corresponding references.*]

Resource 325

Description	A notice from the Contractor to a Funder, where the Contractor has the right to terminate its employment under the Contract.
Type	Notice: Contractor to Funder
Provision references	Part 2 of Schedule 5, paragraph 6.1.2
Drafting notes	Paragraph 11 of the Funder Rights requires that notices under this paragraph must be delivered by hand, or sent by Signed For 1st class or Special Delivery Guaranteed post, or where an email address has been stated for the Funder in the Employer's notice under clause 7.8.1, by email to the Funder at such address or email address as they may from time to time notify the Contractor or (if no such address is then current) their last known principal business address or (where a body corporate) their registered or principal office. It may be helpful to provide a copy of the Funder Rights, to assist the Funder.

Reference: Funder Rights (Part 2 of Schedule 5)

We refer to the Funder Rights, included in Part 2 (Third Party Rights for a Funder) of Schedule 5 (Third Party Rights) of the building contract (the 'Contract') dated [*insert date*] made between us (the 'Contractor') and [*insert name of Employer*] (the 'Employer').

[A copy of the Funder Rights is enclosed, for reference.]

In accordance with paragraph 6.1.2 of the Funder Rights, we write to notify you that we have the right under the Contract forthwith to notify the Employer that our employment under the Contract is terminated, on the following basis:

[*Insert explanation as to why the Contractor has the right to notify the Employer that the Contractor's employment under the Contract is terminated.*]

In accordance with paragraph 6.3.1, we will not issue a notice to the Employer to which paragraph 6.1.2 refers before the lapse of 14 days from your receipt of this notice.

Resource 326

Description	A notice from the Contractor to a Funder, where the Contractor has the right to treat the Contract as having been repudiated by the Employer.
Type	Notice: Contractor to Funder
Provision references	Part 2 of Schedule 5, paragraph 6.2
Drafting notes	Paragraph 11 of the Funder Rights requires that notices under this paragraph must be delivered by hand, or sent by Signed For 1st class or Special Delivery Guaranteed post, or where an email address has been stated for the Funder in the Employer's notice under clause 7.8.1, by email to the Funder at such address or email address as they may from time to time notify the Contractor or (if no such address is then current) their last known principal business address or (where a body corporate) their registered or principal office. It may be helpful to provide a copy of the Funder Rights, to assist the Funder.

Reference: Funder Rights (Part 2 of Schedule 5)

We refer to the Funder Rights, included in Part 2 (Third Party Rights for a Funder) of Schedule 5 (Third Party Rights) of the building contract (the 'Contract') dated [*insert date*] made between us (the 'Contractor') and [*insert name of Employer*] (the 'Employer').

[A copy of the Funder Rights is enclosed, for reference.]

In accordance with paragraph 6.2 of the Funder Rights, we write to notify you that we intend to notify the Employer that we are entitled to treat the Contract as having been repudiated by the Employer, on the following basis:

[*Insert explanation as to why the Contractor considers it is entitled to treat the Contract as having been repudiated by the Employer.*]

In accordance with paragraph 6.3.2, we will not notify the Employer that we are treating the Contract as having been repudiated by the Employer as referred to in paragraph 6.2 before the lapse of 14 days from your receipt of this notice.

Resource 327

Description	A notice from the Contractor to a Funder, if Professional Indemnity insurance ceases to be available.
Type	Notice: Contractor to Funder
Provision references	Part 2 of Schedule 5, paragraph 9
Drafting notes	Paragraph 11 of the Funder Rights requires that notices under this paragraph must be delivered by hand, or sent by Signed For 1st class or Special Delivery Guaranteed post, or where an email address has been stated for the Funder in the Employer's notice under clause 7.8.1, by email to the Funder at such address or email address as they may from time to time notify the Contractor or (if no such address is then current) their last known principal business address or (where a body corporate) their registered or principal office. **Timing:** Paragraph 9 of the Funder Rights requires the Contractor to give this notice 'immediately' upon such insurance ceasing to be available at commercially reasonable rates or terms. It may be helpful to provide a copy of the Funder Rights, to assist the Funder.

Reference: Funder Rights; Professional Indemnity insurance (Part 2 of Schedule 5)

We refer to the Funder Rights, included in Part 2 (Third Party Rights for a Funder) of Schedule 5 (Third Party Rights) of the building contract (the 'Contract') dated [*insert date*] made between us (the 'Contractor') and [*insert name of Employer*] (the 'Employer').

[A copy of the Funder Rights is enclosed, for reference.]

In accordance with paragraph 9 of the Funder Rights, we write to notify you that the Professional Indemnity insurance required under the Contract (on the terms and for the period referred to in clause 6.15 (Professional Indemnity Insurance: Obligation to insure) and its related Contract Particulars) has ceased to be available at commercially reasonable rates or terms.

[We enclose a letter from our insurance broker to that effect.]

We propose to arrange a meeting as soon as possible, to discuss the means of best protecting our respective positions in the absence of such insurance.

Resource 328

Description	The provision by the Contractor to a Funder of documentary evidence of Professional Indemnity insurance.
Type	Correspondence: Contractor to Funder
Provision references	Part 2 of Schedule 5, paragraph 9

Reference: Funder Rights; Professional Indemnity insurance (Part 2 of Schedule 5)

We write further to your request dated [*insert date*], made in accordance with paragraph 9 of the Funder Rights, included in Part 2 (Third Party Rights for a Funder) of Schedule 5 (Third Party Rights) of the building contract (the 'Contract') dated [*insert date*] made between us (the 'Contractor') and [*insert name of Employer*] (the 'Employer').

We enclose documentary evidence to demonstrate that the Professional Indemnity policy required under paragraph 9 of the Funder Rights is being maintained.

Resource 329

Description	A monthly progress report to be prepared by the Contractor.
Type	Contractor's progress report
Provision references	n/a

Below are the suggested outline contents for a monthly progress report, to be written by the Contractor:

- Reporting period
- Executive Summary
- Health and Safety / CDM Regulations / Building Regulations
  - Construction Phase Plan
  - Building Safety Act 2022
  - CDM Regulations: Principal Designer
  - CDM Regulations: Principal Contractor
  - Building Regulations: Principal Designer
  - Building Regulations: Principal Contractor
- Programme and progress
  - Activities completed in the previous reporting period (or an alternative duration)
  - Activities forecast to be completed in the next reporting period (or an alternative duration)
  - Anticipated dates(s) of practical completion
  - Delays to the progress of the Works
- Risk register

- Design
  - Design Submission Procedure
  - Inadequacies, divergences, and discrepancies
  - Statutory Requirements
  - Samples
  - As-built drawings
- Commercial
  - Adjustments to the Contract Sum
  - Interim Payments
  - Employer's instructions
  - Provisional Sums
  - Sub-contracting and procurement
  - Collateral warranties and third party rights
  - Listed Items
  - Cashflow forecast
- Quality Assurance

Other potential sections:

- Statutory Providers
- Third party agreements
- Work not forming part of the Contract
- Instructions or actions required from the Employer

Resource 330

Description	A meeting agenda template for use by the Contractor and the Employer.
Type	Meeting agenda
Provision references	n/a
Drafting notes	Meeting agendas (along with any relevant documents) should be issued as early as possible before the relevant meeting.

Below is a meeting agenda template:

- Project: [*insert name of project*]
- Meeting: [*insert description of meeting*]
- Location: [*insert location, whether in person or online*]
- Date: [*insert date*]
- Time: [*insert start and finish time of the meeting*]
- Invited: [*insert list of those invited to the meeting*]
- Distribution: [*insert distribution list for the agenda, to include those parties who are not invited to the meeting, but who should be kept informed*]
- Agenda: [*insert agenda*]
- Attachments: [*insert description of any relevant documents*]

Resource 331

Description	A meeting minutes template for use by the Contractor and the Employer.
Type	Meeting minutes
Provision references	n/a
Drafting notes	Meeting minutes (along with any relevant documents) should be issued as soon as possible after the relevant meeting.

Below are template meeting minutes:

- Project: [*insert name of project*]
- Meeting: [*insert description of meeting*]
- Location: [*insert location, whether in person or online*]
- Date: [*insert date*]
- Time: [*insert start and finish time of the meeting*]
- Attended: [*insert list of those who attended the meeting*]
- Apologies: [*insert list of those who were invited to the meeting, but did not attend*]
- Distribution: [*insert distribution list for the minutes, to include those parties who are not invited to the meeting, but who should be kept informed*]
- Minutes: [*insert minutes, along with all relevant actions, deadlines, targets, etc.*]
- Attachments: [*insert description of any relevant documents*]

Resource 332

Description	A memorandum template for use by the Contractor and the Employer.
Type	Memorandum
Provision references	n/a

Below is a memorandum template, whether or internal or external use:

- Project: [*insert name of project*]
- Subject: [*insert description of the memorandum*]
- From: [*insert name or party*]
- To: [*insert names or parties*]
- Author: [*insert name of memorandum author*]
- Date: [*insert date*]
- Memorandum: [*insert main body of memorandum*]

Resource 333

Description	A site diary template for use by the Contractor.
Type	Site Diary
Provision references	n/a

Below is a site diary template, for use by the Contractor:

- Project: [*insert name of project*]
- Date: [*insert date*]
- Author: [*insert author of site diary entry*]

Design

- Design Submission Procedure: [*insert details of any Contractor's Design Documents submitted or returned*]
- Approvals or consents: [*insert details or any approvals or consents requested or received*]

Procurement

- Sub-contracting: [*insert update of any sub-contracts or appointments entered into*]
- Materials or goods: [*insert update of any materials or goods ordered or delivered to site or off site*]

Progress of the Works

- Completion Date: [*insert relevant Completion Date*]
- Summary of work carried out today: [*insert summary*]
- Problems / delays: [*insert description of any problems or delays encountered that day*]
- Quality Assurance: [*insert update of any quality assurance undertaken that day*]

Health and Safety

- Concerns and remedial actions: [*insert description of any health and safety concerns, and remedial actions undertaken*]
- Accidents: [*insert update*]

Site conditions

- Nearest Met Office weather station: [*insert location*]
- Weather: [*insert brief description*]
- Temperatures: [*insert maximum and minimum temperature*]
- Ground conditions: [*insert brief description*]

Meetings

[*Insert details of any key meetings on site that day*]

Workforce

- Contractor: [*insert details of the Contractor's staff on site*]
- Contractor's Persons: [*insert details of Contractor's Persons (sub-contractors, consultants, etc.) on site*]

Visitors

[*Insert details of any visitors to the site that day*]

Resource 334

Description	A project calendar template for use by the Contractor and the Employer.
Type	Project calendar
Provision references	n/a

Below is a list of key dates to include in the project calendar, which should be updated regularly:

- the Base Date
- Dates of Possession
- Dates of Completion of Sections
- Completion Dates (as adjusted)
- anticipated dates of practical completion
- dates of Practical Completion Statements / Section Completion Statements
- dates of Non-Completion Notices
- Relevant Dates, relating to partial possession by the Employer
- commencement and expiry of Rectification Periods
- dates of Notices of Completion of Making Good
- Interim Valuation Dates, dates of Interim Payment Applications, due dates, and final dates for payment
- dates relating to the Final Statement / Employer's Final Statement
- dates of insurance renewals
- the date of expiry of performance bonds and parent company guarantees
- dates for release of Retention
- Public Holidays (namely Christmas Day, Good Friday, or a day which under the Banking and Financial Dealings Act 1971 is a bank holiday)

Resource 335

Description	An instruction template for use by the Employer
Type	Instruction
Provision references	n/a
Drafting notes	Whilst in certain circumstances it may be appropriate to include the Employer's estimate of the cost of a particular instruction, any estimated effect on the anticipated dates of practical completion should not be included, to avoid any conflict with the extension of time mechanism provided for within clause 2.23 to clause 2.26 (Adjustment of Completion Date).

Below is a template instruction for use by the Employer:

- Reference: [*insert reference*]
- Date: [*insert date*]
- Conditions reference: [*insert relevant provision within the Conditions, for example, clause 3.9 (Instructions requiring Changes)*]
- Instruction: [*insert details of the instruction*]
- Cost estimate / agreed cost (exclusive of VAT): [*insert amount, and state whether it is the Employer's estimate, or the cost agreed by the Parties*]

Reference

Further Reading

JCT Publications

In addition to the Contract itself, the following JCT publications[1] may also be helpful for the Parties:

- Design and Build Contract Guide 2024 (DB/G 2024)
- Design and Build Contract 2024 Tracked Change Document
- Design and Build Contract 2024 Model Form for the Rights Particulars
- Design and Build Contract 2024 Fluctuations Options A, B and C
- Contractor Collateral Warranty for (i) Purchasers and Tenants (CWa/P&T) and (ii) a Funder (CWa/F)
- Sub-Contractor Collateral Warranty for (i) a Purchaser or Tenant (SCWa/P&T), (ii) a Funder (SCWa/F), and (iii) the Employer (SCWa/E)
- Design and Build Sub-Contract Agreement 2024 (DBSub/A 2024) and JCT Design and Build Sub-Contract Conditions 2024 (DBSub/C 2024)
- Short Form of Sub-Contract (ShortSub)
- Sub-subcontract (SubSub)
- Contract Administration Model Forms (DB 2024 Admin)
- Framework Agreement (FA)
- Formula Rules 2024 (FR 2024)
- JCT 2024 edition of the Construction Industry Model Arbitration Rules (CIMAR)
- Adjudication Agreement (Adj) and JCT Adjudication Agreement (Named Adjudicator) (Adj/N)
- Dispute Adjudication Board Documentation (DAB)
- 2019 Practice Note: BIM and JCT Contracts
- JCT Guidance on The Building Safety Act 2022[2]

Books

Below is a list of some of the classic texts on construction law, and contract law more generally:

- Bailey, J. (2024) *Construction Law*. 4th edition. London: London Publishing Partnership.
- Beale, H.G. (ed.) (2023) *Chitty on Contracts*. 35th edition. London: Sweet & Maxwell.
- Dennys, N., Clay, R., Atkin Chambers (eds.) (2019) *Hudson's Building and Engineering Contracts*. 14th edition. London: Sweet & Maxwell.

[1] Available at https://www.jctltd.co.uk

[2] Available at https://www.jctltd.co.uk/docs/The-Building-Safety-Act-2022.pdf

- Furst, S., Ramsey, V. (eds.) (2020) *Keating on Construction Contracts*. 11th edition. London: Sweet & Maxwell.
- Peel, E. (2020) *Treitel: The Law of Contract*. 15th edition. London: Sweet & Maxwell.
- Uff, J. (2021) *Construction Law*. 13th edition. London: Sweet & Maxwell.

Articles

Articles on construction law are regularly published by the following:

- *Construction Law Journal*. London: Sweet & Maxwell (https://www.sweetandmaxwell.co.uk/Product/Construction-Law/Construction-Law-Journal/Journal/30791391)
- The Society of Construction Law (https://www.scl.org.uk/)
- The Adjudication Society (https://www.adjudication.org/)

Online Resources

Below are a number of further online resources which may prove helpful:

- Cabinet Office (2022) *The Construction Playbook*. Version 1.1. Available at: https://www.gov.uk/government/publications/the-construction-playbook
- Construction Productivity Taskforce (2022) *Trust and Productivity: The Private Sector Construction Playbook*. Available at: https://media.bethebusiness.com/documents/5195_Construction_Brochure_SinglePgs.pdf
- Construction Leadership Council guidance, available at: https://www.constructionleadershipcouncil.co.uk
- Health and Safety Executive guidance, available at: https://www.hse.gov.uk/guidance/index.htm
- Society of Construction Law (2017) *Delay and Disruption Protocol*. 2nd edition. Available at: https://www.scl.org.uk/sites/default/files/documents/SCL_Delay_Protocol_2nd_Edition_Final.pdf
- RISCAuthority, Fire Protection Association, and Contractors Legal Group (2023) *Fire Prevention on Construction Sites: The Joint Code of Practice on the Protection from Fire of Construction Sites and Buildings Undergoing Renovation*. 10th edition, incorporating amendment 1. Gloucestershire: Fire Protection Association. Available at: https://www.thefpa.co.uk/advice-and-guidance/free-documents?q=joint%20code%20of%20practice

Adjudication and Arbitration

Below is the list of nominating bodies for both adjudication and arbitration, identified in the entries against the references to clause 9.3 (Adjudication) and clause 9.4 (Arbitration) in the Contract Particulars.

- Royal Institute of British Architects (https://www.architecture.com)
- The Royal Institution of Chartered Surveyors (https://www.rics.org)
- Chartered Institute of Arbitrators (https://www.ciarb.org)

Legislation

The table below identifies the key provisions in the Contract which relate to specific legislation (rather than simply the 'Statutory Requirements').

References to any legislation throughout Part 1 and Part 2 is to such legislation as amended, and in force from time to time.

Legislation	Provision reference
Arbitration Act 1996	clause 9.8 (Arbitration: Appeal – questions of law), clause 9.9 (Arbitration: Arbitration Act 1996)
Banking and Financial Dealings Act 1971	clause 1.1 (Definitions), definition of '*Public Holiday*'
Bankruptcy (Scotland) Act 2016	clause 8.1 (Meaning of insolvency)
Bribery Act 2010	clause 8.6 (Corruption and regulation 73(1)(b) of the PC Regulations)
Building Regulations 2010 ('Building Regulations')	Article 7 (Building Regulations – Principal Designer and Principal Contractor), clause 2.7.2 (Contract Documents), clause 3.4.2.3 (Conditions of sub-contracting), clause 3.16 (CDM Regulations and Part 2A of the Building Regulations)
Building Regulations etc. (Amendment) (England) Regulations 2023	Article 7 (Building Regulations – Principal Designer and Principal Contractor)
Building Safety Act 2022	clause 3.16 (CDM Regulations and Part 2A of the Building Regulations)
Companies Act 2006	clause 8.1.4.4 (Meaning of insolvency)
'Construction Act' 1996	See the Housing Grants, Construction and Regeneration Act 1996
Construction (Design and Management) Regulations 2015 ('CDM Regulations')	Article 6 (CDM Regulations – Principal Designer and Principal Contractor), clause 1.1 (Definitions), definition of '*Construction Phase Plan*', clause 2.7.2 (Contract Documents), clause 2.37 (As-built Drawings), clause 3.4.2.3 (Conditions of sub-contracting), clause 3.16 (CDM Regulations and Part 2A of the Building Regulations)
Contracts (Rights of Third Parties) Act 1999	clause 1.6 (Contracts (Rights of Third Parties) Act 1999)
Corporate Insolvency and Governance Act 2020	clause 8.1 (Meaning of insolvency)
Defective Premises Act 1972	clause 2.17 (Design Work – liabilities and limitation)

(*Continued*)

Legislation	Provision reference
Finance Act 2004	Fourth Recital, clause 4.5 (Construction Industry Scheme (CIS))
Freedom of Information Act 2000	Supplemental Provision 8 (Transparency)
Health and Safety (Consultation with Employees) Regulations 1996	Supplemental Provision 5, paragraph 5.2.4 (Health and safety)
Housing Grants, Construction and Regeneration Act 1996	Article 8 (Adjudication), clause 9.3 (Adjudication)
Insolvency Act 1986	clause 8.1 (Meaning of insolvency)
Local Government Act 1972	clause 8.6 (Corruption and regulation 73(1)(b) of the PC Regulations)
Management of Health and Safety at Work Regulations 1999	Supplemental Provision 5, paragraph 5.2.3 (Health and safety)
Public Contracts Regulations 2015 ('PC Regulations')	clause 8.6 (Corruption and regulation 73(1)(b) of the PC Regulations), clause 8.11 (Termination by either Party and regulation 73(1)(a) of the PC Regulations), Supplemental Provision 9 (The Public Contracts Regulations 2015)
Scheme for Construction Contracts (England and Wales) Regulations 1998 ('Scheme')	clause 9.3 (Adjudication)

Cases

Below is a list of cases which relate to specific provisions in the Contract:

- *FES Limited v HFD Construction Group Limited* [2024] CSOH 20 (see clause 4.20 (Loss and Expense: Notification and ascertainment)
- *Impresa Castelli SpA v Cola Holdings Ltd* [2002] EWHC 1363 (TCC) (see clause 2.5 (Early use by Employer), and clause 2.30 to clause 2.34 (Partial Possession by Employer))
- *Providence Building Services Limited v Hexagon Housing Association Limited* [2024] EWCA Civ 962 (see clause 8.9 (Termination by Contractor: Default by Employer))
- *Re Compound Photonics Group Ltd [2022] EWCA Civ 1371* (see Article 3 (Collaborative working))
- *Triple Point Technology Inc v PTT Public Company Ltd* [2021] UKSC 29 (see clause 2.29.5 (Payment or allowance of liquidated damages))
- *West Faulkner Associates v London Borough of Newham* [1994] 71 BLR 1 (see clause 2.3 (Date of Possession – progress), and clause 8.4.1.2 (Termination by Employer: Default by Contractor))

Resource Index

Reference	Description	Resource type	Provision references	Page
1	The nomination by the Employer of a replacement Employer's Agent.	Correspondence: Employer to Contractor	Article 4 (Employer's Agent)	408
2	A revision by the Employer to the Employer's Agent's authority to act on its behalf.	Notice: Employer to Contractor	Article 4 (Employer's Agent)	408
3	The replacement of the Employer's nominee or Contractor's nominee, relating to clause 9.1.	Notice: Contractor or Employer to the other Party	Contract Particulars, clause 9.1	409
4	A notice of a conflict within the Contract Documents.	Notice: Contractor or Employer to the other Party	clause 1.3	410
5	A communication protocol template to govern the exchange of notices, communications, and documents between the Parties.	Communication Protocol: Employer and Contractor	clause 1.7.2	410
6	A notice to the other Party confirming a change of email address.	Notice: Contractor or Employer to the other Party	clause 1.7.4	412
7	The confirmation of an oral communication, in the case of an emergency.	Confirmation: Contractor or Employer to the other Party	clause 1.7.5	412
8	A reminder that consents or approvals must not be unreasonably delayed or withheld.	Reminder: Employer and Contractor	clause 1.10	413
9	The provision by the Contractor to the Employer of approvals received by the Contractor, in connection with the Statutory Requirements.	Correspondence: Contractor to Employer	clause 2.1.3	414

(*Continued*)

(*Continued*)

Reference	Description	Resource type	Provision references	Page
20	The provision by the Contractor to the Employer of reasonable proof that materials and goods are compliant.	Reply: Contractor to Employer	clause 2.2.5	419
21	Confirmation by the Contractor to the Employer of steps taken to encourage Contractor's Persons to be CSCS registered cardholders.	Confirmation: Contractor to Employer	clause 2.2.6	420
22	Confirmation by the Employer that it has given possession of the site (or part of it) to the Contractor.	Confirmation: Employer to Contractor	clause 2.3	420
23	Confirmation by the Contractor that it has taken possession of the site (or part of it) from the Employer.	Confirmation: Contractor to Employer	clause 2.3	421
24	A notice from the Employer that it is deferring giving possession of the site (or part of it) to the Contractor.	Notice: Employer to Contractor	clause 2.4	422
25	A request by the Employer for the Contractor's consent for early use.	Request: Employer to Contractor	clause 2.5.1	422
26	Consent given by the Contractor to the Employer for early use.	Request for consent given: Contractor to Employer	clause 2.5.1	423
27	Consent withheld by the Contractor to the Employer for early use.	Request for consent withheld: Contractor to Employer	clause 2.5.1	423
28	A request by the Contractor or Employer for confirmation from the insurers under the Works Insurance Policy (and in relation to the Existing Structures) that early use will not prejudice the insurance.	Request: Contractor or Employer to insurer	clause 2.5.1	424
29	A notice from the Contractor to the Employer of an additional insurance premium for early use.	Notice: Contractor to Employer	clause 2.5.2	424

(Continued)

Reference	Description	Resource type	Provision references	Page
30	An acceptance by the Employer of an additional insurance premium for early use.	Confirmation: Employer to Contractor	clause 2.5.2	425
31	A reminder from the Employer to the Contractor of work not forming part of the Contract.	Reminder: Employer to Contractor	clause 2.6.1	425
32	Permission given by the Contractor to the Employer for the execution of work not forming part of the Contract.	Confirmation: Contractor to Employer	clause 2.6.1	426
33	A request by the Employer for the Contractor's consent for the Employer to arrange for the execution of work not forming part of the Contract.	Request: Employer to Contractor	clause 2.6.2	427
34	Consent given by the Contractor for the Employer to arrange for the execution of work not forming part of the Contract.	Request for consent given: Contractor to Employer	clause 2.6.2	427
35	Consent withheld by the Contractor for the Employer to arrange for the execution of work not forming part of the Contract.	Request for consent withheld: Contractor to Employer	clause 2.6.2	428
36	A request by the Contractor to the Employer to inspect the Contract Documents.	Request: Contractor to Employer	clause 2.7.1	428
37	The provision by the Employer to the Contractor of a certified copy of the Contract Documents, pre-construction information under the CDM Regulations, and building information under the Building Regulations.	Correspondence: Employer to Contractor	clause 2.7.2	428
38	A reminder by the Contractor to the Employer's Agent of its entitlement to access copies of the Contract Documents and the Contractor's Design Documents.	Correspondence: Contractor to Employer's Agent	clause 2.7.3	429

(*Continued*)

Reference	Description	Resource type	Provision references	Page
39	Confirmation from the Contractor to the Employer of the provision of Contractor's Design Documents.	Correspondence: Contractor to Employer	clause 2.8	429
40	A reminder from the Employer to the Contractor of its obligation to comply with the Design Submission Procedure.	Reminder: Employer to Contractor	clause 2.8	430
41	Confirmation by the Employer to the Contractor of the boundaries of the site.	Confirmation: Employer to Contractor	clause 2.9	431
42	A notice from the Contractor to the Employer of a divergence between the Employer's Requirements and the definition of the site boundary, and a request for an instruction.	Notice: Contractor to Employer	clause 2.10.2	431
43	A notice from the Employer to the Contractor of a divergence between the Employer's Requirements and the definition of the site boundary, with an accompanying instruction.	Notice: Employer to Contractor	clause 2.10.2	432
44	A notice from the Contractor to the Employer of an inadequacy in the Employer's Requirements, which is dealt with in the Contractor's Proposals.	Notice: Contractor to Employer	clause 2.12, clause 2.13	432
45	A notice from the Contractor to the Employer of an inadequacy in the Employer's Requirements (which is not dealt with in the Contractor's Proposals), and a request for an instruction.	Notice: Contractor to Employer	clause 2.12, clause 2.13	433
46	An instruction from the Employer following a notice of inadequacy in the design in the Employer's Requirements, under clause 2.12.	Instruction: Employer to Contractor	clause 2.13	433

(*Continued*)

Reference	Description	Resource type	Provision references	Page
47	A notice from the Contractor to the Employer of a discrepancy or divergence in or between the Contractor's Proposals and/or other Contractor's Design Documents.	Notice: Contractor to Employer	clause 2.13, clause 2.14.1	434
48	An instruction from the Employer following a notice of discrepancy or divergence in or between the Contractor's Proposals and/or other Contractor's Design Documents.	Instruction: Employer to Contractor	clause 2.13, clause 2.14.1	434
49	A notice from the Contractor to the Employer of a discrepancy in or between the Employer's Requirements and/or instructions.	Notice: Contractor to Employer	clause 2.13, clause 2.14.2	435
50	An instruction from the Employer following a notice of a discrepancy in or between the Employer's Requirements and/or instructions.	Instruction: Employer to Contractor	clause 2.13, clause 2.14.2	436
51	A notice from the Contractor to the Employer of a divergence between the Statutory Requirements and the Employer's Requirements (including any Change), the Contractor's Proposals, or other Contractor's Design Documents, and a request for the Employer's consent to proceed with the Contractor's proposed amendment to remove the divergence.	Notice: Contractor to Employer	clause 2.15.1	436
52	A notice from the Employer to the Contractor of a divergence between the Statutory Requirements and the Employer's Requirements (including any Change), the Contractor's Proposals, or other Contractor's Design Documents, and a request for a proposed amendment to remove the divergence.	Notice: Employer to Contractor	clause 2.15.1	437

(Continued)

Reference	Description	Resource type	Provision references	Page
53	Consent given by the Employer to the Contractor, for the Contractor's proposed amendment, following a notice of a divergence between the Statutory Requirements and the Employer's Requirements (including any Change), the Contractor's Proposals, or other Contractor's Design Documents.	Consent given: Employer to Contractor	clause 2.15.1	438
54	Consent withheld by the Employer to the Contractor, for the Contractor's proposed amendment, following a notice of a divergence between the Statutory Requirements and the Employer's Requirements (including any Change), the Contractor's Proposals, or other Contractor's Design Documents.	Consent withheld: Employer to Contractor	clause 2.15.1	438
55	Confirmation by the Contractor to the Employer of a change in the Statutory Requirements after the Base Date which necessitates an alteration or modification to the Works, and a request for an instruction.	Confirmation and request: Contractor to Employer	clause 2.15.2.1	439
56	Confirmation by the Employer to the Contractor of a change in the Statutory Requirements after the Base Date which necessitates an alteration or modification to the Works, and an instruction.	Confirmation and instruction: Employer to Contractor	clause 2.15.2.1	439
57	Confirmation from the Contractor to the Employer of an amendment to the Contractor's Proposals for conformity with the Development Control Requirements, and a request for an instruction.	Confirmation: Contractor to Employer	clause 2.15.2.2	440

(*Continued*)

Reference	Description	Resource type	Provision references	Page
58	Confirmation from the Employer to the Contractor of an amendment to the Contractor's Proposals for conformity with the Development Control Requirements, and an instruction.	Confirmation and instruction: Employer to Contractor	clause 2.15.2.2	441
59	Confirmation from the Contractor to the Employer of an amendment to the Employer's Requirements to conform with the Statutory Requirements, and a request for an instruction.	Confirmation and request for instruction: Contractor to Employer	clause 2.15.2.3	442
60	Confirmation from the Employer to the Contractor of an amendment to the Employer's Requirements to conform with the Statutory Requirements, and an instruction.	Confirmation and instruction: Employer to Contractor	clause 2.15.2.3	442
61	A tracker template for recording inadequacies, divergences, and discrepancies under clause 2.11 to clause 2.15.	Tracker: Contractor and Employer	clause 2.11 to clause 2.15	443
62	A notice from the Contractor to the Employer of an emergency compliance with the Statutory Requirements.	Notice: Contractor to Employer	clause 2.16.2	442
63	A notice from the Contractor to the Employer of a potential infringement of patent rights.	Notice: Contractor to Employer	clause 2.20.2	444
64	A notice from the Employer to the Contractor of a potential infringement of patent rights.	Notice: Employer to Contractor	clause 2.20.2	444
65	Confirmation or withdrawal of an instruction by the Employer, relating to a potential infringement of patent rights.	Confirmation: Employer to Contractor	clause 2.20.2	445

(Continued)

Reference	Description	Resource type	Provision references	Page
66	A request from the Contractor to the Employer for consent to remove Site Materials from storage on or adjacent to the Works.	Request: Contractor to Employer	clause 2.21	445
67	Consent given by the Employer for the Contractor to remove Site Materials from storage on or adjacent to the Works.	Consent given: Employer to Contractor	clause 2.21	446
68	Consent withheld by the Employer for the Contractor to remove Site Materials from storage on or adjacent to the Works.	Consent withheld: Employer to Contractor	clause 2.21	446
69	A tracker template for recording Listed Items.	Tracker: Contractor and Employer	clause 2.22	447
70	A notice from the Contractor to the Employer of a delay to the progress of the Works, and particulars of its expected effects.	Notice: Contractor to Employer	clause 2.24	447
71	A notice from the Contractor to the Employer of a material change in the estimated delay, or in other particulars.	Notice: Contractor to Employer	clause 2.24.3	448
72	A notice from the Employer to the Contractor requesting further information to enable the Employer to reach a decision under clause 2.25.1.	Notice: Employer to Contractor	clause 2.24.4	449
73	The provision by the Contractor to the Employer of further information, further to the Employer's notice under clause 2.24.4.	Correspondence: Contractor to Employer	clause 2.24.4	450
74	A reminder from the Contractor to the Employer of its obligation to notify the Contractor of its decision under clause 2.25.1 as soon as reasonably practicable.	Reminder: Contractor to Employer	clause 2.25.2, clause 2.24	450

(Continued)

Reference	Description	Resource type	Provision references	Page
75	A notice from the Employer to the Contractor, giving an extension of time.	Notice: Employer to Contractor	clause 2.25.1	451
76	A notice from the Employer to the Contractor, not giving an extension of time.	Notice: Employer to Contractor	clause 2.25.1	453
77	A notice from the Employer to the Contractor of the final review of the Completion Dates.	Notice: Employer to Contractor	clause 2.25.5	454
78	A reminder from the Employer to the Contractor of its obligation to use best endeavours to prevent delay.	Reminder: Employer to Contractor	clause 2.25.6.1	455
79	A reminder from the Employer to the Contractor to do all that may reasonably be required to the satisfaction of the Employer to proceed with the Works.	Reminder: Employer to Contractor	clause 2.25.6.2	455
80	A checklist for the Contractor for drafting delay notices and providing further particulars.	Checklist: Contractor	clause 2.24, clause 2.25	456
81	A checklist for the Employer for giving extensions of time.	Checklist: Employer	clause 2.25	457
82	A tracker template for delay notices, for both the Contractor and the Employer.	Tracker: Contractor and Employer	clause 2.24, clause 2.25	458
83	A reminder from the Contractor to the Employer of the Contractor's anticipated date of practical completion.	Reminder: Contractor to Employer	clause 2.27	459
84	A Practical Completion Statement template, from the Employer to the Contractor.	Employer to Contractor: Practical Completion Statement	clause 2.27	460
85	A Section Completion Statement template, from the Employer to the Contractor.	Employer to Contractor: Section Completion Statement	clause 2.27	460

(Continued)

Reference	Description	Resource type	Provision references	Page
86	A checklist for the requirements of practical completion.	Checklist: Contractor and Employer	clause 2.27	461
87	A Non-Completion Notice template, from the Employer to the Contractor.	Notice: Employer to Contractor	clause 2.28	462
88	A further Non-Completion Notice template from the Employer to the Contractor, if a new Completion Date was fixed after the issue of the initial Non-Completion Notice.	Notice: Employer to Contractor	clause 2.28	463
89	A notice from the Employer to the Contractor that the Employer may require payment of, or may withhold or deduct, liquidated damages.	Notice: Employer to Contractor	clause 2.29.1.2	464
90	A notice from the Employer to the Contractor that the Employer requires payment of, or will withhold or deduct, liquidated damages.	Notice: Employer to Contractor	clause 2.29.1	464
91	A reminder from the Contractor to the Employer to repay liquidated damages following an extension of time.	Reminder: Contractor to Employer	clause 2.29.3	465
92	Correspondence from the Employer to the Contractor that the Employer no longer may require payment of, or may withhold or deduct, liquidated damages.	Correspondence: Employer to Contractor	clause 2.29.4	466
93	A checklist for the deduction of liquidated damages by the Employer.	Checklist: Employer	clause 2.29	466
94	A request from the Employer for the Contractor's consent for the Employer to take partial possession.	Request: Employer to Contractor	clause 2.30	467
95	Consent given by the Contractor for the Employer to take partial possession.	Consent given: Contractor to Employer	clause 2.30	468

(*Continued*)

Reference	Description	Resource type	Provision references	Page
96	Consent withheld by the Contractor for the Employer to take partial possession.	Consent withheld: Contractor to Employer	clause 2.30	468
97	A notice from the Contractor to the Employer of partial possession.	Notice: Contractor to Employer	clause 2.30	469
98	A Notice of Completion of Making Good template for a Relevant Part, from the Employer to the Contractor.	Notice: Employer to Contractor	clause 2.32	469
99	A tracker template for recording partial possessions by the Employer.	Tracker: Contractor and Employer	clause 2.30 to clause 2.34	470
100	An instruction by the Employer of a schedule of defects, shrinkages, and other faults in the Works, to be made good by the Contractor.	Instruction: Employer to Contractor	clause 2.35.1	470
101	An instruction from the Employer requiring a particular defect, shrinkage, or other fault in the Works to be made good by the Contractor.	Instruction: Employer to Contractor	clause 2.35.2	471
102	An instruction from the Employer requiring a particular defect, shrinkage, or other fault in the Works not to be made good by the Contractor.	Instruction: Employer to Contractor	clause 2.35.2	472
103	A reminder from the Contractor to the Employer of its obligation to issue a Notice of Completion of Making Good.	Reminder: Contractor to Employer	clause 2.36	472
104	A Notice of Completion of Making Good template, from the Employer to the Contractor.	Notice: Employer to Contractor	clause 2.36	473
105	A tracker template for the making good of defects, shrinkages, or other faults in the Works.	Tracker: Contractor and Employer	clause 2.35, clause 2.36	473

(*Continued*)

Reference	Description	Resource type	Provision references	Page
106	A request from the Employer to the Contractor for further documents to show or describe the Works as built, or relate to the maintenance and operation of them, or their installations.	Request for further documents: Employer to Contractor	clause 2.37	474
107	A tracker template for recording the collation of Contractor's Design Documents, as-built drawings, and documents for operation and maintenance.	Tracker: Contractor and Employer	clause 2.37	474
108	A request from the Employer to the Contractor for access for the Employer's Agent and the Employer to the Contractor's workshops and other premises.	Request: Employer to Contractor	clause 3.1	475
109	A request from the Contractor to the Employer for approval to appoint a Site Manager.	Request: Contractor to Employer	clause 3.2	475
110	A request from the Contractor to the Employer for approval to replace the Site Manager.	Request: Contractor to Employer	clause 3.2	475
111	Approval given by the Employer for the Contractor to appoint a Site Manager (or a replacement Site Manager).	Approval given: Employer to Contractor	clause 3.2	476
112	Approval withheld by the Employer for the Contractor to appoint a Site Manager (or a replacement Site Manager).	Approval withheld: Employer to Contractor	clause 3.2	476
113	A request from the Contractor to the Employer for consent to sub-contract the whole or any part of the Works or the design for the Works.	Request: Contractor to Employer	clause 3.3.1	477

(*Continued*)

Reference	Description	Resource type	Provision references	Page
114	Consent given by the Employer for the Contractor to sub-contract the whole or any part of the Works or the design for the Works.	Consent given: Employer to Contractor	clause 3.3.1	477
115	Consent withheld by the Employer for the Contractor to sub-contract the whole or any part of the Works or the design for the Works.	Consent withheld: Employer to Contractor	clause 3.3.1	478
116	A tracker template for the appointment of sub-contractors in accordance with clause 3.3 and clause 3.4.	Tracker: Contractor and Employer	clause 3.3, clause 3.4	478
117	A notice from the Contractor to the Employer of a reasonable objection to an Employer's instruction.	Notice: Contractor to Employer	clause 3.5.1	479
118	A notice from the Employer to the Contractor requiring compliance with an instruction.	Notice: Employer to Contractor	clause 3.6	480
119	A reply from the Contractor to the Employer following a notice from the Employer requiring compliance with an instruction.	Reply: Contractor to Employer	clause 3.6	480
120	A further notice from the Employer to the Contractor, following the Employer's initial notice under clause 3.6, that the Employer will employ and pay other persons.	Notice: Employer to Contractor	clause 3.6	481
121	A further notice from the Employer to the Contractor, following the Employer's initial notice under clause 3.6, that the Employer has employed and paid other persons.	Notice: Employer to Contractor	clause 3.6	481
122	Confirmation by the Contractor to the Employer of an instruction given otherwise than in writing.	Confirmation: Contractor to Employer	clause 3.7.1	482

(*Continued*)

Reference	Description	Resource type	Provision references	Page
123	A notice from the Employer to the Contractor dissenting from the Contractor's confirmation of an instruction given otherwise than in writing.	Notice: Employer to Contractor	clause 3.7.1	482
124	A notice from the Employer to the Contractor confirming the terms of an instruction given otherwise than in writing.	Confirmation: Employer to Contractor	clause 3.7.2	483
125	A notice from the Employer to the Contractor retrospectively confirming the terms of an instruction given otherwise than in writing.	Confirmation: Employer to Contractor	clause 3.7.3	483
126	A request from the Contractor to the Employer to notify which of the Conditions empowers the Employer to issue a particular instruction.	Request: Contractor to Employer	clause 3.8	484
127	A notice from the Employer to the Contractor of the provision of the Conditions which empowers the Employer to issue a particular instruction.	Notice: Employer to Contractor	clause 3.8	484
128	A request from the Employer for the Contractor's consent for the Employer to issue an instruction requiring a Change which is, or makes necessary, an alteration or modification in the design of the Works.	Request: Employer to Contractor	clause 3.9.1	485
129	Consent given by the Contractor for the Employer to issue an instruction requiring a Change which is, or makes necessary, an alteration or modification in the design of the Works.	Consent given: Contractor to Employer	clause 3.9.1	485

(*Continued*)

Reference	Description	Resource type	Provision references	Page
130	Consent withheld by the Contractor for the Employer to issue an instruction requiring a Change which is, or makes necessary, an alteration or modification in the design of the Works.	Consent withheld: Contractor to Employer	clause 3.9.1	486
131	A notice from the Contractor to the Employer of an objection to an instruction, in its capacity as Principal Designer or Principal Contractor.	Notice: Contractor to Employer	clause 3.9.4	486
132	A varied instruction from the Employer to the Contractor, following the Contractor's notice under clause 3.9.4.	Instruction: Employer to Contractor	clause 3.9.4	487
133	Confirmation by the Contractor to the Employer of the varied instruction, following the Contractor's notice under clause 3.9.4.	Confirmation: Contractor to Employer	clause 3.9.4	487
134	An instruction from the Employer to postpone any work to be executed under the Contract.	Instruction: Employer to Contractor	clause 3.10	488
135	Confirmation by the Contractor of an Employer's instruction regarding the postponement of any work to be executed under the Contract.	Confirmation: Contractor to Employer	clause 3.10	488
136	An instruction from the Employer regarding the expenditure of Provisional Sums.	Instruction: Employer to Contractor	clause 3.11	489
137	A Provisional Sums tracker template.	Tracker: Contractor and Employer	clause 3.11	489
138	An instruction from the Employer to open up work for inspection or testing.	Instruction: Employer to Contractor	clause 3.12	490

(*Continued*)

Reference	Description	Resource type	Provision references	Page
139	Confirmation from the Employer to the Contractor that the costs of opening up or inspection will be added to the Contract Sum, where the materials, goods or work were in accordance with the Contract.	Confirmation: Employer to Contractor	clause 3.12	490
140	Confirmation from the Employer to the Contractor that the costs of opening up or inspection will not be added to the Contract Sum, where the materials, goods or work were not in accordance with the Contract.	Confirmation: Employer to Contractor	clause 3.12	491
141	An instruction from the Employer to remove from the site all or any work, materials or goods not in accordance with the Contract.	Instruction: Employer to Contractor	clause 3.13.1	491
142	Confirmation from the Contractor of compliance with the Employer's instruction to remove from the site all or any work, materials, or goods not in accordance with the Contract.	Confirmation: Contractor to Employer	clause 3.13.1	492
143	An instruction from the Employer, as a reasonably necessary consequence of any instruction issued under clause 3.13.1.	Instruction: Employer to Contractor	clause 3.13.2	492
144	Confirmation from the Contractor of compliance with the Employer's instructions issued as a reasonably necessary consequence of any instruction under clause 3.13.1.	Confirmation: Contractor to Employer	clause 3.13.2	493

(Continued)

Reference	Description	Resource type	Provision references	Page
145	An instruction from the Employer to open up for inspection or to test the likelihood or extent of any further similar non-compliance.	Instruction: Employer to Contractor	clause 3.13.3	493
146	Confirmation by the Contractor of compliance with the Employer's instruction made under clause 3.13.3 and clause 3.12.	Confirmation: Contractor to Employer	clause 3.13.3	494
147	An instruction from the Employer as a result of the Contractor's failure to comply with clause 2.1.1 to clause 2.1.4.	Instruction: Employer to Contractor	clause 3.14	494
148	Confirmation from the Contractor of compliance with an Employer's instruction issued as a result of the Contractor's failure to comply with clause 2.1.1 to clause 2.1.4.	Confirmation: Contractor to Employer	clause 3.14	495
149	An attendance note template for consultations between the Contractor and the Employer, relating to materials, goods, work, or workmanship not in accordance with the Contract.	Template: Attendance Note	clause 3.12 to clause 3.14	496
150	A notice from the Contractor to the Employer of the discovery and precise location of any fossils, antiquities, or other objects of interest or value.	Notice: Contractor to Employer	clause 3.15.1	496
151	An instruction from the Employer concerning the object reported by the Contractor under clause 3.15.1.	Instruction: Employer to Contractor	clause 3.15.2	497
152	A notice from the Contractor to the Employer of the discovery of any asbestos, contaminated material, or unexploded ordnance.	Notice: Contractor to Employer	clause 3.15.3	497

(*Continued*)

Reference	Description	Resource type	Provision references	Page
153	An instruction from the Employer concerning the material or item reported by the Contractor under clause 3.15.3.	Instruction: Employer to Contractor	clause 3.15.4	498
154	A tracker template for recording Employer's instructions.	Tracker: Contractor and Employer	clause 3.5 to clause 3.15	498
155	A checklist for the Contractor upon receiving an Employer's instruction.	Checklist: Contractor	clause 3.5 to clause 3.15	499
156	The provision from the Contractor to the Employer of the health and safety file.	Correspondence: Contractor to Employer	clause 3.16.2	499
157	A notice from the Employer to the Contractor of the appointment of a replacement Principal Designer or Principal Contractor, for the purposes of either the CDM Regulations or the Building Regulations.	Notice: Employer to Contractor	clause 3.16.4	500
158	A checklist for other amounts to be added to, or deducted from, the Contract Sum, as referred to in clause 4.2.6.	Checklist: Contractor and Employer	clause 4.2.6	500
159	A reminder from the Contractor to the Employer to take adjustments to the Contract Sum into account in the next Interim Payment.	Reminder: Contractor to Employer	clause 4.3	501
160	A reminder from the Contractor to the Employer in respect of a VAT exemption.	Reminder: Contractor to Employer	clause 4.4.2	502
161	Confirmation from the Employer to the Contractor of the Employer becoming a 'contractor' for the purposes of the Construction Industry Scheme.	Confirmation: Employer to Contractor	clause 4.5	502
162	A tracker template to record the reimbursement to the Employer of an advance payment made to the Contractor.	Tracker: Contractor and Employer	clause 4.6	503

(*Continued*)

Reference	Description	Resource type	Provision references	Page
163	An Interim Payment Application template, from the Contractor to the Employer.	Template (Contractor): Interim Payment Application	clause 4.7.3	503
164	A Payment Notice template, from the Employer to the Contractor.	Template: Payment Notice	clause 4.7.5	504
165	A Final Payment Notice template.	Final Payment Notice: Contractor or Employer to the other Party	clause 4.8	505
166	A Pay Less Notice template.	Pay Less Notice: Employer and Contractor	clause 4.9.5	506
167	A notice from the Contractor to the Employer of the Contractor's intention to suspend the performance of its obligations under the Contract.	Notice: Contractor to Employer	clause 4.11.1	507
168	A reply from the Employer to the Contractor's notice of intention to suspend the performance of its obligations under the Contract.	Reply: Employer to Contractor	clause 4.11.1	507
169	Confirmation from the Contractor to the Employer of the Contractor's intention to suspend the performance of its obligations under the Contract.	Confirmation: Contractor to Employer	clause 4.11.1	508
170	An application from the Contractor to the Employer for payment of costs and expenses incurred as a result of the Contractor's suspension of the performance of its obligations under the Contract.	Application: Contractor to Employer	clause 4.11.3	509

(*Continued*)

Reference	Description	Resource type	Provision references	Page
171	A request from the Employer to the Contractor for details of costs and expenses incurred as a result of the Contractor's suspension of the performance of its obligations under the Contract.	Request: Employer to Contractor	clause 4.11.3	509
172	A checklist for calculating the Gross Valuation in accordance with Alternative A (Stage Payments).	Checklist: Contractor and Employer	clause 4.12	510
173	A checklist for calculating the Gross Valuation in accordance with Alternative B (Periodic Payments).	Checklist: Contractor and Employer	clause 4.13	512
174	A tracker template for recording Interim Payments.	Tracker: Contractor and Employer	clause 4.6 to clause 4.14	513
175	A request from the Contractor for the Employer's approval to provide a bond from a particular surety.	Request: Contractor to Employer	clause 4.15.4	514
176	Approval given from the Employer for the Contractor to provide a bond from a particular surety.	Approval given: Employer to Contractor	clause 4.15.4	514
177	Approval withheld from the Employer for the Contractor to provide a bond from a particular surety.	Approval withheld: Employer to Contractor	clause 4.15.4	515
178	A checklist for the Employer to determine whether the conditions for inclusion of the value of a Listed Item in the Gross Valuation before its delivery to or adjacent to the Works have been met.	Checklist: Contractor and Employer	clause 4.15	515
179	A request from the Contractor to the Employer to place the Retention in a separate bank account.	Request: Contractor to Employer	clause 4.16.2	517

(*Continued*)

Reference	Description	Resource type	Provision references	Page
180	A notice from the Employer to the Contractor that the Retention has been placed in a separate bank account.	Notice: Employer to Contractor	clause 4.16.2	517
181	A request from the Contractor for the Employer's approval to provide a Retention Bond from a particular surety.	Request: Contractor to Employer	clause 4.17.2	517
182	Approval given from the Employer for the Contractor to provide a Retention Bond from a particular surety.	Approval given: Employer to Contractor	clause 4.17.2	518
183	Approval withheld from the Employer for the Contractor to provide a Retention Bond from a particular surety.	Approval withheld: Employer to Contractor	clause 4.17.2	518
184	Confirmation from the Employer of the deduction of Retention by the Employer, if the Contractor fails to provide or maintain the Retention Bond.	Confirmation: Employer to Contractor	clause 4.17.3	519
185	A request from the Employer for the Contractor to confirm how the additional Retention will be accounted for.	Request: Employer to Contractor	clause 4.17.4	519
186	A checklist for the calculation of the Retention which may be deducted.	Checklist: Contractor and Employer	clause 4.18	520
187	A notice from the Contractor to the Employer of loss and expense incurred by the Contractor, arising from a deferment of giving possession of the site.	Notice: Contractor to Employer	clause 4.20.1, clause 4.20.2	521
188	A notice from the Contractor to the Employer of loss and expense incurred by the Contractor, arising from a Relevant Matter.	Notice: Contractor to Employer	clause 4.20.1, clause 4.20.2	522
189	A requirement by the Employer of further information from the Contractor to ascertain loss and expense incurred.	Requirement: Employer to Contractor	clause 4.20.3	523

(*Continued*)

Reference	Description	Resource type	Provision references	Page
190	A notice from the Employer to the Contractor of the ascertained amount of loss and expense incurred.	Notice: Employer to Contractor	clause 4.20.4	523
191	A tracker template for recording applications for reimbursement of loss and expense.	Tracker: Contractor and Employer	clause 4.19 to clause 4.22	524
192	The submission by the Contractor to the Employer of the Final Statement.	Correspondence: Contractor to Employer	clause 4.24.1	524
193	A request by the Employer for further supporting documents required from the Contractor, in respect of the Final Statement.	Request: Employer to Contractor	clause 4.24.1	525
194	A Final Statement template, from the Contractor to the Employer.	Template: Final Statement	clause 4.24.2	525
195	A notice from the Employer to the Contractor of the Employer's intention to issue the Employer's Final Statement.	Notice: Employer to Contractor	clause 4.24.3	526
196	The submission by the Employer to the Contractor of the Employer's Final Statement.	Correspondence: Employer to Contractor	clause 4.24.4	526
197	An Employer's Final Statement template, from the Employer to the Contractor.	Template: Employer's Final Statement	clause 4.24.4	527
198	A notice disputing the Final Statement / Employer's Final Statement.	Notice: Contractor or Employer to the other Party	clause 4.24.6	527
199	A checklist for establishing whether an instruction is a Change.	Checklist: Contractor and Employer	clause 5.1	528
200	A tracker template for recording Changes, work to be treated as a Change, and the expenditure of Provisional Sums.	Tracker: Contractor and Employer	clause 5.2	528
201	A checklist for carrying out a Valuation in accordance with the Valuation Rules.	Checklist: Contractor and Employer	clause 5.4 to clause 5.7	529

(*Continued*)

Reference	Description	Resource type	Provision references	Page
202	A request from the Contractor for the Employer's approval to place the insurance required under clause 6.5.1 with a particular insurer.	Request: Contractor to Employer	clause 6.5.2	531
203	Approval given by the Employer for the Contractor to place the insurance required under clause 6.5.1 with a particular insurer.	Approval given: Employer to Contractor	clause 6.5.2	532
204	Approval withheld by the Employer for the Contractor to place the insurance required under clause 6.5.1 with a particular insurer.	Approval withheld: Employer to Contractor	clause 6.5.2	532
205	The provision by the Contractor to the Employer of the policy for the insurance required under clause 6.5.1.	Correspondence: Contractor to Employer	clause 6.5.2	532
206	An instruction from the Employer not to renew Terrorism Cover.	Instruction: Employer to Contractor	clause 6.10.4	533
207	A notice of the non-availability of Terrorism Cover.	Notice: Contractor or Employer to the other Party	clause 6.11.1	533
208	A notice from the Employer to the Contactor, following the non-availability of Terrorism Cover.	Notice: Employer to Contractor	clause 6.11.2	534
209	An instruction from the Employer for the Contractor to effect and maintain an alternative or additional form of Terrorism Cover.	Instruction: Employer to Contractor	clause 6.11.3	534
210	A request for evidence of insurance.	Request: Contractor or Employer to the other Party	clause 6.12.1	535
211	The provision of evidence of insurance.	Reply: Contractor or Employer to the other Party	clause 6.12.1	535

(*Continued*)

Reference	Description	Resource type	Provision references	Page
212	Confirmation of a failure to provide evidence of insurance.	Correspondence: Contractor or Employer to the other Party	clause 6.12.2	536
213	A notice from the Contractor to the Employer of loss or damage affecting executed work, Site Materials, or the Existing Structures.	Notice: Contractor to Employer	clause 6.13.1	537
214	An authorisation from the Contractor to the insurers to pay insurance monies to the Employer.	Authorisation: Contractor to insurers	clause 6.13.3	537
215	The provision by the Employer to the Contractor of a reinstatement work statement.	Reinstatement work statements: Employer	clause 6.13.5.1	538
216	A notice of termination following material loss or damage to any of the Existing Structures.	Notice: Contractor or Employer to the other Party	clause 6.14	539
217	The invocation of a dispute resolution procedure to determine whether the termination under clause 6.14 is just and equitable.	Correspondence: Contractor or Employer to the other Party	clause 6.14	540
218	A request from the Employer to the Contractor for evidence of professional indemnity insurance.	Request: Employer to Contractor	clause 6.15.3	540
219	The provision by the Contractor to the Employer of evidence of professional indemnity insurance.	Reply: Contractor to Employer	clause 6.15.3	541
220	A notice from the Contractor to the Employer of the increased cost or non-availability of professional indemnity insurance.	Notice: Contractor to Employer	clause 6.16	541
221	A tracker template for insurances required under the Contract, and their renewal.	Tracker: Contractor and Employer	clause 6.1 to clause 6.16	542

(*Continued*)

Reference	Description	Resource type	Provision references	Page
222	The provision of a notice from the insurers of the Remedial Measures required, as a result of a breach of the Joint Fire Code.	Correspondence from insurers: Employer / Contractor	clause 6.19.1	543
223	A reply from the Contractor to the Employer following receipt of the Remedial Measures required by the insurers, as a result of a breach of the Joint Fire Code.	Reply: Contractor to Employer	clause 6.19.1	544
224	Confirmation from the Employer to the Contractor that the Employer will employ and pay others to carry out the Remedial Measures, following the Contractor's failure to proceed with the Remedial Measures.	Notice: Employer to Contractor	clause 6.19.2	544
225	Confirmation from the Employer to the Contractor that the Employer has employed and paid others to carry out the Remedial Measures, following the Contractor's failure to proceed with the Remedial Measures.	Notice: Employer to Contractor	clause 6.19.2	545
226	A request for consent from the other Party to assign the Contract or any rights thereunder.	Consent: Contractor or Employer to the other Party	clause 7.1	545
227	Consent given to assign the Contract or any rights thereunder.	Consent given: Contractor or Employer to the other Party	clause 7.1	546
228	Consent withheld to assign the Contract or any rights thereunder.	Consent withheld: Contractor or Employer to the other Party	clause 7.1	546
229	A request from the Contractor for the Employer's consent to provide a performance bond from a particular surety.	Request: Contractor to Employer	clause 7.3.1	547

(*Continued*)

Reference	Description	Resource type	Provision references	Page
230	Approval given from the Employer for the Contractor to provide a performance bond from a particular surety.	Approval given: Employer to Contractor	clause 7.3.1	547
231	Approval withheld from the Employer for the Contractor to provide a performance bond from a particular surety.	Approval withheld: Employer to Contractor	clause 7.3.1	548
232	A request by the Contractor to the Employer to amend the form of performance bond or parent company guarantee.	Request: Contractor to Employer	clause 7.3	548
233	An agreement by the Employer for the Contractor to change the proposed form of performance bond or parent company guarantee.	Agreement: Employer to Contractor	clause 7.3	549
234	A refusal by the Employer for the Contractor to change the proposed form of performance bond or parent company guarantee.	No agreement: Employer to Contractor	clause 7.3	549
235	A tracker template for recording performance bonds and parent company guarantees, across multiple building contracts.	Tracker: Contractor	clause 7.3	550
236	A notice from the Employer to the Contractor of the vesting of P&T Rights as third party rights in a Purchaser or Tenant.	Notice: Employer to Contractor	clause 7.7.1	550
237	A request from the Contractor to a Purchaser or Tenant to amend or vary clause 7.7 or Part 1 of Schedule 5, relating to P&T Rights.	Request: Contractor to Purchaser or Tenant	clause 7.7.2	551
238	A notice from the Employer to the Contractor of the vesting of Funder Rights as third party rights in a Funder.	Notice: Employer to Contractor	clause 7.8.1	552

(*Continued*)

Reference	Description	Resource type	Provision references	Page
239	A request from the Contractor to the Funder to amend or vary clause 7.8 or Part 2 of Schedule 5, relating to Funder Rights.	Request: Contractor to Funder	clause 7.8.2	552
240	A notice from the Employer to the Contractor to enter into a Collateral Warranty with a Purchaser or Tenant.	Notice: Employer to Contractor	clause 7.9	553
241	A notice from the Employer to the Contractor to enter into a Collateral Warranty with a Funder.	Notice: Employer to Contractor	clause 7.10	553
242	A tracker template to record P&T Rights and Funder Rights, by way of third party rights or collateral warranty.	Tracker: Contractor or Employer	clause 7.7–7.10	554
243	A notice from the Employer to the Contractor requesting third party rights or collateral warranties from a sub-contractor.	Notice: Employer to Contractor	clause 7.11.1.1	554
244	Confirmation or provision by the Contractor to the Employer of third party rights or a collateral warranty from a sub-contractor.	Correspondence: Contractor to Employer	clause 7.11.1.2	555
245	Approval given/withheld to the sub-contractor to amend the form of third party rights or collateral warranty.	Approval: Contractor or Employer to sub-contractor	clause 7.11.2	555
246	A request from the Contractor to a beneficiary to amend, vary, or rescind the sub-contract.	Request: Contractor to beneficiary	clause 7.11.3	556
247	Consent given/withheld by the Employer to the Contractor to amend the sub-contract.	Consent: Employer to Contractor	clause 7.11.3	557
248	A tracker template to record third party rights and collateral warranties from sub-contractors.	Tracker: Contractor and Employer	clause 7.11	557

(*Continued*)

Reference	Description	Resource type	Provision references	Page
249	Confirmation of the reinstatement of the Contractor's employment under the Contract.	Correspondence: Contractor or Employer to the other Party	clause 8.3.2	558
250	A notice from the Employer to the Contractor of a specified default.	Notice: Employer to Contractor	clause 8.4.1	558
251	A reply from the Contractor to the Employer's notice of a specified default.	Reply: Contractor to Employer	clause 8.4.1	559
252	A notice of termination from the Employer to the Contractor, following the continuation of a specified default.	Notice: Employer to Contractor	clause 8.4.2	560
253	A notice of termination from the Employer to the Contractor, following the repetition of a specified default.	Notice: Employer to Contractor	clause 8.4.3	560
254	A notice of termination from the Employer to the Contractor, following the Contractor becoming Insolvent.	Notice: Employer to Contractor	clause 8.5.1	561
255	A notice from the Contractor to the Employer relating to the Contractor's insolvency.	Notice: Contractor to Employer	clause 8.5.2	561
256	A notice of termination from the Employer to the Contractor, relating to corruption and regulation 73(1)(b) of the PC Regulations.	Notice: Employer to Contractor	clause 8.6	562
257	A requirement by the Employer for the Contractor to remove from the Works temporary buildings, plant, tools, equipment, goods and materials belonging to the Contractor.	Requirement: Employer to Contractor	clause 8.7.2.1	562
258	The provision by the Contractor to the Employer of all Contractor's Design Documents then prepared.	Correspondence: Contractor to Employer	clause 8.7.2.2	563

(*Continued*)

Reference	Description	Resource type	Provision references	Page
259	A requirement by the Employer for the Contractor to assign agreements for the supply of materials or goods, and/or for the execution of any work, for the purposes of the Contract.	Requirement: Employer to Contractor	clause 8.7.2.3	563
260	A notice from the Employer to the Contractor of the completion of the Works and the making good of defects in them	Notice: Employer to Contractor	clause 8.7.4	564
261	A notice from the Employer to the Contractor of the Employer's decision not to complete the Works.	Notice: Employer to Contractor	clause 8.8.1	564
262	A notice from the Contractor to the Employer of a specified default.	Notice: Contractor to Employer	clause 8.9.1	565
263	A notice from the Contractor to the Employer of a specified suspension event.	Notice: Contractor to Employer	clause 8.9.2	565
264	A reply from the Employer to the Contractor's notice of a specified default or a specified suspension event.	Reply: Employer to Contractor	clause 8.9.1, clause 8.9.2	566
265	A notice of termination from the Contractor to the Employer, following the continuation of a specified default or a specified suspension event.	Notice: Contractor to Employer	clause 8.9.3	567
266	A notice of termination from the Contractor to the Employer, following the repetition of a specified default or a specified suspension event.	Notice: Contractor to Employer	clause 8.9.4	568
267	A notice of termination from the Contractor to the Employer, following the Employer becoming Insolvent.	Notice: Contractor to Employer	clause 8.10.1	569
268	A notice from the Employer to the Contractor relating to the Employer's insolvency.	Notice: Employer to Contractor	clause 8.10.2	569

(*Continued*)

Reference	Description	Resource type	Provision references	Page
269	A notice of a potential termination of the Contractor's employment under the Contract, following a suspension of the carrying out of the uncompleted Works for a continuous period.	Notice: Contractor or Employer to the other Party	clause 8.11.1	570
270	A notice of termination of the Contractor's employment under the Contract, following a suspension of the carrying out of the uncompleted Works for a continuous period.	Notice: Contractor or Employer to the other Party	clause 8.11.1	572
271	A notice of termination from the Employer to the Contractor, relating to regulation 73(1) of the PC Regulations.	Notice: Employer to Contractor	clause 8.11.3	572
272	The provision of documents relating to clause 2.37 (As-built Drawings) from the Contractor to the Employer.	Correspondence: Contractor to Employer	clause 8.12.2	573
273	The provision of all documents necessary for calculating the amounts referred to in clause 8.12.5, from the Contractor to the Employer.	Correspondence: Contractor to Employer	clause 8.12.3	573
274	The provision of an account of the amounts referred to in clause 8.12.5, from the Employer to the Contractor.	Account: Employer to Contractor	clause 8.12.3	574
275	A template Employer's Termination Payment Notice, from the Employer to the Contractor.	Template: Employer's Termination Payment Notice	clause 8.13.2	575
276	A template Contractor's Termination Payment Notice, from the Contractor to the Employer.	Template: Contractor's Termination Payment Notice	clause 8.13.5.1	577
277	A notice of a matter which appears likely to give rise to a dispute or difference.	Notice: Contractor or Employer to the other Party	clause 9.1	579

(Continued)

Reference	Description	Resource type	Provision references	Page
278	A request to mediate a difference or dispute between the Parties.	Request: Contractor or Employer to the other Party	clause 9.2	580
279	The submission by the Contractor of Contractor's Design Documents.	Submission: Contractor to Employer	Schedule 1, paragraph 1	580
280	The return by the Employer of Contractor's Design Documents.	Return: Employer to Contractor	Schedule 1, paragraph 2	581
281	Confirmation by the Contractor to the Employer of Contractor's Design Documents regarded as marked 'A'.	Confirmation: Contractor to Employer	Schedule 1, paragraph 3	582
282	A reply from the Contractor to the Employer to marked Contractor's Design Documents.	Reply: Contractor to Employer	Schedule 1, paragraph 5	582
283	The resubmission by the Contractor of a Contractor's Design Document marked 'B'.	Resubmission: Contractor to Employer	Schedule 1, paragraph 5.2	583
284	The resubmission by the Contractor of Contractor's Design Document marked 'C'.	Resubmission: Contractor to Employer	Schedule 1, paragraph 5.3	583
285	A notice from the Contractor to the Employer of a disagreement with the Employer's comment on a Contractor's Design Document.	Notice: Contractor to Employer	Schedule 1, paragraph 7	584
286	A reply from the Employer to the Contractor's notice of disagreement with the Employer's comment on a Contractor's Design Document.	Confirmation: Employer to Contractor	Schedule 1, paragraph 7	585
287	A tracker template for recording Contractor's Design Documents.	Tracker: Contractor and Employer	Schedule 1	585

(*Continued*)

Reference	Description	Resource type	Provision references	Page
288	A notice from the Contractor to the Employer of the Contractor's inability to enter into a sub-contract with a Named Sub-Contractor.	Notice: Contractor to Employer	Supplemental Provision 1, paragraph 1.1.2	586
289	An instruction from the Employer following a notice from the Contractor of the Contractor's inability to enter into a sub-contract with a Named Sub-Contractor.	Instruction: Employer to Contractor	Supplemental Provision 1, paragraph 1.1.2	587
290	A notice from the Contractor to the Employer of the potential termination of a Named Sub-Contractor's employment under their sub-contract.	Notice: Contractor to Employer	Supplemental Provision 1, paragraph 1.3.1	587
291	A request from the Contractor for the Employer's consent to terminate a Named Sub-Contractor's employment under their sub-contract.	Request for consent: Contractor to Employer	Supplemental Provision 1, paragraph 1.3.2	588
292	Consent given by the Employer for the Contractor to terminate a Named Sub-Contractor's employment under their sub-contract.	Consent given: Employer to Contractor	Supplemental Provision 1, paragraph 1.3.2	588
293	Consent withheld by the Employer for the Contractor to terminate a Named Sub-Contractor's employment under their sub-contract.	Consent withheld: Employer to Contractor	Supplemental Provision 1, paragraph 1.3.2	589
294	The provision of estimates by the Contractor to the Employer of the effects of a Change instruction.	Estimates: Contractor to Employer	Supplemental Provision 2, paragraph 2.2	589
295	Confirmation from the Employer to the Contractor that estimates are not required.	Correspondence: Employer to Contractor	Supplemental Provision 2, paragraph 2.2.1	590
296	An objection from the Contractor to the Employer to the provision of estimates.	Objection: Contractor to Employer	Supplemental Provision 2, paragraph 2.2.2	591

(*Continued*)

Reference	Description	Resource type	Provision references	Page
297	Confirmation or withdrawal by the Employer of an instruction to the Contractor.	Confirmation / withdrawal of instruction: Employer to Contractor	Supplemental Provision 2, paragraph 2.5	591
298	A tracker template for instructions and estimates relating to Supplemental Provision 2.	Tracker: Contractor and Employer	Supplemental Provision 2	592
299	The provision by the Contractor to the Employer of an estimate of the direct loss and expense incurred by the Contractor.	Estimate: Contractor to Employer	Supplemental Provision 3, paragraph 3.2	592
300	The provision by the Contractor to the Employer of further estimates of the direct loss and expense incurred by the Contractor.	Further estimates: Contractor to Employer	Supplemental Provision 3, paragraph 3.3	593
301	A request by the Employer to the Contractor for further information and details relating to the Contractor's estimate of the direct loss and expense incurred by the Contractor.	Request: Employer to Contractor	Supplemental Provision 3, paragraph 3.4	594
302	A notice from the Employer to the Contractor relating to the Contractor's estimate of the direct loss and expense incurred by the Contractor.	Notice: Employer to Contractor	Supplemental Provision 3, paragraph 3.4	594
303	A tracker template for estimates of direct loss and expense incurred by the Contractor, relating to Supplemental Provision 3.	Tracker: Contractor and Employer	Supplemental Provision 3	595
304	A request by the Employer for the Contractor to provide an Acceleration Quotation.	Request: Employer to Contractor	Supplemental Provision 4, paragraph 4.1.1	595

(*Continued*)

Reference	Description	Resource type	Provision references	Page
305	The provision by the Contractor to the Employer of an Acceleration Quotation, or an explanation why it would be impracticable to achieve practical completion earlier than the Completion Date.	Acceleration Quotation: Contractor to Employer	Supplemental Provision 4, paragraph 4.1.1	596
306	An Acceleration Quotation template, from the Contractor to the Employer.	Template: Acceleration Quotation	Supplemental Provision 4, paragraph 4.1.1	596
307	A request by the Employer for the Contractor to issue revised proposals, in respect of an Acceleration Quotation.	Request: Employer to Contractor	Supplemental Provision 4, paragraph 4.1.3	597
308	A notice from the Employer to the Contractor of an adjustment to the periods referred to in Supplemental Provision 4.	Notice: Employer to Contractor	Supplemental Provision 4, paragraph 4.2.3	598
309	A Confirmed Acceptance from the Employer to the Contractor, by way of an instruction.	Instruction: Employer to Contractor	Supplemental Provision 4, paragraph 4.3	598
310	A checklist of the Contractor's duties in relation to health and safety.	Checklist: Contractor	Supplemental Provision 5	599
311	A proposed change to the Works by the Contractor, for the Employer's benefit.	Proposed change: Contractor to Employer	Supplemental Provision 6, paragraph 6.1	599
312	An instruction by the Employer of a change proposed by the Contractor.	Instruction: Employer to Contractor	Supplemental Provision 6, paragraph 6.3	600
313	A tracker template for proposals made by the Contractor, relating to Supplemental Provision 6.	Tracker: Contractor and Employer	Supplemental Provision 6	600
314	The provision by the Contractor to the Employer of information for performance indicators and monitoring.	Provision of information: Contractor to Employer	Supplemental Provision 7, paragraph 7.2	601

(*Continued*)

Reference	Description	Resource type	Provision references	Page
315	Informing the Contractor that targets for performance indicators may not be met.	Correspondence: Employer to Contractor	Supplemental Provision 7, paragraph 7.3	601
316	Informing the Contractor of a request made in accordance with the Freedom of Information Act 2000, in relation to the Contract.	Correspondence: Employer to Contractor	Supplemental Provision 8, paragraph 8.2	601
317	The provision by the Contractor to the Employer of the Joint Names Policy.	Correspondence: Contractor to Employer	Schedule 3, paragraph A.1	602
318	Confirmation by the Contractor to the Employer of the use of the Contractor's annual policy, as an alternative to a Joint Names Policy.	Correspondence: Contractor to Employer	Schedule 3, paragraph A.2	602
319	The provision by the Employer to the Contractor of the Joint Names Policy.	Correspondence: Employer to Contractor	Schedule 3, paragraph B.1	603
320	The provision by the Employer to the Contractor of the Joint Names Policy for Specified Perils.	Correspondence: Employer to Contractor	Schedule 3, paragraph C.1	604
321	The provision by the Employer to the Contractor of the Joint Names Policy for All Risks.	Correspondence: Employer to Contractor	Schedule 3, paragraph C.2	604
322	A notice from the Contractor to a Purchaser or Tenant, if Professional Indemnity insurance ceases to be available.	Notice: Contractor to Purchaser or Tenant	Part 1 of Schedule 5, paragraph 5	605
323	The provision by the Contractor to a Purchaser or Tenant of documentary evidence of Professional Indemnity insurance.	Correspondence: Contractor to Purchaser or Tenant	Part 1 of Schedule 5, paragraph 5	606
324	A notice from the Contractor to a Funder, relating to step-in rights, and prior to the Contractor being entitled to terminate its employment under the Contract.	Notice: Contractor to Funder	Part 2 of Schedule 5, paragraph 6.1.1	606

(*Continued*)

Reference	Description	Resource type	Provision references	Page
325	A notice from the Contractor to a Funder, where the Contractor has the right to terminate its employment under the Contract.	Notice: Contractor to Funder	Part 2 of Schedule 5, paragraph 6.1.2	607
326	A notice from the Contractor to a Funder, where the Contractor has the right to treat the Contract as having been repudiated by the Employer.	Notice: Contractor to Funder	Part 2 of Schedule 5, paragraph 6.2	608
327	A notice from the Contractor to a Funder, if Professional Indemnity insurance ceases to be available.	Notice: Contractor to Funder	Part 2 of Schedule 5, paragraph 9	609
328	The provision by the Contractor to a Funder of documentary evidence of Professional Indemnity insurance.	Correspondence: Contractor to Funder	Part 2 of Schedule 5, paragraph 9	610
329	A monthly progress report to be prepared by the Contractor.	Template: Contractor's progress report	n/a	610
330	A meeting agenda template for use by the Contractor and the Employer.	Template: Meeting agenda	n/a	611
331	A meeting minutes template for use by the Contractor and the Employer.	Template: Meeting minutes	n/a	612
332	A memorandum template for use by the Contractor and the Employer.	Template: Memorandum	n/a	612
333	A site diary template for use by the Contractor.	Site Diary	n/a	613
334	A project calendar template for use by the Contractor and the Employer.	Project calendar	n/a	614
335	An instruction template for use by the Employer.	Instruction	n/a	615

Subject Index

The subject index below links each subject with their corresponding key provisions in the Contract.

The defined terms from the Contract are identified in **bold.**

(*Continued*)

(*Continued*)

(*Continued*)

(*Continued*)

(*Continued*)

(*Continued*)

(*Continued*)

(*Continued*)

(*Continued*)

(*Continued*)

(Continued)

(*Continued*)

(*Continued*)

(*Continued*)

(*Continued*)

(*Continued*)

(*Continued*)

(*Continued*)